MICROBICIDES FOR THE PROTECTION OF MATERIALS

A HANDBOOK

MICROBICIDES FOR THE PROTECTION OF MATERIALS

A HANDBOOK

by

WILFRIED PAULUS

President of the International Biodeterioration Research Group

CHAPMAN & HALL

London · Glasgow · New York · Tokyo · Melbourne · Madras

Published by Chapman & Hall, 2-6 Boundary Row, London SE1 8HN

Chapman & Hall, 2-6 Boundary Row, London SE1 8HN, UK

Blackie Academic & Professional, Wester Cleddens Road, Bishopbriggs, Glasgow G64 2NZ, UK

Chapman & Hall Inc., One Penn Plaza, 41st Floor, New York, NY10019, USA

Chapman & Hall Japan, Thomson Publishing Japan, Hirakawacho Nemoto Building, 6F, 1-7-11 Hirakawa-cho, Chiyoda-ku, Tokyo 102, Japan

Chapman & Hall Australia, Thomas Nelson Australia, 102 Dodds Street, South Melbourne, Victoria 3205, Australia

Chapman & Hall India, R. Seshadri, 32 Second Main Road, CIT East, Madras 600 035, India

First edition 1993

Printed in Great Britain by Galliard (Printers) Ltd, Great Yarmouth

ISBN 0 412 53450 9

A catalogue record for this book is available from the British Library

Library of Congress Cataloging-in-Publication data

Paulus, Wilfried
Microbicides for the protection of materials: a handbook / by Wilfried Paulus
p. cm
Includes bibliographical references and index.
ISBN 0-412-53450-9
1. Disinfection and disenfectants—Handbooks, manuals, etc.
2. Industrial hygiene—Handbooks, manuals, etc. 3. Materials—Health aspects—Handbooks, manuals, etc. I. Title.
RA765.P38 1993
620.1′1223—dc20 92-31479
CIP

∞ Printed on permanent acid-free text paper, manufactured in accordance with ANSI/NISO Z39.48-1992 (Permanence of Paper).

To my wife
whose tolerance made this book a reality

Preface

This book is chiefly intended for those who are using microbicides for the protection of materials. Another purpose is to inform teachers and students working on biodeterioration and to show today's technical standard to those engaged in R & D activities in the microbicide field.

When trying to classify, or to subclassify, material-protecting microbicides according to their mode of action, e.g. as membrane-active and electrophilic-active ingredients, it turned out that a clear assignment was not always possible. For that reason the author has resorted to chemistry's principle of classifying according to groups of substances (e.g. alcohols, aldehydes, ketones, acids, esters, amides, etc.), thus providing the first necessary information about the microbicides' properties. The description of the various groups of substances includes, whenever possible, an outline of the mode and mechanism of action of the active ingredients involved.

The effective use of microbicides presupposes knowledge of their characteristics. That is why the microbicides' chemico-physical properties, their toxicity, ecotoxicity, effectiveness, and effective spectrum are described in greater detail.

As mentioned before, the characteristics of microbicides play an important role. They have to be suited to the intended application to avoid detrimental effects on the properties and the quality of the material to be protected; also production processes in which microbicides are used to avoid disturbances by microbial action must not be disturbed by the presence of those microbicides.

Microbicides which discolour materials or cause unpleasant smells, which delay the drying of coatings or tend to spoil the surface quality of coatings have proved unsuitable for certain applications. The same is true for microbicides which decompose when the materials are processed. In processing water, surface-active and corrosive microbicides tend to disturb the normal cycle of production. The effective concentration of preservatives for the in-can/in-tank protection of aqueous functional fluids must not cause any change in the viscosity of the liquids. The requisite microbicide concentration depends on the resistance of the microbes to be inhibited. The range of effectiveness is another important factor to be considered. It is necessary to choose microbicides with the proper effective spectrum; a full range covering all species of microbes is not always a must.

The present book is intended to inform about all those issues. In addition, the chapter 'Applications' describes the object of using microbicides and the problems that may arise in the various fields of application, with proposals for suitable microbicides.

The chapter 'Effectiveness — mode and mechanism of action' was written to improve the reader's understanding of microbicides and their applications.

The assistance of those involved in the development and production of microbicides for the protection of materials is earnestly solicited in the correction of possible errors in the text; in the collection of descriptions of microbicides that may have been omitted from the text; in supplying more detailed descriptions of such microbicides as are described incompletely; and in furnishing complete descriptions of newly developed microbicides, or in directing the attention of the author to publications of such newly described microbicides.

ACKNOWLEDGEMENTS

The author desires to express his appreciation and thanks to Miss C. Malz who gave invaluable technical assistance in the preparation of the book. Also the patience and tolerance of the publishers in the face of repeated delays on the part of the author must be acknowledged.

ERRATA

Page 59, the right-hand side of the chemical structure should read: $-(CH_2-O)_x-H$

Page 156, half-way down the page, line beginning LD_{50} percutaneous, dosage should read: >5000 mg/kg rat

Page 337, the right-hand side of the chemical equation should show a double bond between N and S/S in the same manner as the bond between N and S/SH on the left-hand side

Page 440, neither of the two chemical structures should indicate a shared bond between all of the components of the hexagonal constructs. This is also true for the structure shown for Trichloro melamin on page 441

Microbicides for the Protection of Materials. By Wilfried Paulus. Published in 1993 by Chapman & Hall, London.
ISBN 0 412 53450 9

Contents

PART IV

PART I

Introduction

Antimicrobial substances will be referred to in this book as 'microbicides' in as far as they kill micro-organisms and as 'microbistats' in as far as they inhibit the multiplication of micro-organisms. Whether the action of a substance is microbicidal or microbistatic generally depends on the concentration at which the substance is used.

Microbicides belong to the 'biocides'. The important distinctions between the various types of biocide, which can be listed as follows, are far too often disregarded.

- Biocides → Bios
  - Microbicides → Microbes
    - Bactericides → Bacteria
    - Fungicides → Fungi
    - Algicides → Algae
  - Insecticides → Insects
  - Acaricides → Mites
  - Molluscicides → Molluscs
  - Herbicides → Plants
  - ... → ...
  - ... → ...
  - ... → ...

The experts are fully conversant with these distinctions and take them for granted. Whenever appropriate, however, the same distinctions should be used more widely — not simply to prevent confusion, but also so that the general public is constantly reminded that modern active ingredients and those undergoing development are not intended to attack the entire 'bios', but only a limited part of it. Their effects — in other words — are specific; microbicides, for example, are active only against micro-organisms.

Micro-organisms in the sense of this book are bacteria, yeasts, fungi, algae and lichens. The last-mentioned are double organisms consisting of an alga species and a fungus species that vegetate in symbiotic association (symbiosis). The alga uses carbon dioxide, light, water and trace elements to photosynthesize organic material, on which the fungus feeds. For its part, the mycelium of the fungus stores water, which enables the alga to continue living in times of drought.

Microbial cells were the first living cells on earth. They appeared about three and a half thousand million years ago as the starting point of the evolution of life on our planet. Microfossils that have been found in flint and certain other very fine sediments are regarded as direct proof of such early life — of cyanobacteria, for example, which probably formed a blue-green slime in the littoral regions of lakes close to volcanoes.

Micro-organisms, whose cells measure just a few micrometres (thousandths of a millimetre) and contain 75–85% of water, have since become ubiquitous. Wherever moisture and nutrients are available they reproduce explosively — exponentially, that is. Thus,

$$m_1 = m_0 \cdot e^{\mu t}$$

where

m_1 = number of microbes/ml at point of time t,
m_0 = the number at point of time 0,
μ = specific growth rate.

A gram of fertile arable soil contains some 2–5 thousand million microbes. Waste water — and likewise, indeed, fluids used for technical purposes (e.g. lubricoolants) — contain between 1 and 10 million microbes per millilitre, while even high-quality drinking water has up to 100 microbes per millilitre.

All organic matter existing on earth is decomposable by micro-organisms, which use it as their source of nutrition. Those known as mineralizers are dependent on oxygen as their source of energy. The end-products of decomposition processes in which oxygen is consumed, which are known as mineralization processes, are inorganic compounds and carbon dioxide, which, together with water and light, are used by plants for photosynthesis. The organic matter produced by photosynthesis is the principal nutrient of the great majority of living organisms and the prerequisite of the existence of life.

Where oxygen is absent there are microbe species which obtain their energy from different sources. Methane bacteria get it by reducing carbon dioxide to methane and are thus responsible for the formation of marsh gas or natural gas. Sulphate-reducing bacteria obtain their energy by reducing sulphate to hydrogen sulphide under unaerobic conditions, causing many objectionable odours.

As in alcoholic fermentation, the production of sauerkraut, the ripening of cheese, the baking of bread, the production of penicillin, the purification of waste water at biological treatment plants and the production of biogas from agricultural waste micro-organisms may be extremely useful, and even indispensable, to Man. Yet in other cases they are harmful or highly dangerous: by causing infectious diseases, by forming poisonous or carcinogenic metabolites and wherever — in addition to decomposing wastes — they attack valuable materials, disturb production processes or impair the quality of products. According to estimates, microbial decomposition destroys materials to the value of at least one hundred thousand million dollars several times over every year.

It was not until the 19th century that the causes of fermentation, decay and infectious diseases were investigated and the way opened to measures directed specifically against the harmful micro-organisms. Pasteur, the pioneer in this field, whose discoveries were made possible by Leeuwenhoek's invention of the microscope in the 17th century, initiated the war against harmful micro-organisms with a physical method: he killed them with heat. Chemical methods — the use of substances that destroy or inhibit microbes — soon followed. In 1867 Lister made use for the first time of the antiseptic properties of 'carbolic acid' (phenol) to kill bacteria on medical instruments, dressings and wounds.

As will be explained in Part III, virtually every class of chemical substances contains many microbicides (and microbistats). Their use in or on inanimate objects or their surroundings is referred to as the protection of materials against biodeterioration. Since micro-organisms occur everywhere, their destructive activity is basically unavoidable. Nevertheless materials can be protected with the aid of compounds that prevent the multiplication of microbes at the relevant sites, either by killing them or by inhibiting their development. Microbicides for material protection include preservatives for in-tank/in-can protection of aqueous functional fluids (e.g. polymer emulsions, paints, adhesives, mineral slurries, metalworking fluids, cosmetics, pharmaceutical and food products), actives for the antimicrobial treatment of widely differing materials (e.g. paint films, plastics, leather, textiles, wood), active ingredients for algicides and slimicides, and also those that form the basis of disinfectants that are used on equipment or the inanimate environment (e.g. in hospitals, households, animal stables, the food industry).

The demand for microbicides by chemical-technical industries led at first to the use of such oxidizing agents as chlorine and hydrogen peroxide, which, though intrinsically very effective, are unstable. Other chemicals followed, but the requirements that had to be met were still modest, comprising little more — in the case of microbicides used to protect materials — than the need:

- to have a strong microbicidal effect and broad spectrum of activity;
- for the greatest possible stability; persistence;
- for economy in use.

This short list, still remarkable for its simplicity, led inevitably to the organometallic compounds and polyhalogenated phenol derivatives — active ingredients which are both highly effective and highly persistent and which may indeed be termed 'biocides', since they are harmful not only to micro-organisms, but also to plants, insects, snails, fish and other organisms — in short, to all forms of life.

In the meantime, however, a new catalogue of requirements has evolved — or, rather, the earlier ones have been modified to meet the needs of present-day civilization as seen in its heightened sense of environmental responsibility and desire for quality of life.

The characteristics demanded of microbicides for the protection of materials today are as follows:

- the possession of degrees of stability and spectra of activity graduated according to the application;

— very low toxicity;
— very low ecotoxity;
— economy in use.

The user no longer insists on persistence at any price, but is content with limited persistence graduated to the application: for stability or instability as dictated by the circumstances.

Often furthermore, there is no need to use a microbicide which, like pentachlorophenol, combines fungicidal, bactericidal, algicidal and insecticidal effects, since a satisfactory result can be obtained with a product that controls only one kind of organism. If this product is also effective at very low concentrations, so much the better. In other words, microbicides must now be tailored to the intended use and to the microbe species to be controlled.

Already there exists so much knowledge on the structure and functioning of the various microbe species, as well as on the mechanisms of activity of antimicrobial compounds, that is no longer a utopian demand, provided that the principal of synergetics is followed. According to this principle, appropriate interaction of individual entities or the combined action of suitable components leads to new structures and mechanisms of activity.

Are we looking too far ahead? Let us see how much progress has been made in that direction. Pressure for replacement of organomercury and organotin compounds and the more highly chlorinated phenol derivatives is growing; but these microbicides are still manufactured and used, though on a decreasing scale. Their replacement to meet a new list of requirements entails growing diversification because the available range of microbicides must be adapted to the many different processes for which these compounds are used. Those chosen must ensure that industrial processes are not disturbed by microbial activity, and they must be effective against many species of microbes that are resistant in different degrees. So the idea that destruction of materials and disturbance of processes by microbes can be prevented with a mere handful of preservatives will remain an illusion. The trend towards greater diversification is also gaining strength from demands made in particular instances of microbicide application — relating to biodegradability and toxicity, for example. But the obstacles to the development of new and 'progressive' microbicides for material protection that arise from complex and labour-intensive bureaucratic registration procedures are delaying desirable changes. One result of the consequent disincentive to innovation is that manufacturers are seeking to broaden the application of those progressive microbicides that have already overcome these hindrances. For this purpose they are either developing additional formulations or combinations of individual microbicides with extended activity spectra, or exploiting synergisms that enable ingredients to be used at lower concentrations.

One should not overlook the fact, however, that the use of microbicides for material protection represents a significant contribution to environmental protection.

Let us see what this means in practice. Well preserved functional fluids — metalworking fluids, for instance — can be used longer than unpreserved fluids and therefore appear less frequently as wastes that are difficult to dispose of and place a burden on the environment. Greater durability of products arising from the use of microbicides also means that replacements are needed less often, with the result that the burdens imposed on the environment by industry may be reduced and that valuable raw materials of limited availability may be spared. This fact will undoubtedly keep the search for new microbicides going. If antimicrobial compounds that are effective at extremely low concentrations can be found, their discovery and development will be justified even if they are comparatively costly. Finally, growing knowledge of the relationships between antimicrobial effect and chemical composition and structure, and — in this context — the opportunities for computer-assisted tailoring of microbicides, will keep the cost of development, which could otherwise become prohibitive, within reasonable limits.

If microbicides for the protection of materials are to be used appropriately and without risk to humans and the environment, the user, too, should be as familiar as possible with their property patterns, aware of the toxicological and chemicophysical properties of the active ingredients and, lastly, know the concentrations at which they are effective under practical conditions. It is hoped that the present book will contribute to this goal.

PART II

Effectiveness — Mode of Action

Part II will be devoted mainly to explaining the technical terms used in Part III in connection with the antimicrobial effects and modes of action of microbicides. It will also consider how chemical constitution and antimicrobial effect are related. However, the subject matter will be covered only in sufficient detail to explain what microbicides are and how they can be used to protect materials. Readers who would like to know more should consult the specialist literature, e.g. the books by Corbett *et al.* (1984), Franklin & Snow (1989) and Russel & Chopra (1990).

As mentioned in Part I, microbicides are used for material protection to ensure that micro-organisms, which are present elsewhere, are unable to multiply in materials and in process fluids. Bacteria are often divided into those which are stainable with iodine/crystal violet (gram-positive bacteria) and those that cannot be stained in this way (gram-negative bacteria). It is important to know that the differences in stainability arise from differences in the morphology and composition of the bacterial cell wall, which surrounds and protects the cytoplasm and cytoplasmic membrane (see Fig. 1). By virtue of its composition and structure, the cell wall, unlike the cytoplasmic membrane, is able to keep the shape of the bacterial cell unchanged, i.e. it withstands the high osmotic pressure exerted by the cytoplasm within the cell. The fundamental difference between gram-positive and gram-negative bacteria is that the latter have a so-called outer membrane, which is not found in gram-positive bacteria.

An antimicrobial effect can be defined as an interaction between an active substance and specific targets of the microbial cell. On their way to these targets microbicides encounter a variety of structures that are responsible for the different degrees to which individual microbe species resist one and the same microbicide — as explained above in connection with gram-positive and gram-negative bacteria. Experience has shown that gram-negative bacteria (those additionally protected by an outer membrane), especially Pseudomonades, are, for example, more resistant than gram-positive bacteria; these are, in other words, more easily controlled. Nevertheless, the molecules of active substances are also able to pass through the outer cell membrane by way of water-filled pores (so-called porins), provided this is not prevented by the size, charge and lipophilic properties of the

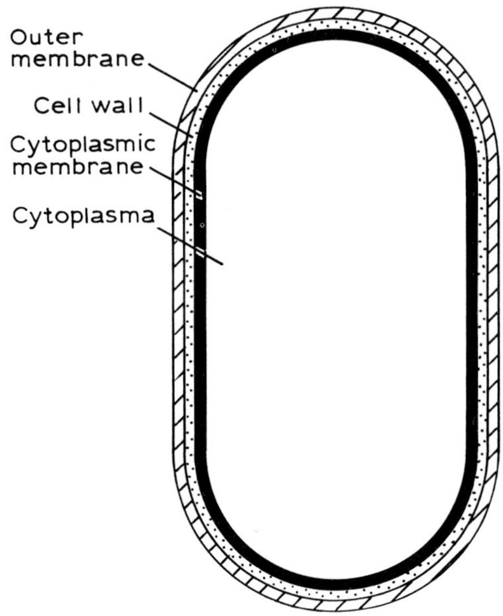

Fig. 1. Simplified, schematic drawing of a gram-negative bacterium cell.

molecule. If a substance is soluble in water and its molecular weight below 600, the likelihood of its being able to pass through porin channels is high. It is also possible for active substances to penetrate the outer membrane after having initially destabilized and disorganized it. Finally the microbicides reach their primary sites of action on the cytoplasmic membrane or within the cytoplasm and start to exert their effects. These may initiate secondary or subsequent effects, which may then cause the actual microbistatic or microbicidal effect, from which it is evident that the mode of action of a given microbicide may be too complex to permit clear description.

The processes described — transportation of the microbicide to the surface of the microbe cell, adsorption, diffusion, penetration and interaction at the target site, are not instantaneous. The time they take, the killing time, may well differ from microbicide to microbicide. The differences also depend on the mode of action, as well as on the chemical constitution and chemico-physical properties of the microbicide. As the latter characteristics may be very different among microbicides that share the same mode of action, one cannot say that a given mode of action corresponds to a short or long killing time. It will be understood that killing times decrease as the microbicide concentration and temperature rise; but here, too, the influences depend on the microbicide concerned.

The effects of microbicides are not in all cases influenced by the pH value. In some cases, however, the pH level affects the microbicidal action decisively and must therefore be taken into consideration when the microbicides concerned are used. This will be considered in more detail in Part III when the respective groups of compounds are described. They include the antimicrobially effective acids (III.6), the phenol derivatives (III.5), which may be considered as belonging to the same category, and the esters of *p*-hydroxy-benzoic acid (III.6), which are active mainly in the undissociated form. As the dissociation of acids increases as pH values rise, their antimicrobial effects become weaker at the same time. Where certain acids are concerned, a characteristic constant important in this connection is the pK_a value. This is the pH value at which 50% of an acid is present in the undissociated form (see Table 65). Likewise highly dependent on the pH value is the antimicrobial effect of glutaraldehyde (III.2.3), which, as an electrophilically active substance, is able to react with nucleophilic cell constituents, e.g. with components bearing amino and SH groups. The intensity of this interaction increases with the pH value; glutaraldehyde is therefore most effective in alkaline media, e.g. at pH 8·5.

Increasing pH values also affect the surfaces of the microbe cells: they become more negatively charged. This, in turn, naturally increases the affinity of cationogenic, membrane-active substances, to the cell surface and hence the intensity of their effects on it. Examples are provided by quaternary ammonium compounds (III.16.1), biguanides (III.16.2) and dibenzamidines (III.10).

The ambient medium impairs the effectiveness of microbicides also if its constituents include those capable of interacting with a microbicide in competition with the constituents of the microbe cell. This is true of electrophilically active microbicides in general as far as the ambient medium contains nucleophilically active constituents with which the microbicide can react in competition with the corresponding cell constituents. It is also true, however, of membrane-active microbicides if adsorption of the microbicide on 'organic matter' competes with the adsorptive processes on the cytoplasmic membrane or if such microbicides, e.g. phenol derivatives, become incorporated in micelles that are formed in certain media at levels above the critical micelle concentration with the result that the incorporated active substance molecules are no longer available for the antimicrobial effect (see III.16, Fig. 34).

Membrane-active microbicides include:

Alcohols (III.1)
Acids (III.6)
Phenol derivatives (III.5)
Salicyl anilides and carbanilides (III.8)
Dibenzamidines (III.10)
Surface active agents (III.16).

All of them damage the inner cell membrane. As, however, they belong to different groups of compounds they do so in different ways. The cytoplasmic membrane is semi-permeable, a property which enables it, together with several important enzymes, to control the passage of essential matter into and out of the cytoplasm. It is thus a highly sensitive target for microbicides. Damage is manifested as leakage of intracellular constituents, cell lysis, dissipation of the electrochemical potential of protons and inhibition of important membrane-associated enzymes.

Alcohols are regarded as membrane disrupters; they induce protoplast (germ cell) lysis.

Acids may dissipate pH across the cell membrane and induce leakage of intracellular constituents.

The reversible adsorption of phenols at the cytoplasmic membrane damages the permeability barrier, thus causing leakage of intracellular substances from the metabolic pool. In this phase the effect is microbistatic and can be terminated by dilution. Relatively high phenol concentrations, however, lead to penetration and deterioration of the cell wall, accompanied by destruction of the microbe cell (microbicidal effect).

Active salicylanilides and carbanilides (III.8) affect the membrane permeabilities in various ways, thus exerting mainly microbistatic effects.

Because of their cationic nature, dibenzamidines (III.10), like the antimicrobially effective biguanidines (III.16.2) and, of course, quaternary ammonium salts (III.16.1), are attracted by the negatively charged surface of the microbe cell and adsorbed very strongly by constituents containing phosphate. This impairs the functioning of the outer cell membrane, thus enabling the cationogenic active substances to penetrate to the cytoplasmic membrane, where they reduce its permeability until it is no longer able to function as such. The microbicidal action then starts with the precipitation of intracellular constituents. Antimicrobially active anionic detergents are like QACs, surface active agents. In contrast to cationogenic microbicides, anionogenic detergents interact with the cell membrane protein (though not until a relatively high concentration is reached), so that the cell membrane breaks down. In this way anionic surface active agents may make it easier for other microbicides to penetrate the cell, thus intensifying their effects. Finally, mention should be made of the ampholytes (III.16.3), which combine the surface active properties of anionic detergents with the microbicidal effectiveness of cationogenic surfactants.

The already mentioned electrophilically active substances have, as toxophoric constituent, an electrophilic group which is responsible for the antimicrobial effect in as far as it enables these active substances to react with specific nucleophilic entities of the microbial cell (Paulus, 1991). Examples of nucleophilic reaction partners in the cell are amino and thiol groups and amide groups of amino acids, proteins, enzymes and nucleic acids. The reactions are often irreversible and lead to inactivation of the active substance. This represents a fundamental difference between the compounds concerned and membrane-active microbicides, which remain intact and thus available for further antimicrobial action, after dissolution of the cell, for example, when they are no longer adsorbed. Long-lasting activity in practical material protection is thus especially achieved with microbicides, that are membrane-active. Electrophilically active substances, on the other hand, may be termed intrinsically non-persistent. In the meantime it has been universally accepted that the fact that electrophilically active microbicides do not accumulate in the environment outweighs their disadvantages of limited stability and duration of activity.

EXAMPLES OF ELECTROPHILICALLY ACTIVE MICROBICIDES

Aldehydes, formaldehyde releasing compounds (III.2 and III.3)

The electrophilic character of aldehydes results from the electron deficiency at the carbonyl carbon atom (Fig. 2). Among aldehydes, formaldehyde (III.2.1) is the compound with the lowest molecular weight (M_r = 30); it is therefore one with the lowest degree of sterical hindrance and very reactive. So one can assume that formaldehyde does not react very selectively and that by reacting with constituents of the ambient medium it may be consumed relatively quickly. The mode of action of formaldehyde has been investigated by Neely (1963*a,b*), among others; according to this publication formaldehyde is able to block the synthesis of methionine by condensing with the methionine precursor homocysteine to form a thiazane ring.

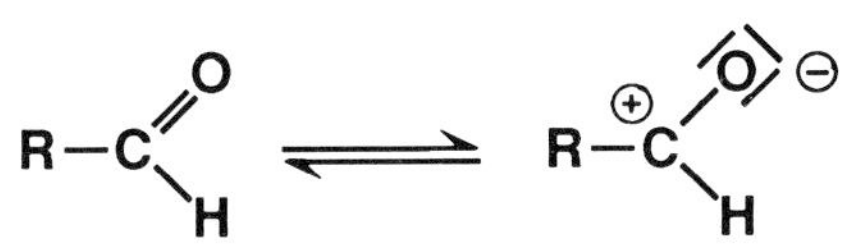

Fig. 2. Electron hole in the carbonyl-carbon atom of aldehydes.

Glutaraldehyde (III.2.3) is less reactive than formaldehyde. As a dialdehyde, however, it has two toxophoric groups, which enable it to couple nucleophilic substrates.

Substances with an activated halogen group in the α-position or/and in the vinyl position to an electronegative group E (III.13)

These electrophilic substances can react with nucleophilic compounds, for example, of the microbial cell, by way of an elimination reaction in the case of microbicides with an activated halogen in the *α*-position to an electronegative group, whereas microbicides with vinyl-activated halogen combine with nucleophiles by way of an addition–elimination reaction (Fig. 3).

Substances bearing a vinyl group in the α,β-position to an electronegative group E

Such compounds are very reactive, i.e. unstable, especially if the molecule does not contain a resonance-increasing group (Fig. 4). The high reactivity of such vinyl compounds leads to a strong antimicrobial effect. This arises because the substances participate in an addition reaction with nucleophilic cell constituents (Fig. 5).

Because, as may be seen, these vinyl compounds have both a carbon atom with electron deficiency and one with electron surplus, thus combining electrophilic

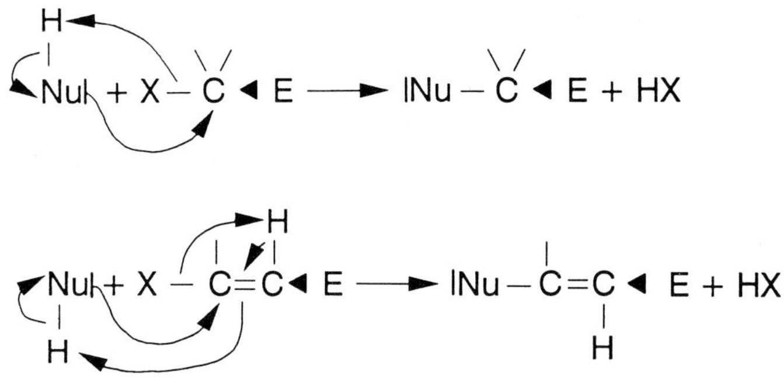

H—Nul = nucleophilic entities of the microbial cell

E = electronegative group

X = halogen

Fig. 3. Mechanism of activity of microbicides with activated halogen.

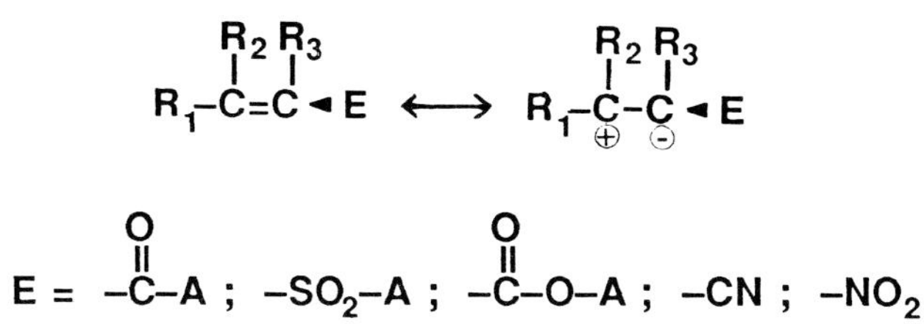

A = a resonance increasing group, e.g. aryl

Fig. 4. Reactive vinyl compound.

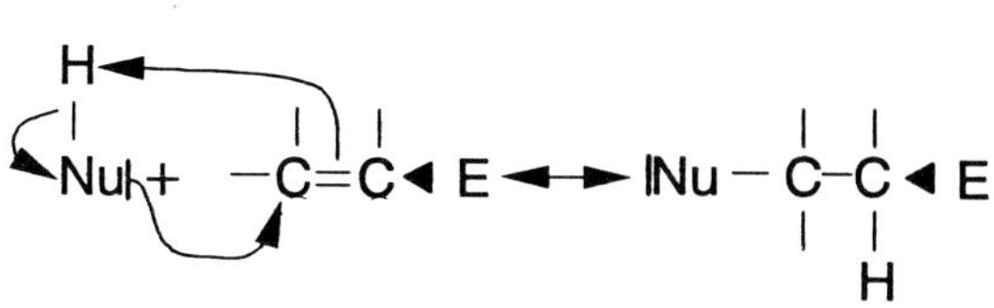

H—Nul = nucleophilic entities of the microbial cell

E = electronegative group

Fig. 5. Mechanism of activity of microbicides with an activated vinyl group.

and nucleophilic properties, they also react with one another. This tendency to polymerize naturally impairs their usefulness as microbicides. However, if a so-called 'leaving group' (LG) is added to the unstable vinyl bonds, stable substances constituting potential vinyl compounds are obtained (Fig. 6). These exhibit a microbicidal effect comparable to that of the vinyl compounds from which they originate (Paulus, 1976).

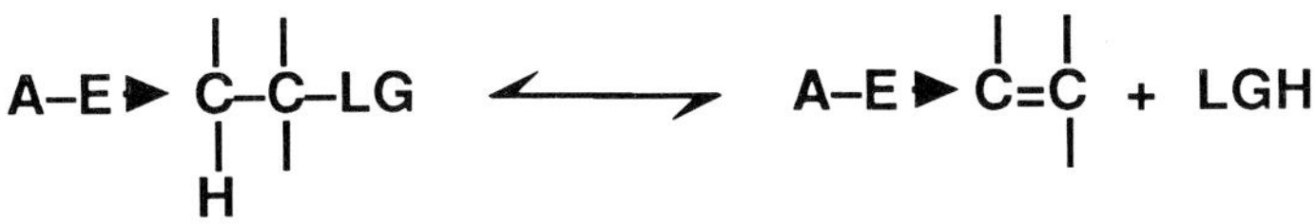

A = resonance-increasing group
E = electronegative group
LG = leaving group, e.g. halogen, N<R R, ⊕N (R)$_3$, S–R, ⊕S (R)$_2$, ⊕P (R)$_3$

Fig. 6. Cleavage of potential microbicidal vinyl compounds.

Cleavage of the potential vinyl compounds (elimination of the leaving group) to produce the active vinyl substances takes place even under physiological conditions (room temperature and approximately neutral pH). In general, the rate of cleavage is many times higher than those of the reactions leading to inactivation of the vinyl compound produced in the cleavage process. This represents an advantage in the practical use of potential vinyl compounds as microbicides. A biokinetic test produced a curve (Fig. 7) illustrating the above statements (Paulus, 1976).

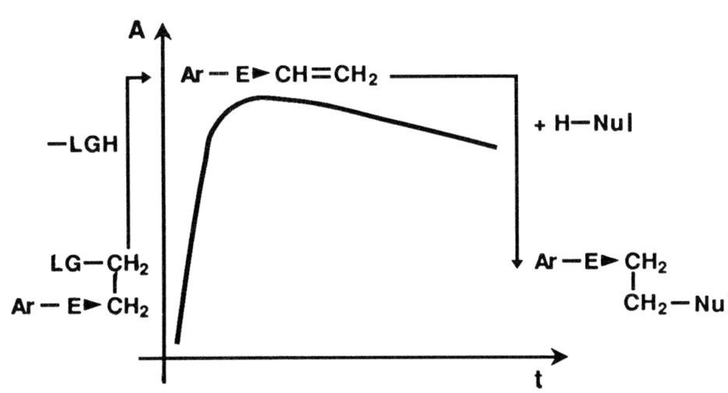

LG = leaving group
E = electronegative group
H–Nul = nucleophilic entities of the microbial cell

Fig. 7. Antimicrobial activity (*A*) of an aqueous solution of a potential aryl vinyl ketone dependent on time (*t*) at pH 6·5.

Microbicides with an activated N-S bond as toxophoric structural element

Activated N-S bonds are found in substances such as those containing the isothiazolinone ring (III.12) and in *N*-haloalkylthio compounds (III.14) (see Fig. 8). The isothiazolinone ring reacts with nucleophilic cell entities such as amino acids, proteins and enzymes, with opening of the ring, thus exerting antimicrobial activity (see Fig. 9). The antimicrobial activity of N-haloalkylthio compounds is likewise based on the ability of the N-S bond to open and react with nucleophilic entities such as SH groups. Such reactions apparently proceed by way of several steps and lead to disulphides. The course of the reaction can be summarized as shown in Fig. 10 for N-trihalomethylthio derivatives.

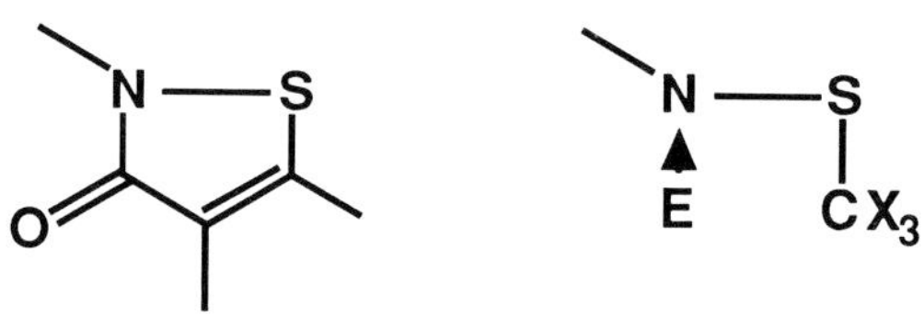

E = electronegative group
X = halogen

Fig. 8. Activated N—S bonds.

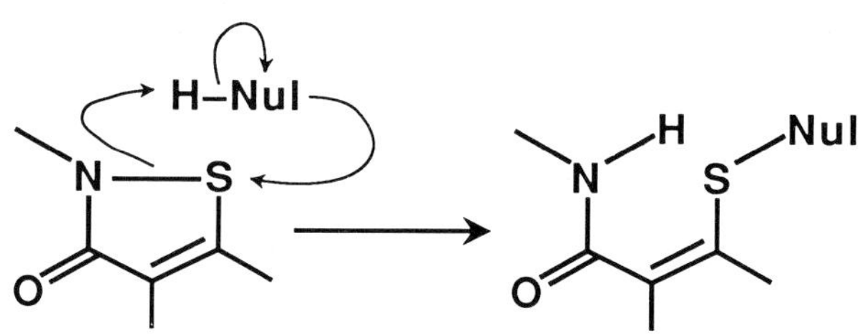

H–Nul = nucleophilic entities of the microbial cell

Fig. 9. Mechanism of activity of the isothiazolinone structure.

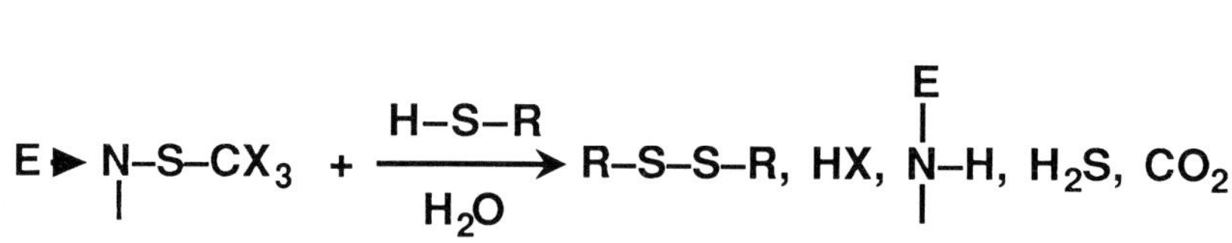

E = electronegative group
X = halogen

Fig. 10. Reaction of *N*-trihalomethylthio derivatives with nucleophilic compounds.

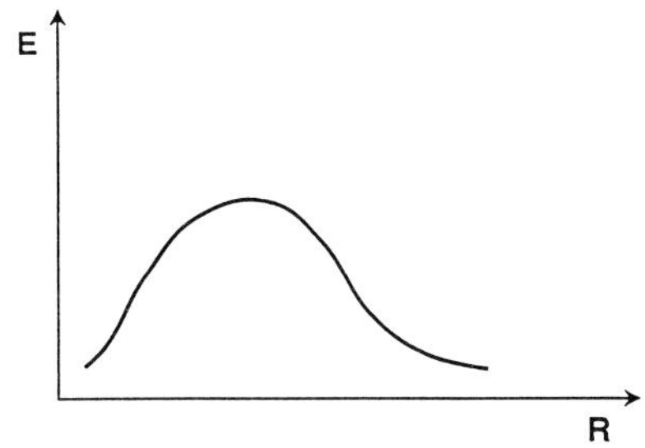

Fig. 11. Connection between effectiveness (E) and reactivity (R) of electrophilically active microbicides.

The effectiveness and efficiency of microbicides result from the interplay of the chemicophysical properties of the active substance molecule, which are determined by the molecule's constitution. Solubility, polarity, ionicity and reactivity are examples of properties that influence effectiveness. Taking the N-trihalomethylthio derivative as examples, Paulus & Kühle (1986) drew attention to an important principle that evidently applies to electrophilically active microbicides in general, namely that the relationship shown in Fig. 11 exists between the antimicrobial effectiveness and reactivity of microbicide molecules. As the reactivity increases, so, too, does the effectiveness—until it peaks at a moderately high reactivity level. Thereafter the antimicrobial activity decreases as the reactivity continues to rise because at this stage competition reactions — interactions of the reactive microbicide molecules with constituents of the surrounding medium — predominate.

Microbicidal organometallic compounds (III. 17)

Organometallic compounds such as phenylmercury acetate (PMA) and tributyltin fluoride (TBTF) are classified as electrophilically active microbicides

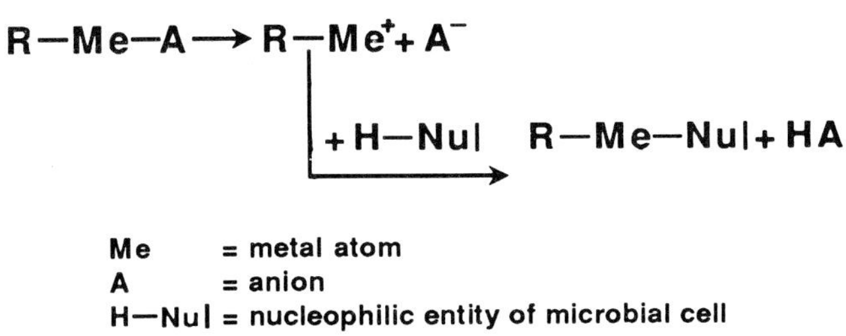

Fig. 12. Reaction of organometallic compounds with nucleophilic sites (H-Nu) of the microbial cell. In this way organometallic compounds interfere with a large number of cell processes. For example, the reaction with essential thiol groups leads to the inhibition of enzymes. However, amines, amides, amino acids and sulphur-free proteins are nucleophilic reaction partners for organometallic compounds, too; the corresponding reactions cause disturbances of many kinds in the metabolism of the microbe cell.

because they can form corresponding cations, e.g. Ph-Hg^+ or $(butyl)_3Sn^+$, as active species, which, as shown in Fig. 12, are able to react with nucleophilic cell constituents.

MICROBICIDAL ACTION OF HEAVY METALS

The prerequisite for the toxic effect of heavy metals is penetration of the cell wall. For this, certain carriers are needed to transport the heavy metals through lipophilic centres in the cell membrane. Oxines, 2-mercapto-pyridine-N-oxides (III.11) and dialkyl-dithiocarbamates (III.9) are examples of compounds capable of complexing heavy metal ions such as Cu^{2+} ions and transporting them into the microbe cell. The carriers named are also toxic, as such, to micro-organisms. But, as a copper carrier, copper-8-hydroxyquinoline is about 100 times more effective than 8-hydroxyquinoline (oxine III.11.4). Once heavy metal ions have overcome the cell wall with the aid of the carrier molecules, they may compete with magnesium, calcium and potassium ions for receptor sites, inhibit enzymes and cause non-specific precipitation of proteins.

MICROBICIDAL ACTIVITY BY CHELATING AGENTS

The aforementioned carriers for heavy metal ions exert their antimicrobial effects in the microbe cell even if they are not accompanied by these ions. As chelating ions they can complex metal cations that are essential for the functioning of the cell metabolism, thus disturbing it. The mode of action of alkylene bisdithiocarbamates, e.g. zineb, nabam (III.9.9), is, however, reported to be fundamentally different from that of the dialkyl dithiocarbamates. It is assumed that metabolites of the alkylene bisdithiocarbamates react with cellular thiol groups (Corbett *et al.*, 1984).

FUNGICIDAL INHIBITORS OF STEROL BIOSYNTHESIS

A new generation of acceptable microbicides exists in the shape of the so-called azole fungicides (III.12). They include fungicides of the triazole type, such as azaconazole (III.12.3), propiconazole (III.12.2) and tebuconazole (III.12.1), fungicides of the imadazole type, such as imazalil (III.12.4), and others also, e.g. tridemorph. Common to all of them is the ability to block the biosynthesis of ergosterol. Ergosterol is the main sterol in many fungus species. It plays an important part in the formation and functioning of the cell membrane. Azole fungicides inhibit the formation of ergosterol by preventing the conversion of lanosterol to ergosterol (Kato, 1986). They do so by inhibiting the enzyme C-14 demethylase, which normally catalyses the demethylation of lanosterol. As the C-14 demethylation is a reaction dependent on cytochrome P-450, Gadher *et al.* (1983) assume that azole fungicides occupy the sterol binding site of the enzyme competitively, for example, in that the meta-N atom of the azole ring enters into a coordinate bond with the protohaem iron of cytochrome P-450. In addition it is possible that the lipophilic part of azole fungicides forms hydrophobic and hydrogen bridge bonds with cytochrome P-450. The final consequence is retardation of fungal growth at lower concentrations of active ingredient. As, however, the azole derivatives also have cationogenic, surface active properties, they act at higher concentrations than as membrane-active microbicides.

We have seen that the microbicidal effect of an active ingredient depends on many factors and that microbicides exert their effects according to various mechanisms. The microbicidal concentration ($C_{\text{eff.}}$) of an active substance is likewise not an absolute value, varying as it does according to the conditions the substances encounters, such as the medium, pH value, temperature and microbe density. Under uniform conditions, however, microbicides can be characterized by $C_{\text{eff.}}$. In practical material protection with the aid of microbicides it must nevertheless be remembered that the microbicide concentration — especially where reactive microbicides are concerned — naturally decreases in time and that, as may be seen in Fig. 13, the steeper the slope α of the straight line, the faster this happens. With reactive microbicides, for example, those that are electrophilically active, α is only a characteristic quantity if it is determined under comparable conditions. The measurements obtained in different media and at different states of the medium (temperature, pH value) vary. Yet, if the slope acquired by α under practical conditions is known, one can predict when the microbicide concentration will fall below $C_{\text{eff.}}$; in other words predictions as to the activity duration ($T_{\text{eff.}}$) of a given microbicide concentration higher than $C_{\text{eff.}}$ are possible.

As already stated, microbicides exert their effects through various mechanisms, which attack different targets, and in some several targets, on and in the microbe cell. Depending on their properties, the microbicides find it easier or less easier to reach the target sites. Furthermore, depending on the microbe

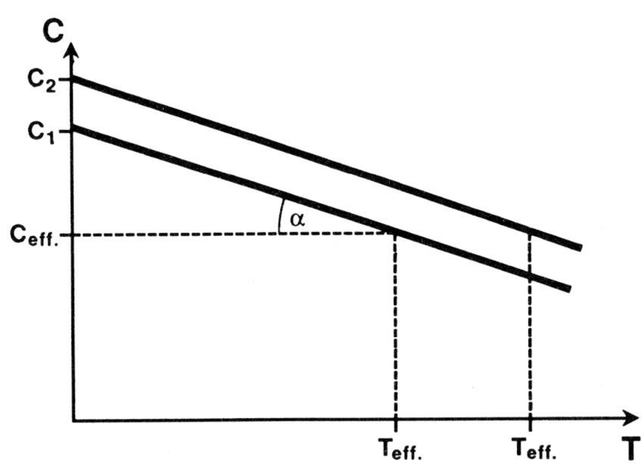

Fig. 13. Dependence of the activity duration ($T_{\text{eff.}}$) of a microbicide, whose reactivity is characterized by the slope α, on the microbicide concentration C_1 or C_2.

species, they encounter a variety of hindrances. The outer membrane of gram-negative bacteria, which makes the target sites inaccessible to certain microbicides, has been mentioned already. Vegetative cells of certain microbes (sporogenic species) are capable of producing spores which, when mature, are set free after lysis of the mother cell. They have an extremely resistant external envelope that enables them to resist both chemical and physical influences. Spores may be described as a form of latent life in which microbes can withstand great heat, and also — for almost unlimited periods — dryness. The resistance of spores to chemicals such as organic solvents and many microbicides is remarkable.

In an appropriate medium, and under conditions more favourable to them, spores will germinate, i.e. pass from dormancy to a metabolically active state. Finally a cell capable of dividing grows, vegetative cells are formed and multiplication and sporulation occur again.

Although the spore envelope is enormously resistant to certain conditions and virtually metabolically inert, sporicidal microbicides exist, i.e. those that damage the spores so much that they can no longer be activated and germinated (Cook & Pierson, 1983). Sporicidally active microbicides are generally highly reactive substances, examples being aldehydes (III.2), especially glutaraldehyde and formaldehyde, which, as electrophilically active substances, attack the spore both at the surface and, in consequence of their penetrating power, internally (Russel, 1983). Oxidizing agents (III.19) — chlorine, hypochlorites and other chlorine releasing agents, for example — and also peroxides, cause external and internal oxidative changes in spores, by which these are destroyed (Bloomfield & Uso, 1985). Alkylating agents of high reactivity, such as ethylene oxides and β-propiolactone, also exert sporicidal effects — ethylene oxide as a gaseous active substance and β-propiolactone as a liquid-phase and vapour-phase sporicide.

Many microbicides are sporistatic in that they inhibit spore germination and outgrowth without killing the spore. They include membrane-active substances, such as alcohols (III.1), phenol derivatives (III.5), esters of *p*-hydroxy-benzoic acid (III.6) and quaternary ammonium salts (III.16), and, at low concentrations, also the above-mentioned aldehydes, formaldehyde and glutaraldehyde, which at higher concentrations are sporicidal. The sporistatic effects of the microbicides just mentioned, especially of those that are membrane-active, are reversible, however, because the binding of the microbicides to the spore surface is relatively weak: it can be negated by diluting the surrounding medium or washing the spores.

The individual sensitivity and resistance of microbes to microbicides, which depend on the composition and structure of the outer cell layers, are termed intrinsic resistance in contradistinction to acquired resistance, which is a consequence of the selection pressure exerted on a microbe population in the presence of microbicides. The appearance of resistant strains within a microbe population that was initially sensitive to an applied microbicide is a consequence of mutations and selection of the resistant mutants.

Acquired resistance is important in chemotherapy with antibiotics, where it causes serious difficulties. In plant protection with systemic active substances difficulties are also caused by acquired resistance through the appearance and selection of strains resistant to the applied microbicide concentrations which are limited by the plant's system. In chemotherapy with antibiotics it is the system of humans or animals which limits the concentrations of active ingredients. As corresponding limitations of microbicide concentrations in material protection are not generally necessary, acquired resistance has practically no importance here. Quite the opposite is true of intrinsic resistance; this must be taken into account whenever microbicides are used for material protection.

Intrinsic resistance is possessed by, for example, gram-negative bacteria because the outer membrane prevents a large number of microbicides from reaching the target sites in the cell. It is not simply that active substance molecules with molecular weights above 600 cannot pass through the porin channels of the outer membrane (many microbicides have lower molecular weights). The ability of the outer membrane to function as an exclusion barrier is also explained by the hydrophilicity of the cell surface of many gram-negative bacteria. This prevents the passage of hydrophobic active substances through the aqueous pore system into the cell. The minimum inhibition concentrations of various phenol derivatives with molecular weights below 600, but differences in degree of hydrophobicity, make this clear (see Table 1). On the other hand, improved lipoid solubility improves the effectiveness of microbicides towards the gram-positive bacterium *Staphylococcus aureus*.

The waxy cell wall of mycobacteria is an example of a hydrophobic barrier to hydrophillic microbicides.

The minimum inhibition concentrations in Table 1 also show that, by virtue of intrinsic resistance, *Escherichia coli*, is less sensitive than the gram-positive bacterium *Staphylococcus aureus*, but more sensitive than Pseudomonades. The phenomenal resistance of *Pseudomanas aeruginosa* to quaternary ammonium salts (III.16) enables this non-sporing bacterium to survive in solutions of QACs. This intrinsic QAC resistance is attributed to the special composition of the

Table 1. Sensitivities of Gram-negative and Gram-positive Bacteria to Phenol Derivatives Arranged in a Succession of Increasing Degrees of Hydrophobicity — Minimum Inhibition Concentrations (mg/litre) in Nutrient Agar

Phenol derivative	M_r	*Test organisms*			
		Pseudomonas aeruginosa	Pseudomonas fluorescens	Escherichia coli	Staphylococcus aureus
2-methyl-4-chlorophenol (III.5.3.1)	142·5	800	800	250	200
2-phenyl-phenol (III.5.4.1)	170·1	1500	1500	200	100
Chlorophen (III.5.3.5)	218·5	5000	>5000	3500	20
Dichlorophen (III.6.7.3)	269·1	>5000	3500	100	5

lipopolysaccharides in the outer membrane; evidently the cationogenic QAC reacts with anionic groups of the lipopolysaccharide, whereby it is inactivated.

The ability of bacteria to form biofilms by means of exopolysaccharide glycocalyx polymers may also be looked upon as a form of intrinsic resistance to microbicides. These polymers cause the bacteria to adhere irreversibly to surfaces. For certain microbicides the biofilm represents a penetration barrier within which the microorganisms are protected. In some instances biofilm formation explains the failure of microbicides which at certain concentrations have definitely proved effective against the microbes in free cell cultures.

The resistance of certain *Pseudomonas putida* strains to formaldehyde and formaldehyde-releasing compounds is a case of intrinsic resistance attributable to the presence of the constitutive enzyme formaldehyde dismutase in these microbes (Adroer *et al.*, 1990). The stoichiometric dismutation of formaldehyde to methanol and formic acid, known to chemists as Cannizaro's reaction, occurs under physiological conditions in microbe cells in which formaldehyde dismutase is present. Furthermore, if formaldehyde is supplied, this promotes in the cells the formation of formaldehyde dehydrogenase, an enzyme that catalyses the conversion of formaldehyde to formic acid (Kato *et al.*, 1984).

PART III

Substance classes: Properties* — Effectiveness — Applications

**Sources for toxicity data, if not otherwise indicated:*
Registry of toxic effects of chemical substances (1990). Department of Health and Human Services, Cincinnati, USA.
Lexikon der Hilfsstoffe für Pharmazie, Kosmetik und angrenzende Gebiete *(1989), Ed. H. P. Fiedler, Editio Cantor, Aulendorf, Germany.*
Product information and safety data sheets of microbicide suppliers.
All indications are given in good faith and conscience. This also applies for the 'maximum working place concentrations' (MWC), the 'acceptable daily intake' (ADI) values and to the data with regard to ecotoxicity and biodegradability of microbicides which, as far as accessible, were also incorporated.

1 ALCOHOLS

Microbicidal alcohols are, as a general rule, colourless volatile liquids. In numerous fields of application they thus offer the advantage of being effective without leaving residues.

Alcohols rank among the membrane-active antimicrobial agents; they are adsorbed at the cytoplasmic membrane which — as a multi-purpose instrument of the microbe cell — is particularly susceptible to functional disturbances. Membrane proteins form, for example, parts of enzymes that catalyse the transformation/decomposition of substances so that metabolites important for the cells' vegetation may form and pass the cellular membrane. The denaturation of membrane proteins by alcohols necessarily disturbs those processes and is the main reason for the alcohols' antimicrobial effect, which is limited to vegetative microorganisms. Spores will not be destroyed. The virucidal efficacy is confined to a few viri. Alcohols are characterized by acting at a quick rate but only in comparatively high concentrations, preferably in the presence of water. Absolute ethanol for instance does have a dehydrating effect, but a lower denaturing effect than an aqueous ethanol solution. Gram-negative bacteria are more sensitive to alcohols than Gram-positive ones. By first approximation it can be said that the effectiveness of alcohols increases with their molecular weight. More complex is the mechanism of action of some unsaturated iodinated alcohols.

Whilst lower alcohols are chiefly employed for disinfecting hands and surfaces, higher alcohols lend themselves to use in preservatives — in cosmetics and pharmaceuticals rather than in industrial fluids.

1.1 Ethanol — Ethyl alcohol

CH_3-CH_2-OH C_2H_6O M_r 46·07
CAS-No. 64-17-5

Chemical and physical properties

Appearance	Clear, colourless, burnable fluid
Density at 20°C	0·79 g/ml
Viscosity at 20°C	1·2 mPa s
Surface tension at 20°C	22·8 dyn/cm
Vapour pressure at 20°C	59 hPa
Refractive index (25°C)	n_D 1·3595
Solidification point	–114·5°C
Boiling point	78·5°C
Flash point	18·9°C (100%) 21·1°C (95%)
Ignition temperature	425°C
Solubility	Miscible with water, alcohols, acetone, ether, trichloromethane, benzene
Stability	Volatile, hygroscopic

Toxicity data

LD_{50} oral	13 700 mg/kg rat 5500 mg/kg guinea pig 9500 mg/kg rabbit
LD_{50} intravenous	4200 mg/kg rat 2300 mg/kg guinea pig
MWC	1000 ml/m^3 (1900 mg/m^3)

Antimicrobial effectiveness/applications

Table 2. Minimum Inhibition Concentrations (MIC) of Ethanol in Nutrient Agar

Test organism	*MIC (%)*
Escherichia coli	11
Pseudomonas aeruginosa	7·5
Aspergillus niger	6
Rhizopus nigricans	6

Table 2a. The Killing Action (Exposure Time in Seconds) of Various Concentrations of Ethanol for Various Test Organisms (Wallhäußer, 1984)

Test organism	*Ethanol concentration (%)*			
	60	*70*	*80*	*95*
Staphylococcus aureus	15	15	10	
Staphylococcus epidermidis	30	30		
Streptococcus pyogenes			90	
Escherichia coli	60	30	30	
Serratia marcescens		10		
Salmonella typhosa		10		
Pseudomonas aeruginosa		10		
Mycobacterium tuberculosum	60	30	30	30
Spores of *Trichophyton gyps.*			30 min	

The presence of water is essential for the antimicrobial effectiveness of ethanol. Concentrations of 60–70% ethanol in water exhibit the strongest killing action (Price, 1950). Non-sporulating bacteria are rapidly killed by such concentrations, while ethanol at all concentrations does not affect bacterial spores (Russel, 1971).

Ethanol is mainly used in disinfectants for surfaces, instruments and for the skin, e.g. in hand disinfectants. Sometimes ethanol is applied as a preservative, e.g. in pharmaceutical and cosmetic products, where odourless and non-irritating preservatives are preferred.

1.2 *n*-Propanol — Propyl alcohol

CH_3-CH_2-CH_2-OH C_3H_8O M_r 60·1
CAS-No. 71–23–8

Chemical and physical properties

Appearance	Clear, colourless, burnable fluid with a slight alcoholic odour
Density at 20°C	0·8035 g/ml
Viscosity at 20°C	2·2 mPa s
Surface tension at 20°C	23·8 dyn/cm
Vapour pressure at 20°C	18·7 hPa
Refractive index (20°C)	n_D 1·3853
Solidification point	−127°C
Boiling point	97·4°C
Flash point	22°C
Ignition temperature	405°C
Solubility	Miscible with water, alcohols, ketones, ether
Stability	Volatile

Toxicity data

LD_{50} oral	5400 mg/kg rat
LD oral	5700 mg/kg man
Inhalation LD for rat	4000 ppm (4 h)
MWC	400 ml/m^3 (980 mg/m^3)
LC_{50} for fish	4560 mg/litre

Antimicrobial effectiveness/applications

Compared with ethanol, *n*-propanol is effective at lower but still relatively high concentrations, e.g. 50–60% for hand disinfection. The presence of water is essential for the efficacy.

n-Propanol is a suitable active ingredient in disinfectants, e.g. for the hands; it may also be used as a preservative for cosmetic products.

1.3 Isopropanol — Isopropyl alcohol

CH_3-CH-CH_3 (with OH on the central carbon) C_3H_8O M_r 60·1
CAS-No. 67-63-0

Chemical and physical properties

Appearance	Clear, colourless, flammable fluid with an alcoholic odour
Density at 20°C	0·785 g/ml
Viscosity at 20°C	2·27 mPa s
Surface tension at 20°C	22·6 dyn/cm
Vapour pressure at 20°C	43 hPa
Refractive index (20°C)	n_D 1·3771
Solidification point	−89°C
Boiling point	82·4°C
Flash point	22·2°C
Ignition temperature	425°C
Solubility	Miscible with water, alcohols, ketones, ether
Stability	Volatile

Toxicity data

LD_{50} oral	5850 mg/kg rat
LD oral	8600 mg/kg man
MWC	480 ml/m^3 (980 mg/m^3)
LC_{50} for fish	8970 mg/litre

Antimicrobial effectiveness/applications

The presence of water is essential for the activity of isopropanol, too. Most effective are concentrations of approx. 50%. Isopropanol and *n*-propanol are the highest alcohols which are miscible with water. The sporicidal effectiveness of isopropanol is poor.

Isopropanol may be used in hand disinfectants at concentrations of 60–70% and as a preservative for cosmetics.

1.4 Phenyl methanol — Benzyl alcohol

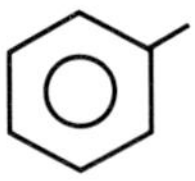

CH_2-OH C_7H_8O M_r 108·14
CAS No. 100-51-6

Chemical and physical properties

Appearance	Clear, colourless fluid with an aromatic odour
Density at 20°C	1·5 g/ml
Surface tension at 20°C	38·8 dyn/cm
Vapour pressure at 20°C	0·13 hPa
Refractive index (21·9°C)	n_D 1·5373
Solidification point	−15·3°C
Boiling point	205°C
Flash point	94°C
Ignition temperature	435°C
Solubility	In water: 38 g/litre; soluble in alcohols, ketones, ether, aromatic solvents, oils
Stability	Slow oxidation in contact with oxygen to benzaldehyde

Toxicity data

LD_{50} oral	1230 mg/kg rat
	1580 mg/kg mouse
	1940 mg/kg rabbit
LD oral	500 mg/kg man

Benzyl alcohol is also percutaneously toxic (Wollmann *et al.*, 1963).

Antimicrobial effectiveness/applications

Benzyl alcohol exhibits a broad spectrum of effectiveness which covers bacteria, yeasts and moulds. However, the activity against moulds is more intense than against bacteria.

Benzyl alcohol is listed in the EC positive list of preservatives for cosmetic products (maximum concentration for application: 10 000 mg/litre). The activity of benzyl alcohol is not very much affected by the pH and the composition of the medium to be protected. As an auxiliary solvent with antimicrobial efficacy benzyl alcohol is used in preservative compositions for industrial fluids (Paulus *et al.*, 1970*a*). A well-known preservative for cosmetics and industrial fluids is benzyl alcohol mono(poly)hemiformal (Paulus, 1976) which is a formaldehyde releasing compound and therefore listed in Section 3.1.1.

Table 3. Minimum Inhibition Concentrations (MIC) of Benzyl Alcohol in Nutrient Agar

Test organism	*MIC (mg/litre)*
Aerobacter aerogenes	5000
Escherichia coli	4000
Pseudomonas aeruginosa	3000
Pseudomonas fluorescens	6000
Staphylococcus aureus	5000
Formaldehyde resistant bacteria	5000
Candida albicans	3500
Candida crusei	6000
Aureobasidium pullulans	1500
Chaetomium globosum	2500
Trichoderma viride	4500
Trichophyton mentagrophytes	3000

1.5 (2,4-Dichloro-)phenyl methanol — 2,4-Dichloro benzyl alcohol

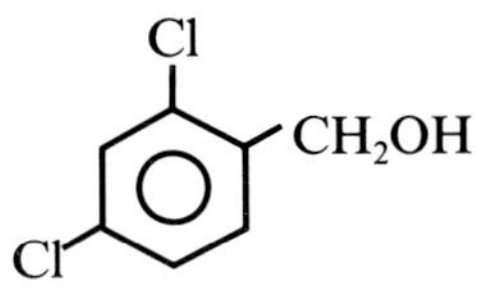

$C_7H_6Cl_2O$

M_r 176·95
CAS-No. 1777-82-8

Chemical and physical properties

Appearance	White to yellow crystals
Melting point	55–58°C
Boiling point at 20 hPa	150°C
Flash point	>110°C
Solubility	In water: approx. 1 g/litre; soluble in alcohols, ketones, ether
Stability	Oxidisable to 2,4-dichloro benzaldehyde

Toxicity data

LD_{50} oral	2300 mg/kg mouse
LD_{50} subcutaneous	1700 mg/kg mouse

2,4-Dichloro benzyl alcohol does not cause skin irritation and sensitization, is not mutagenic and not teratogenic (Wallhäußer, 1984).

Antimicrobial effectiveness/applications

Table 4. Minimum Inhibition Concentrations (MIC) of 2,4-Dichloro Benzyl Alcohol in Nutrient Agar

Test organism	*MIC (mg/litre)*
Escherichia coli	500
Pseudomonas aeruginosa	1500
Staphylococcus aureus	500
Aspergillus niger	500
Chaetomium globosum	200
Penicillium glaucum	500
Candida albicans	500

2,4-Dichloro benzyl alcohol is much more effective than benzyl alcohol (see Table 3) and additionally exhibits a very equalized spectrum of effectiveness.

The solubility of 2,4-dichloro benzyl alcohol in water is poor; in aqueous systems it tends to migrate into the organic phase. The use of 2,4-dichloro benzyl alcohol as a preservative therefore is very limited. Nevertheless it is listed in the EC positive list of preservatives for cosmetic products (maximum concentration for the applications: 0·15%). Its main application, however, is in hand disinfectants.

1.6 2-Phenyl ethanol — Phenylethyl alcohol

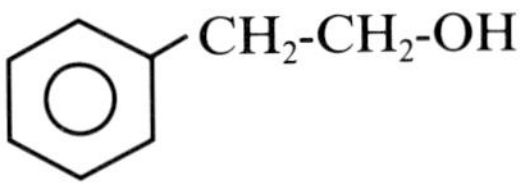

$C_8H_{10}O$ M_r 122·17
CAS-No. 60-12-8

Chemical and physical properties

Appearance	Clear, colourless fluid with the odour of roses
Density at 20°C	1·020 g/ml
Refractive index (20°C)	n_D 1·532
Solidification point	−27°C
Boiling point	215–217°C
Solubility	In water: approx. 20 g/litre; miscible with alcohols, ketones, ether, soluble in fatty oils, poor solubility in mineral oils
Stability	Sensitive to oxidizing agents

Toxicity data

LD_{50} oral	1790 mg/kg rat
LD_{50} cutaneous	5–10 ml/kg guinea pig

Antimicrobial effectiveness/applications

Table 5. Minimum Inhibition Concentrations (MIC) of 2-Phenyl Ethanol in Nutrient Agar

Test organism	*MIC (mg/litre)*
Escherichia coli	2500
Pseudomonas aeruginosa	3500
Staphylococcus aureus	4750
Aspergillus niger	2750
Chaetomium globosum	2000
Penicillium glaucum	2750
Candida albicans	2500

As the effectiveness against bacteria is weak, weaker than the activity against moulds, 2-phenyl ethanol is in general applied in combination with other microbicides, e.g. with *p*-hydroxy-benzoic acid esters (Section 6.1.10), quaternary ammonium compounds (Section 16·1), *p*-chloro-*m*-cresol (Section 5.3.1) for the preservation of cosmetic and pharmaceutical products. It is especially effective in acidic media.

1.7 2-Phenoxy ethanol — Phenoxyethyl alcohol

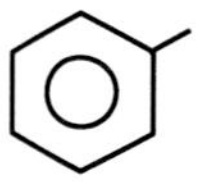
O-CH_2-CH_2-OH $C_8H_{10}O_2$ M_r 138·17
CAS-No. 122-99-6

Chemical and physical properties

Appearance	Clear, colourless, slightly viscous fluid with a pleasant odour
Density at 20°C	1·107 g/ml
Refractive index (20°C)	n_D 1·538
Melting point	11–13°C
Boiling point at 14 hPa	134–135°C
Solubility	In water: approx. 26 g/litre; miscible with alcohols, ketones, ether

Toxicity data

LD_{50} oral	1300 mg/kg rat

Does not cause skin irritation.

Antimicrobial effectiveness/applications

Table 6. Minimum Inhibition Concentrations (MIC) of 2-Phenoxy Ethanol in Nutrient Agar

Test organism	*MIC (mg/litre)*
Escherichia coli	4000
Pseudomonas aeruginosa	4000
Staphylococcus aureus	2000
Candida albicans	4000
Aspergillus niger	4000
Chaetomium globosum	2000
Penicillium glaucum	4000
Rhizopus nigricans	3000

2-Phenoxy ethanol is mentioned in the EC positive list of preservatives for cosmetic products (maximum authorized concentration: 1%) In general it is applied in combination with other preservatives, such as *p*-hydroxy benzoic acid esters (Section 6.1.10), 1,2-dibromo-2,4-dicyanobutane (Section 15.11) and others. More effective is the 2-phenoxy ethanol hemiformal (Section 3.1.3).

1.8 1-Phenoxypropan-2-ol

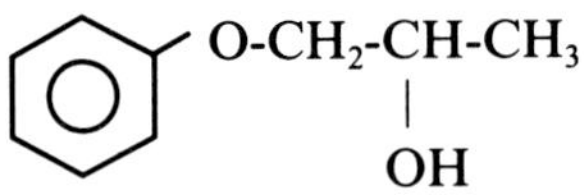

$C_9H_{12}O_2$ M_r 152·2
CAS-No. 770-35-4

Chemical and physical properties

Appearance	Clear, colourless fluid
Boiling point	233°C
Density at 20°C	1·062 g/ml
Refractive index (20°C)	n_D 1·52397
Solubility	Practically insoluble in water, soluble in polar and non-polar organic solvents

Toxicity data

According to the indications of Fiedler (1989) the substance is not irritant to the skin and mucous membranes.

Antimicrobial effectiveness/applications

Because of its broad spectrum of effectiveness and its favourable dermatological properties 1-phenoxypropan-2-ol is mentioned in the EC list of preservatives provisionally allowed for the protection of cosmetics with a maximum authorized concentration of 1% and for rinse-off products only.

1.9 3-Iodopropargyl alcohol — 3-Iodo-2-propin-1-ol (IPA)

I-C≡C-CH_2-OH	C_3H_3IO	M_r 165·97 CAS-No. 1725-82-2

Chemical and physical properties

Appearance	Yellowish solid of an unpleasant odour
Melting point	41–42°C
Solubility	In water 1·5 g/litre; highly soluble in polar and aromatic solvents
Stability	Not stable in hot water: partial transformation to 1,1,2-triiodo-propene(1)-3-ol which exhibits strong antimicrobial activity, too

Toxicity data

LD_{50} oral	140–170 mg/kg rat

Irritant and corrosive to skin and mucosa.
Non-mutagenic (Salmonella microsome test).

Antimicrobial effectiveness/applications

IPA exhibits an extremely broad activity spectrum covering fungi, yeasts and bacteria. However, IPA is approximately 10 times more effective against fungi and yeasts than against bacteria (see Table 7).

3-Iodopropargyl derivatives (Sections 9.1–9.3 and 3.1.5), e.g. 3-iodopropargyl-N-butylcarbamate are used mainly in fungicidal finishing of materials; but the less complex compound, IPA, although very effective, is unsuitable for that, being volatile and easily leached. IPA has also acquired no appreciable importance as a preservative. Its high price in comparison with other preservatives, an irritating effect on the skin, troublesome odour and a toxicity profile which is, on the whole, unfavourable, have prevented this.

Table 7. Minimum Inhibition Concentrations (MIC) of IPA in Nutrient Agar

Test organism	*MIC (mg/litre)*
Alternaria alternata	5
Aspergillus niger	5
Aureobasidium pullulans	5
Chaetomium globosum	5
Cladosporium cladosporioides	2
Coniophora puteana	1
Lentinus tigrinus	5
Penicillium glaucum	5
Polyporus versicolor	2
Sclerophoma pityophila	2
Trichoderma viride	10
Candida albicans	20
Candida krusei	20
Rhodotorula mucilaginosa	10
Rhodotorula glutinis	15
Sporobolomyces roseus	15
Torula rubra	10
Escherichia coli	100
Formaldehyde resistant bacteria	15
Staphylococcus aureus	150

1.10 2,3,3-Tri-iodallyl alcohol (TIAA)

$I_2C{\equiv}CI\text{-}CH_2\text{-}OH$ $C_3H_3I_3O$ M_r 435·77
CAS-No. 42778-72-3

Chemical and physical properties

Appearance	Pale yellowish to greyish brown crystalline powder
Melting point	150–152°C
Solubility	Highly soluble in dimethyl sulphoxide, dimethyl formamide, dioxane, cyclohexane; moderately soluble in methanol, ethanol, acetone, ethyl acetate, isopropanol; slightly soluble in ethylene glycol, isobutanol; hardly soluble in water

Toxicity data

LD_{50} oral >5000 mg/kg mouse

Antimicrobial effectiveness/applications

TIAA performs as a broad spectrum microbicide which may be used in anti-mould and anti-sapstain agents and in wood preservatives because of its extraordinary efficacy against wood destroying fungi (Lee *et al.*, 1990).

Table 8. Minimum Inhibition Concentrations (MIC) of Tri-iodallyl Alcohol in Nutrient Agar

Test organism	*MIC (mg/litre)*
Alternaria alternata	1
Aspergillus niger	3·5
Aureobasidium pullulans	1
Chaetomium globosum	7·5
Coniophora puteana	0·5
Lentinus tigrinus	1
Penicillium glaucum	2
Polyporus versicolor	2
Sclerophoma pityophila	2
Trichoderma viride	10
Escherichia coli	50
Staphylococcus aureus	10
Slime bacteria[a]	2·5

[a] Described by Kato & Fukumura (1962).

2 ALDEHYDES

Aldehydes belong to the group of electrophilic active agents, which due to the electron deficiency at the carbonyl carbon atom can react with nucleophilic cell entities and thus exert antimicrobial activity (Fig. 14).

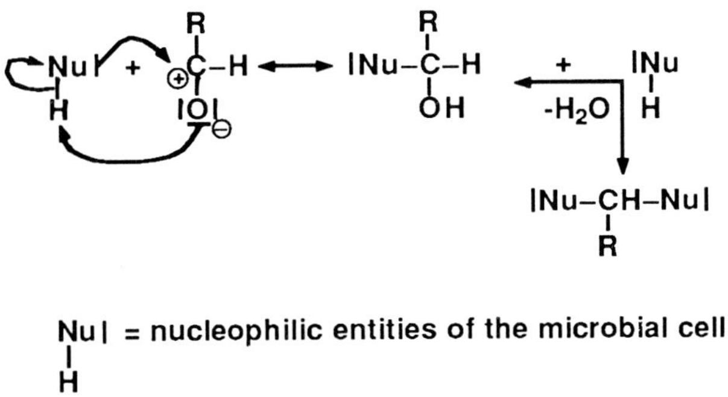

Fig. 14. Mechanism of activity of aldehydes.

Examples of nucleophilic reaction partners in the cell are amino and thiol groups, as well as the amide groups of amino acids or proteins (Fig. 15). These in turn are components of enzymes, which are inactivated by the reaction of their nucleophilic groups with aldehydes.

Dialdehydes have two toxophoric groups which are responsible for the antimicrobial effect, i.e. they can react with nucleophilic centres of the microbe cell; naturally there is a possibility of manifold connecting reactions. Formaldehyde

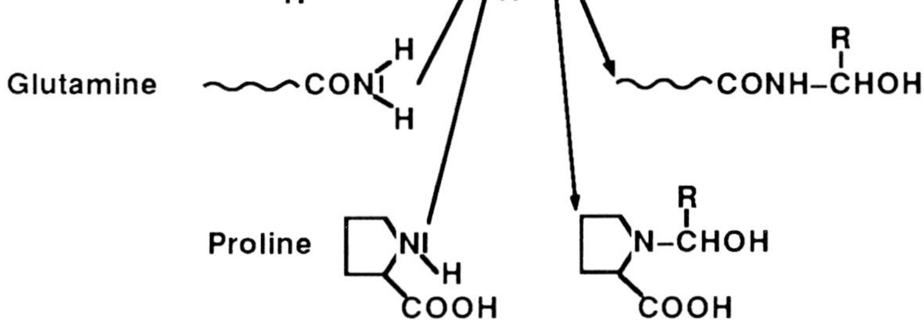

Fig. 15. Reaction of aldehydes with amino acids.

is the most reactive and thus most effective among monoaldehydes. Glutaric aldehyde is the most effective dialdehyde. Other dialdehydes such as glyoxal (see Section 2.2), malonic aldehyde, succinic aldehyde and adipic aldehyde have not attained appreciable significance as microbicides although possessing sporicidal effects. Higher aliphatic aldehydes with molecular weights higher than that of the adipic aldehyde display no noticeable microbicidal effect. Aromatic mono- and dialdehydes, too, exhibit microbicidal effectiveness. Up to now these aldehydes have not gained importance in practice, although several authors emphasize their utility as preservatives and active ingredients in sterilizing and disinfecting compositions (Bruckner *et al.*, 1988; Maddox, 1988; Sharma, 1989).

Because of the mechanism mentioned above the aldehydes with microbicidal effect are distinguished by broad spectra of effectiveness which also cover viri and spores. Since only few sporicides are found among microbicides, the sporicidal effect of aldehydes, particularly that of glutaric aldehyde is of special interest. Aldehydes are above all used for disinfecting purposes; but also for preservation, then chiefly in the form of so-called aldehyde releasing compounds (see Sections 3 and 4).

2.1 Formaldehyde — Methanal

$O{=}CH_2$	CH_2O	M_r 30·03 CAS-No. 50-00-0

Chemical and physical properties

Appearance	Reactive, colourless, inflammable gas of pungent odour; the gas obeys the ideal gas laws
Solidification point	−118°C
Boiling point	−19°C
Solubility	Soluble in H_2O, forming formalin, soluble in alcohols, forming hemiformals, soluble in polar solvents
Stability	The dry gas tends to polymerize slowly; as a reductive agent formaldehyde is easily oxidized, e.g. by hydrogen peroxide, iodine, sodium permanganate; it reacts with ammonia and proteins under inactivation; the influence of alkali leads to disproportionation of formaldehyde to methanol and formic acid

2.1(a) Paraformaldehyde — Polyoxymethylene

$HO \cdot (CH_2)_n \cdot OH$ $\quad n = 8 - 100$

Paraformaldehyde is a mixture of the polymerization products of formaldehyde, containing 91–99% formaldehyde and small amounts of H_2O which are added to chains of formaldehyde, thus disrupting the polymerization.

Common grades include flaked, powdered and granular white crystalline materials with the distinct odour of formaldehyde. The water solubility of the polymerizates decreases with increasing degree of polymerization. Paraformaldehyde depolymerizes especially in acidic but in alkaline media, too. Depolymerization of paraformaldehyde also takes place in polar solvents, very rapidly if heat is involved and traces of an alkaline material, e.g. potassium carbonate are added.

A typical grade of paraformaldehyde is characterized as follows:

CH_2O content	95%
Melting point	163–165°C
Flash point	71°C

2.1(b) 1,3,5-Trioxane — Trioxymethylene

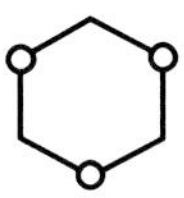

$C_3H_6O_3$ — M_r 90·08
CAS-No. 110-88-3

Trioxane is the cyclic trimer of formaldehyde which originates from gaseous formaldehyde in the form of volatile needles.

Melting point	63°C
Flash point	45°C
Solubility	Soluble in water and polar solvents

2.1(c) Formalin

The addition of H_2O to formaldehyde leads to formalin containing 30–45% by weight formaldehyde. The 37% standard is characterized as follows:

Appearance	Clear, colourless liquid with the pungent odour of formaldehyde
CH_2O content	37%
Refractive index at 20°C	n_D 1·3765
Density at 20°C	1·083 g/ml
Flash point	56°C
Solubility	Miscible with water and alcohols

Contains 10–15% methanol to avoid polymerization

Toxicity data for formaldehyde

LD_{50} oral	800 mg/kg rat
LD_{50} subcutaneous	300 mg/kg mouse
MWC	0·5 ml/m^3 (0·6 mg/m^3)

Formaldehyde is a strong irritant agent for the skin and mucous membranes. However, the presence of formaldehyde is already registered at concentrations of approx. 0·1–1·0 ppm by the very sensitive sense of smell.

Antimicrobial effectiveness/applications

The powerful antimicrobial properties of formaldehyde were first demonstrated in 1886 (Loew).

Looking at the MIC in Table 9 more closely it becomes obvious that bacteria are better covered by the antimicrobial efficacy of formaldehyde than fungi and yeasts. In addition, formaldehyde also has sporicidal and virucidal effects. This extremely broad range of effectiveness is a result of the distinct chemical reactivity of the formaldehyde which, as already mentioned before, belongs to the electrophilic active agents. When using formaldehyde, e.g. as a preservation agent, this reactivity may naturally also have an adverse effect. The coincidence of formaldehyde with ammonia thus leads to an inactivation due to the formation of hexamethylenetetramine (Section 3.3.1). The reaction with proteins also results in inactivation; at the same time disturbing coagulates are being formed by this reaction. On the other hand, formaldehyde is — without loss in effectiveness — compatible with anionic, cationic and non-ionic detergents and effective — largely independent of the pH value (optimum pH value: 3–10).

Formalin and paraformaldehyde are significant forms for applications which are advantageously available in a high degree of purity and as colourless products. When evaluating formaldehyde as a preserving agent its reactivity and volatility

Table 9. Minimum Inhibition Concentrations (MIC) of Formaldehyde in Nutrient Agar

Test organism	*MIC (mg/litre)*
Fungi	
Aspergillus niger	300
Aureobasidium pullulans	140
Chaetomium globosum	70
Penicillium glaucum	750
Rhizopus nigricans	200
Trichoderma viride	1200
Yeasts	
Candida albicans	1000
Candida crusei	200
Torula rubra	700
Torula utilis	80
Bacteria	
Aerobacter aerogenes	50
Aeromonas punctata	15
Bacillus subtilis	15
Desulfovibrio desulfuricans	15
Escherichia coli	45
Proteus vulgaris	10
Pseudomonas aeruginosa	60
Pseudomonas fluorescens	75
Staphylococcus aureus	20
Formaldehyde resistant bacteria[a]	1400

[a] Paulus (1976).

are considered a disadvantage, because both impair permanence and compatibility and are moreover responsible for odour nuisances. The volatility is on the other hand a benefit by converting formalin and paraformaldehyde into preservatives with vapour phase effect, e.g. in the head space of cans and tanks. Also attractive from the economic point of view is formaldehyde which, as a multi-purpose microbicide, is today still used in disinfectants and preservation agents.

As a preservative for industrial fluids, formaldehyde (formalin and paraformaldehyde) is preferably used in cosmetic products and in polymer dispersions (natural and synthetic latex) often in combination with other microbicides. Formaldehyde is listed in the EC list of preservatives allowed for the in-can protection of cosmetics (maximum authorized concentration: 0·2%, except for products for oral hygiene: 0·1%, prohibited in aerosol dispensers). Oil, fat and wax emulsions, starch and dextrine glues, adhesive dispersions, pigment and filler slurries, thickening solutions and other aqueous formulations containing no proteins can be preserved by the addition of formaldehyde, too; in order to improve the permanence and the range of effectiveness (synergism) generally a further microbicide, preferably a fungicide, is also applied.

Formalin or paraformaldehyde can also be employed as vapour phase microbicides (fungicides) without residues being left (e.g. in packings).

2.2 Glyoxal — Ethane-1,2-dial

```
O=C-C=O
  | |
  H H
```

$C_2H_2O_2$

M_r 58·04
CAS-No. 107-22-2

Chemical and physical properties

Appearance	White to yellow prisms
Density at 20°C	1·14 g/ml
Refractive index (20·5°C)	n_D 1·3826
Melting point	15°C
Boiling point	51°C
Solubility	More than 40% in water; soluble in alcohols
Stability	Tends to polymerize in contact with water

Toxicity data

LD_{50} oral	2·02 g/kg rat

Antimicrobial effectiveness/applications

Glyoxal exhibits a microbicidal effect only at relatively high concentrations (1–4%). 4000 mg/litre of a 30% solution of glyoxal in water does not inhibit the growth of bacteria, yeast and fungi on a nutrient agar. Because of the poor antimicrobial effectiveness glyoxal is not significantly applied as a preservative for industrial fluids or as an active ingredient in disinfectants.

2.3 Glutaraldehyde — Pentane-1,5-dial

O=C-$(CH_2)_3$-C=O 	| H H	$C_5H_8O_2$	M_r 100·12 CAS-No. 111-30-8

Chemical and physical properties

Appearance	Oily fluid
Boiling point	187–9°C (decomposition) 71–2°C at 13 hPa
Solubility	Soluble in water

Typical properties of aqueous solutions

	Glutaraldehyde concentration	
	25% by wt	*50% by wt*
Density (g/ml at 20°C)	1·066	1·131
Vapour pressure (hPa at 20°C)	21·86	20
Solidification point (°C)	–10	–21
Refractive index (at 20°C)	n_D 1·375	n_D 1·421
Surface tension (dyn/cm at 20°C)	45	48
pH at 25°C	3·1–4·5	3·1–4·5
Stability	Volatile with water vapour, tends to polymerize in water, is inactivated by ammonia and primary amines at neutral and higher pH	

Toxicity data

LD_{50} oral	600 mg/kg rat
Inhalation	Rats survived an 8-h exposure to saturated glutaraldehyde vapour
Sensitization	Possible
Non-mutagenic	
MWC	0·2 ml/m^3 (0·8 mg/m^3)

Antimicrobial effectiveness/applications

Glutaraldehyde was introduced as an antimicrobial substance not before 1962 and has been studied and reviewed extensively in the meantime (Gorman *et al.*, 1980).

Glutaraldehyde may be considered as a 'chemosterilizer' because of its capability to destroy bacteria, fungi and the corresponding spores, tubercle bacilli and viruses, although non-lipid viruses usually are more resistant to the attack of glutaraldehyde than the enveloped lipophilic viruses. The most important property of glutaraldehyde is, however, its sporicidal efficacy, as there are only a few sporicides available. Glutaraldehyde is at a more acceptable contact time 2–8 times more sporicidal than formaldehyde which has additionally several other disadvantages including irritation and lack of penetration. Ease of use, activity in the presence of organic matter (e.g. serum) and rapidity of action — bacteria are killed within 1 min, most viruses are destroyed within 10 min — are further advantages of glutaraldehyde.

The mechanism of microbicidal action of glutaraldehyde is based on its two toxophoric (aldehyde) groups which can interact with microbial cell constituents, e.g. amino and thiol groups of proteins, with ring nitrogen atoms of purine bases. This interaction increases with increasing pH. Thus pH is the most important factor in regulating the activity of glutaraldehyde under typical use conditions. The curves in Fig. 16 show that the rate of kill of the gram-negative bacterium *Escherichia coli* is approximately twenty times faster at pH 8·5 than pH 5. One can assume that in an acidic medium the reactive sites of the cell wall are protonated and therefore protected from an interaction with the toxophoric groups of glutaraldehyde. This explains also the fact, that in the presence of ammonia, primary amines or protein, which can interact with (inactivate) glutaraldehyde,

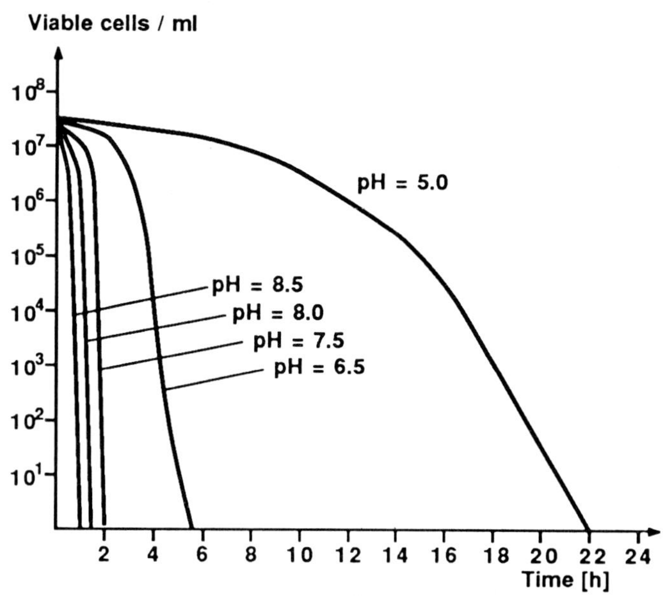

Fig. 16. Rate of kill curves for glutaraldehyde (conc. 44 ppm) at different pH values; temperature 20°C; test organism *Escherichia coli* (source: Union Carbide Corporation, Specialty Chemicals Division, USA).

the latter is significantly effective only at pH 4·5 or less. In the acidic state the effectiveness of glutaraldehyde is presumably caused by the ability of the active ingredient to penetrate the cell wall and to reach internal areas of neutral pH which allows interaction with free internal amino groups; however, the rate of kill is slower (s.a.). In the acidic state, the addition of surfactants permit faster penetration of glutaraldehyde through the cell wall thus increasing the activity. Suitable surfactants for the combination with glutaraldehyde may be cationic, nonionic or anionic agents.

Glutaraldehyde is usually obtained commercially as 2% or 25% or 50% solution of acidic pH. The degree of polymerization is negligible in acidic solution but high at basic pH. Polymerization leads to an extensive loss of aldehyde groups and in consequence to a loss of effectiveness, however, in weeks whereas the microbicidal efficacy is observed within minutes or hours, especially in basic media. Glutaraldehyde therefore is applied more as a disinfecting and sterilizing agent than as a preservative.

Optimum performance involves balancing factors such as pH, temperature matrix and additives in the final product. In practice glutaraldehyde is generally available as a 2% solution to which an activator is added to bring the pH to approximately 8 before application, e.g. disinfection of instruments. The activated solution disinfects at room temperature within 10 min and sterilizes within 10 h. The activated solution has to be discarded 14 days after activation. Another important application for glutaraldehyde solutions is the disinfection of surfaces.

Because of its effectiveness against slime forming microbes (minimum inhibition concentration approx. 2·5 mg/litre) glutaraldehyde may be used in paper mills as a slimicide which does not cause any waste water problems. Especially since paper mills increase the pH in their process water to 7 or even to the slightly alkaline state, they profit from the higher activity of glutaraldehyde in such media.

At concentrations of 0·05–0·2% glutaraldehyde can act as a preservative for the in-can protection of industrial fluids, the pH of which is between 4 and 6 (optimum pH 5). The average lifespan for glutaraldehyde is 6 months at pH 7, however, up to 2 years at pH 4.

In the EC glutaraldehyde has been added to the list of preservatives provisionally allowed in cosmetics. The warning 'contains glutaraldehyde' must appear on labels when concentrations in the finished product exceed 0·05%.

2.4 Acrolein — Propen-2-al

$CH_2{=}CH{-}C{=}O$ (with H on the C)	C_3H_4O	M_r 56·06 CAS-No. 107-02-8

Chemical and physical properties

Appearance	Colourless fluid with a pungent and irritant odour
Density at 20°C	0·8389 g/ml
Vapour pressure at 20°C	296 hPa
Melting point	−87°C
Boiling point	52–52·5°C (decomposition)
Flash point	−25°C
Ignition temperature	240°C
Solubility	26·7 g/100 ml H_2O; miscible with lower alcohols, ketones, benzene, diethyl ether and other common organic solvents
Stability	Reacts with numerous compounds and polymerizes readily (stabilizer, e.g. hydroquinone) — half-life in H_2O at pH 5: 150 h, at pH 7: 120–180 h, at pH 9: 5–40 h

Toxicity data

LD_{50} oral	46 mg/kg rat; 7 mg/kg rabbit
LD_{50} dermal	562 mg/kg rabbit

Powerful skin irritant; lachrymatory effect and irritant action on respiratory organs (2·3 mg/m^3 are already intolerable).

MWC	0·1 ml/m^3 (0·25 mg/m^3) 0·1–0·8 mg/m^3 are registered by the very sensitive sense of smell.
LC_{100} for fish	Approx. 1–5 mg/litre

Antimicrobial effectiveness/applications

There are two toxophoric groups in the acrolein molecule, the activated vinyl group and the aldehyde group; both groups are able to react with nucleophilic cell entities. Due to this fact acrolein is highly reactive and highly effective and may be defined as a biocide rather than a microbicide.

Minimum inhibition concentrations for:

Slime forming organisms	1 mg/litre
Fresh water algae	5–10 mg/litre

Acrolein may be used as a slimicide and algicide in industrial water circuits and injection water for oil rccovery as a non-persistent active ingredient which does not cause waste water problems. It was also recommended for combatting submerging aquatic plants (Austin, 1964). However, because of its powerful irritant action and toxicity, handling of acrolein is very difficult; therefore it has not gained significant importance in applications as a microbicide.

2.5 α-Bromocinnamaldehyde (BCA)

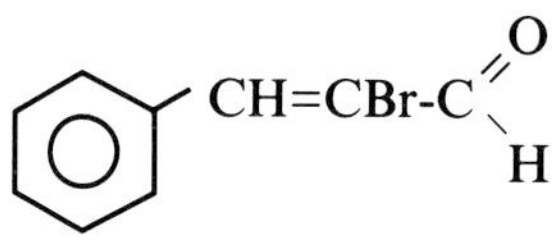

C_9H_7BrO

M_r 211·06
CAS-No. 5443-49-2

Chemical and physical properties

Appearance	Light yellow to brown powder with an odour similar to cinnamon
Bulk density	650–700 g/litre
Melting point	71–72°C
Vapour pressure at 50°C	0·0013 hPa
Solubility	Sparingly soluble in water, soluble in organic solvents such as ethanol, acetone, methylethylketone, benzene, toluene, xylene
Stability	Stable in a pH range of 3–11; decomposition at 230–240°C

Toxicity data

LD_{50} oral	1795 mg/kg mouse
dermal	2220 mg/kg mouse

Irritant to skin and mucosa.

Antimicrobial effectiveness/applications

BCA is a volatile chemical which may be used as a vapour phase microbicide. Small amounts of it are able to inhibit the growth of mould producing fungi of yeasts and bacteria. In containers, machines, parts of telecommunication devices BCA is used as a vapour phase microbicide preventing mould growth during storage and transport of the devices. It is especially useful where articles do not allow the incorporation of a microbicide for protection against biodeterioration.

Table 10. Minimum Inhibition Concentrations (MIC) of BCA in Nutrient Agar

Test organism	*MIC (mg/litre)*
Alternaria alternata	50
Aspergillus niger	100
Aureobasidium pullulans	100
Chaetomium globosum	50
Lentinus tigrinus	5
Penicillium glaucum	35
Sclerophoma pityophila	100
Trichoderma viride	100
Escherichia coli	35
Pseudomonas aeruginosa	500
Staphylococcus aureus	35

2.6 3,5-Dichloro-4-hydroxybenzaldehyde (DCHB)

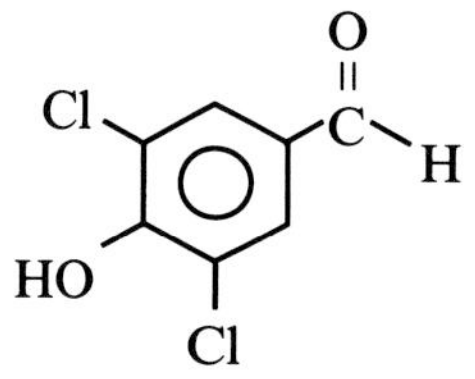

$C_7H_4Cl_2O_2$

M_r 191·02
CAS-No. 2314-36-5

Chemical and physical properties

Appearance	Crystalline solid with a characteristic odour
Melting point	156°C
Solubility	Highly soluble in ethanol, propylene glycol, ether, acetic acid, less soluble in benzene, chloroform, sparingly soluble in water

Toxicity data

LD_{50} oral	>5 g/kg rat
dermal	>2 g/kg rat (exposure: 24 h)

Not irritant to skin and mucous membranes.
Ames test negative.

Antimicrobial effectiveness/applications

Because of its broad spectrum of activity which covers a wide range of bacteria, yeasts and moulds, DCHB may be a suitable preservative for the in-can protection of industrial fluids, e.g. detergent concentrates and surfactant-based products; addition rates: 0·1–0·15%. Chelating agents, such as EDTA, support the efficacy of DCHB so that lower concentrations can be applied, e.g. 0·05–0·1% (Maddox, 1988). Assuming that further toxicological evaluations will deliver satisfactory results, DCHB can be used as a preservative in cosmetics.

Table 11. Minimum Inhibition Concentrations (MIC) of a 30% Solution of DCHB in Propylene Glycol According to Maddox (1988)

Test organism	*MIC (mg/litre)*
Pseudomonas aeruginosa-isolate	1500
Pseudomonas putida-isolate	1500
Pseudomonas fluorescens-isolate	700
Pseudomonas fluorescens NCTC 10038	70
Klebsiella oxytoca-isolate	700
Klebsiella pneumoniae-isolate	1500
Proteus vulgaris NCTC 4175	700
Citrobacter freundii-isolate	150
Salmonella entiridis NCTC 5765	700
Escherichia coli-isolate	700
Staphylococcus aureus NCTC 4163	700
Streptococcus mutans NCTC 10449	1500
Staphylococcus epidermidis	1500
Candida albicans NCYC 597	2300
Saccharomyces cerevisiae NCYC 200	2300
Penicillium expansum	2300
Penicillium notatum	1500
Cladosporium herbarum	700

2.7 2-Hydroxy-1-naphthaldehyde (HNA)

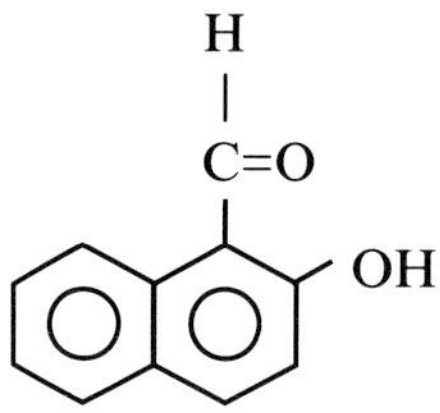

$C_{11}H_8O_2$

M_r 172·19
CAS-No. 708-06-5

Chemical and physical properties

Appearance	Crystalline solid
Melting point	82–85°C
Boiling point	192°C at 36 hPa
Solubility	Practically insoluble in water, soluble in alcohol, ether, petro-ether, highly soluble in alkaline solutions

Antimicrobial effectiveness/applications

HNA is especially effective against mould producing fungi; it offers an extraordinary broad and equal spectrum of activity. Sharma (1989) has determined the minimum inhibition concentrations for 40 different species of fungi and found that none of the species tolerated more than 60 mg/litre of HNA. At the inhibitory concentrations of HNA the secretion of extracellular enzymes (amylase, cellulase, protease lipase) is suppressed (= 100% inhibition). According to the findings of Sharma (1989) the morphological changes of test organism *Aspergillus flavus* were comparable to those produced by chlorinated phenols, e.g. *p*-chloro-*m*-cresol, demonstrating that HNA belongs to the membrane-active substances. The presence of the aldehyde groups makes HNA, in addition, an electrophilic agent and increases its toxicity.

Because of its broad spectrum of effectiveness, its insolubility in water and its stability it is proposed to use HNA as a long-term protectant against biodeterioration of materials, e.g. leather, footwear, cotton, textiles, paper.

3 FORMALDEHYDE RELEASING COMPOUNDS

Formaldehyde as such is often too volatile and too reactive to be used as a microbicide for the protection of materials or in disinfectants. It additionally produces unwelcome side-effects, and has an insufficient balanced range of activity. One therefore has looked for formaldehyde releasing compounds which do not exhibit the disadvantageous formaldehyde effects but maintain or even improve the antimicrobial action of formaldehyde.

The first step towards formaldehyde releasing compounds generally consists in the process of hydroxymethylation, that is, introduction of the hydroxymethyl group into such molecules as possess active hydrogen atoms will allow a reversible reaction with formaldehyde, the series ranging from hydrol to the amino acid taurine (Fig. 17).

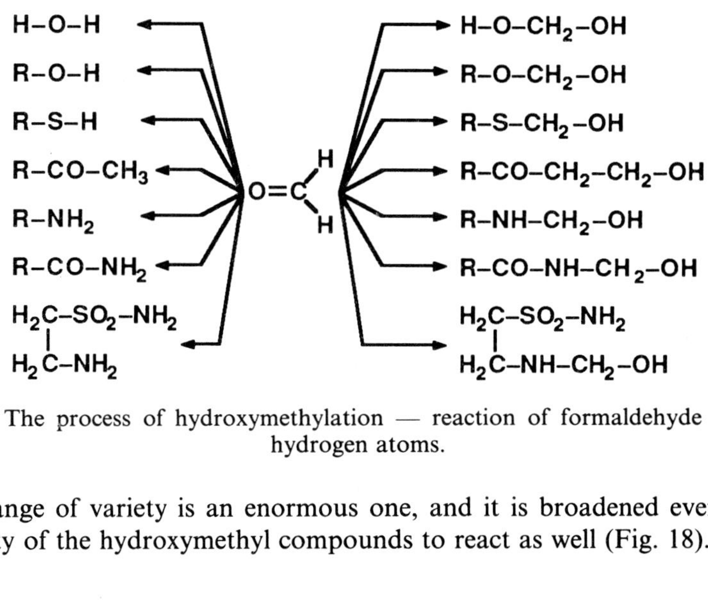

Fig. 17. The process of hydroxymethylation — reaction of formaldehyde with active hydrogen atoms.

The range of variety is an enormous one, and it is broadened even more by the ability of the hydroxymethyl compounds to react as well (Fig. 18).

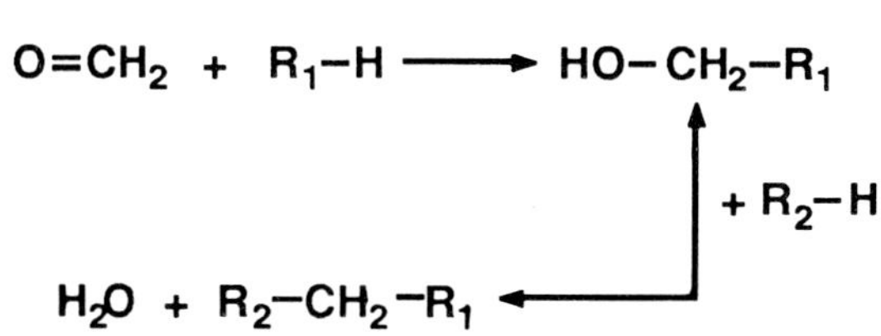

Fig. 18. Reversible formation of hydroxymethyl/methylene compounds.

Formaldehyde releasing compounds can be solids or liquids, water soluble or oil soluble, odourless, alkaline, neutral, or slightly acidic. They open up applications to the active agent, formaldehyde, which would otherwise be closed to it due to its unfavourable properties.

3.1 *O*-Hydroxymethyl compounds (hemiformals) and formals

The reaction of formaldehyde with alcohols (R-OH), easily taking place under neutral or weakly alkaline conditions, leads to the formation of hemiformals which are relatively heat resistant and in equilibrium with the starting products. However, the equilibrium is widely shifted to the side of the hemiformal; free formaldehyde is detectable in mere traces. Under acidic conditions the reaction goes on to produce formals.

If formaldehyde acts upon the more acidic phenolic hydroxy group, the reaction products obtained are unstable and cannot be isolated as the equilibrium is widely shifted to the side of the starting products. However, for practical applications ones takes advantage of the antimicrobial activities of such mixtures which are based on the effective spectra of the mixing partners usually complementing each other advantageously; a remarkable synergism is often also noticed. One has to keep in mind, however, that under certain conditions, e.g. at pH values above 9–9·5, the reaction of formaldehyde with phenols leads to nucleus-hydroxymethylated phenols to which the formaldehyde naturally is fixed irreversibly, thus being prevented from producing antimicrobial effects (see Fig. 19). Additionally the introduction of the hydroxymethyl groups into the phenyl ring has an adverse effect on the antimicrobial activity of phenolic compounds.

In general the antimicrobial efficacy of the hemiformals is in conformity with their formaldehyde contents (exception: benzylalcoholhemiformal; see Section 3.1.2).

This is not true for the formals which do not exhibit remarkable antimicrobial activity. Exception: (3-iodopropargyl)-(4-chlorophenyl)formal (see Section 3.1.5)

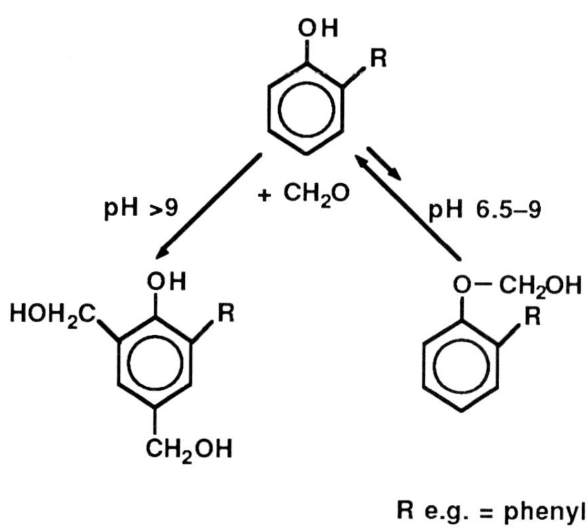

Fig. 19. Reaction of phenol derivative with formaldehyde.

the antimicrobial activity of which is not based on the formaldehyde incorporated into the molecule. It can pass for a rule that formaldehyde releasing compounds are effective if their formaldehyde contents can be analysed in diluted aqueous solutions using the method described by Tannenbaum & Bricker (1951) (reaction of the compounds with phenylhydrazine and potassium hexacyanoferrate). However, in case the formaldehyde content of a compound is not detectable by the Tannenbaum method, it may be effective in spite of that. Examples of such compounds are certain *N*-hydroxymethyl amides (see Section 3.3).

The importance of hemiformals as preservatives for practical applications is based on their differing and superior chemico-physical properties as compared with formalin.

3.1.1 *n*-Butanolhemiformal

$CH_3\text{-}(CH_2)_2\text{-}CH_2\text{-}O\text{-}CH_2\text{-}OH$ $C_5H_{12}O_2$ M_r 104·15
CAS-No. 3085-35-6

Chemical and physical properties

Appearance	Colourless fluid with the odour of formaldehyde
CH_2O-content	28·8%
Solubility	In water approx. 100 g/litre; miscible with alcohols
Stability	Decomposition by heating to the boiling point

Antimicrobial effectiveness/applications

The action of *n*-butanolhemiformal corresponds to its formaldehyde content. Minimum inhibition concentrations:

300 mg/litre for *Escherichia coli*
4000 mg/litre for formaldehyde resistant bacteria (Paulus, 1976).

n-Butanolhemiformal may be used as a preservative for water based products with the advantage that the affinity of formaldehyde to *n*-butanol is stronger than in formalin to water. Additionally *n*-butanolhemiformal can serve as a formulation aid (e.g. solvent) with antimicrobial effectiveness, for example, when a microbicide has to be formulated to a preservative.

3.1.2 Benzylalcoholhemiformal (BHF)

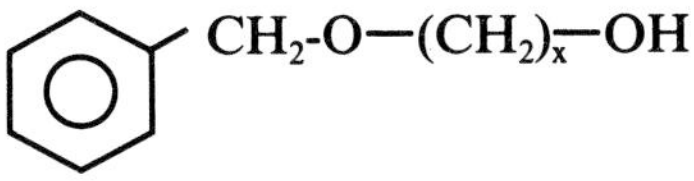

x = 1·5 M_r 153
CAS-No. 14548-60-8

More precisely, the BHF meant here is benzylalcoholmono(poly)hemiformal as it is the reaction product of 1 mol benzylalcohol with 1·5 mol formaldehyde.

Chemical and physical properties

Appearance	Water-clear liquid with a characteristic and slight odour of formaldehyde
CH_2O content	29%
Density at 20°C	1·11 g/ml
Viscosity at 20°C	16·3 mPa s
Vapour pressure at 20°C	21 hPa
Solidification point	<-60°C
Boiling point	105°C (decomposition)
Flash point	Approx. 95°C
Ignition temperature	400°C
Solubility	In water at 20°C: 25 g/litre; soluble in organic solvents
Reaction in water	Neutral
Ionicity	Non-ionic
Stability	Releases formaldehyde when heated and in water based solutions

Toxicity data

LD_{50} oral	1700 mg/kg rat
LD_{50} dermal	>1000 mg/kg rat (exposition: 7 days)
LC_{50} inhaled	>0·5 mg/litre for rats (exposition: 4 h)

Concentrated BHF causes skin irritation, a 0·2% dilution in water does not.

LC_0 for *Leuciscus idus* 20 mg/litre (exposition: 48 h)

BHF is easily biodegraded; activated sludge organisms in sewage plants are not disturbed in their activity by BHF concentrations up to 500 mg/litre.

Antimicrobial effectiveness/applications

Table 12. Minimum Inhibition Concentrations (MIC) of BHF in Nutrient Agar

Test organism	*MIC (mg/litre)*
Aerobacter aerogenes	150
Bacillus mycoides	100
Bacillus punctatus	50
Bacillus subtilis	50
Bacterium vulgare	50
Desulfovibrio desulfuricans	50
Escherichia coli	100
Formaldehyde resistant bacteria[a]	2000
Pseudomonas aeruginosa	150
Pseudomonas fluorescens	130
Staphylococcus aureus	70
Alternaria alternata	500
Aspergillus flavus Link	240
Aspergillus niger	900
Aspergillus terreus	1000
Aureobasidium pullulans	700
Chaetomium globosum Kunze	300
Penicillium glaucum	1500
Rhizopus nigricans	600
Sclerophoma pityophila	500
Trichoderma viride	2000
Candida albicans	1000
Candida crusei	500
Torula utilis	400

[a] Paulus (1976). The isolate is considered to be a strain of *Pseudomonas putida*. However, it is atypical in not producing acid from xylose.

According to the MIC in Table 12 BHF has a higher degree of efficiency than one would conclude from its formaldehyde content. A careful examination of the spectrum of effectiveness reveals the phenomenon of synergism which makes itself felt in the case of bacteria, fungi and yeasts (Paulus, 1976). Additionally BHF offers the advantage that its efficacy is independent of pH value and is developed fully even in the presence of ionic or non-ionic agents. On the other hand the disinfecting power of BHF is slow compared to phenolic active ingredients or to quaternary ammonium compounds (Table 13).

The MIC in Table 12 show that the antimicrobial effectiveness of BHF against bacteria is stronger than against fungi and yeasts. When looking for possibilities to overcome this deficiency which is characteristic for most of the formaldehyde releasing compounds it is self suggesting to use phenolic microbicides as combination partners, which are known to have an excellent fungicidal effect.

Table 13. Disinfecting Concentrations (%) of BHF and Mixtures of BHF with Phenolic Microbicides, Determined by the Suspension Method after 10 min

Active ingredient	*Test organism*		
	Staphylococcus aureus	Escherichia coli	Pseudomonas aeruginosa
BHF	1·50	1·00	1·50
BHF/2-phenyl-phenol (4/1)	0·15	0·15	0·25
BHF/3-methyl-4-chlorophenol (4/1)	0·25	0·15	0·25
BHF/2,2'-dihydroxy-5,5'-dichloro-diphenylmethane (4/1)	0·04	0·07	0·25

A remarkable synergism which especially extends to fungi, but also to yeasts, is noticed, when BHF is combined with microbicidal phenol derivatives (Paulus *et al.*, 1970*b*). This synergism is so striking that even additions of 15–20% phenol derivatives to BHF are sufficient to broaden the spectrum of activity to the desired extent. However, phenols are known to be able to react with formaldehyde and irreversible nucleus hydroxymethylations may occur under specific conditions. In this way the formaldehyde releasing compounds are deprived of microbicidal formaldehyde. In addition, the phenol derivatives hydroxymethylated in the phenyl ring are less effective than the corresponding starting products. However, these reactions, which reduce the antimicrobial effect of the mixing partners, can be suppressed as long as pH levels between 6 and 9, and max. 9·5 are available (Paulus, 1980).

Thanks to its excellent properties — low toxicity, good skin compatibility, negligible environmental toxicity, colourlessness, neutral reaction, efficacy independent of pH and unaffected by detergents or emulsifying agents — BHF is successfully employed in a variety of industrial fluids of widely differing compositions. Examples: cosmetics, polymer emulsions, pigment slurries, thickening solutions, emulsion paints, adhesive dispersions, metal working fluids, glues, biopolymers, concrete additives, wax emulsions, polishes and other water-based formulations. BHF is provisionally registered in the EC list of preservatives for cosmetics with a maximum authorized concentration of 0·2%.

In disinfectants BHF is used in combination with other active ingredients, preferably with phenol derivatives (see Table 13) or quaternary ammonium compounds (Section 16.1). In 'chemical WCs' BHF has both a disinfecting and a deodorizing effect.

3.1.3 2-Phenoxy ethanol hemiformal

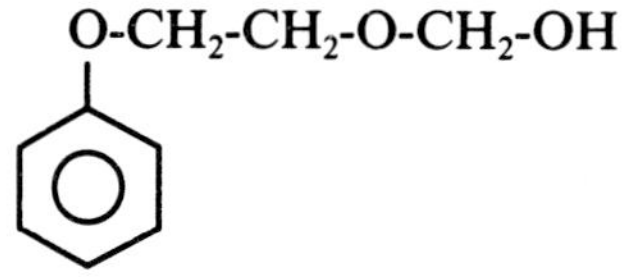

$C_9H_{12}O_3$

M_r 168·17
CAS-No. 41965-76-8

Chemical and physical properties

Appearance	Clear, colourless, slightly viscous fluid with a slight odour of formaldehyde
CH_2O content	17·8%
Density at 20°C	1·113 g/ml
Solubility	In water approx. 30 g/litre; miscible with alcohols, ketones

Antimicrobial effectiveness/applications

Table 14. Minimum Inhibition Concentrations (MIC) of 2-Phenoxy Ethanol Hemiformal in Nutrient Agar

Test organism	*MIC (mg/litre)*
Escherichia coli	500
Pseudomonas aeruginosa	1000
Staphylococcus aureus	500
Candida albicans	1000
Aspergillus niger	1500
Chaetomium globosum	500
Penicillium glaucum	2000

As is demonstrated in Table 14 2-phenoxy-ethanol hemiformal is much more effective than 2-phenoxy ethanol itself (see Section 1.7). It may be used alone or in combination with other active ingredients, especially those which exhibit a more fungicidal effect, for the in-can protection of cosmetic products, and also as a preservative for other industrial fluids. Although 2-phenoxy ethanol hemiformal is not mentioned in the EC positive list of preservatives for cosmetic products, there should be no inhibition for this application as it consists of 2-phenoxy ethanol and formaldehyde (addition product) which are registered in the a.m. EC list. Additionally, 2-phenoxy ethanol hemiformal may be regarded as a solvent with antimicrobial effectiveness which proves useful for the formulation of preservatives.

3.1.4 Ethylene glycolhemiformals and ethyleneformal

The addition of formaldehyde to ethylene glycol leads to mono- and bishemiformal of ethylene glycol with microbicidal efficacy (Fig. 20). Under acidic conditions water is eliminated from the EG hemiformal and ethyleneformal is formed, the formaldehyde content of which is not detectable with the Tannenbaum method. Accordingly 1,3-dioxolane does not exhibit substantial antimicrobial effectiveness.

Chemical and physical properties

	EG mono-hemiformal	EG bis-hemiformal	Ethylene-formal
Appearance	Liquid with the odour of CH_2O, colourless	Liquid with the odour of CH_2O, colourless	Liquid with aromatic odour, colourless
CH_2O content	32·6%	42·2%	40·5%
Density (g/ml at 20°C)	1·17	1·12	1·064
Refractive index (20°C)			1·401
Solidification point			−95°C
Boiling point			75–76°C
Stability	Release CH_2O when heated and in water based solutions		Does not release CH_2O under mild conditions
Solubility	Miscible with water, alcohols, ketones		

Antimicrobial effectiveness/applications

Table 15. Minimum Inhibition Concentrations (MIC) of EG Hemiformals and Ethyleneformal

Test organism	*MIC (mg/litre)*		
	EG monohemiformal	*EG bishemiformal*	*Ethyleneformal*
Escherichia coli	500	350	>5000
Pseudomonas aeruginosa	1500	1000	>5000
Staphylococcus aureus	500	200	>5000
Candida albicans	500	200	>5000
Aspergillus niger	1000	500	>5000
Chaetomium globosum	500	200	>5000
Penicillium glaucum	2000	1000	>5000

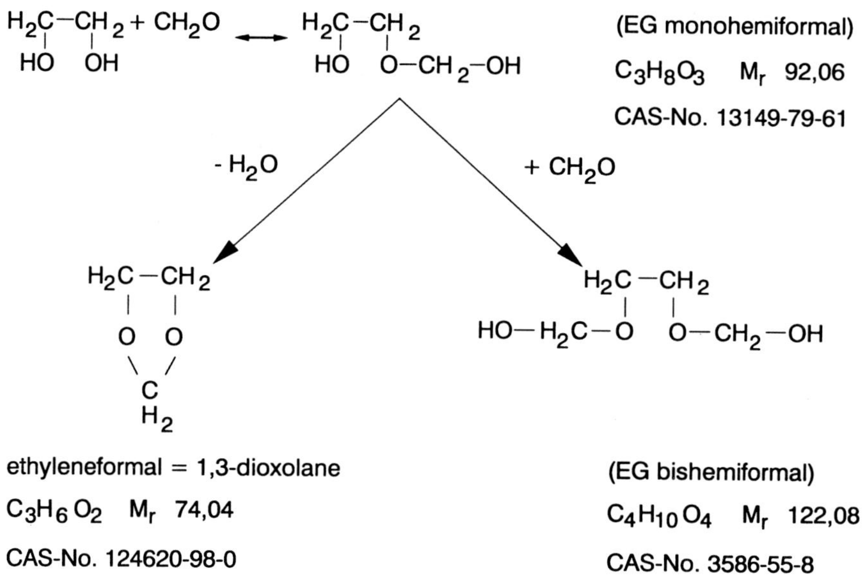

Fig. 20. The reaction of ethylene glycol (EG) with formaldehyde.

According to their composition the EG hemiformals are especially effective against bacteria and therefore useful for the in-can protection of a large variety of industrial fluids, mainly together with other active ingredients, e.g. fungicides. EG hemiformals are also of importance as formulations aids in preservatives.

3.1.5 (3-Iodopropargyl)-(4-chlorophenyl)formal

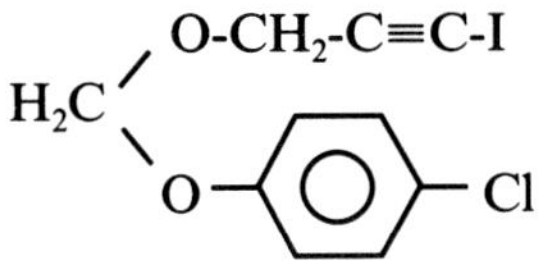

$C_{10}H_8ClIO_2$ M_r 322·54
CAS-No. 29772-02-9

Chemical and physical properties

Appearance	Pale brown liquid with an unpleasant odour
CH_2O content	9·3%
Density at 25°C	1·572 g/ml
Vapour pressure at 25°C	$<10^{-3}$ hPa
Refractive index (25°C)	n_D 1·583
Viscosity at 25°C	16 mPa s
Boiling point at 0·27 hPa	130°C
Flash point	164°C
Solubility	Sparingly soluble in water, easily soluble in organic solvents
Stability	Hydrolyses under severe conditions to the starting compounds: 3-iodopropargylalcohol, 4-chlorophenol and formaldehyde

Toxicity data

LD_{50} oral	1250 mg/kg rat
dermal	>2000 mg/kg rat

Irritant to skin and mucosa; skin irritation was not observed in tests with formulations containing less than 10% of the active ingredient (patch test, exposure: 24 h).

Fish toxicity: LC_0 against Killi fish: 1·5 mg/litre

Antimicrobial effectiveness/applications

(3-Iodopropargyl)-(4-chlorophenyl)formal is a powerful fungicide with a broad spectrum of activity but only a few species of bacteria are covered by the efficacy of the compound (see Table 16). In view of the MIC it is apparent that formaldehyde (being mainly effective against bacteria) does not take a significant part in the activity of the formal. More relation can be found to 3-iodopropargyl alcohol (see Section 1.9) which can generate from the compound.

Table 16. Minimum Inhibition Concentrations (MIC) of (3-Iodopropargyl)-(4-chlorophenyl)formal in Nutrient Agar

Test organism	*MIC (mg/litre)*
Alternaria alternata	25
Aureobasidium pullulans	25
Aspergillus niger	5
Chaetomium globosum	15
Coniophora puteana	0·5
Lentinus tigrinus	5
Penicillium glaucum	3·5
Cladosporium cladosporioides	1·5
Sclerophoma pityophila	7·5
Trichoderma viride	50
Escherichia coli	>1000
Pseudomonas aeruginosa	>1000
Staphylococcus aureus	20

Formulations of the active ingredient are recommended for the protection of wood, leather, textiles, paints, paper. Because of the gaps in the activity spectrum it is not suitable to prevent bacterial deterioration in aqueous functional fluids. The compound also exhibits slimicidal and algicidal effectiveness.

3.2 C-methylols

Activated hydrogen atoms, prepared to react with formaldehyde to yield C-methol compounds, one finds in nitro-hydrocarbons at the carbon atom adjacent to the nitro group, in ketones at the carbon atom in α-position towards the carbonyl group, in certain phenol derivatives. Whereas the C-methylols deriving from ketones and phenol derivatives generally do not exhibit antimicrobial activity the C-methylols deriving from ineffective nitro-hydrocarbons do so, although the formaldehyde combined in these substances cannot be determined by the Tannenbaum method. As the C-methylolated nitro-hydrocarbons are able to react with further formaldehyde, amines and amides, microbicides having a wide range of chemical and physical properties can be synthesized within this class of substances. The antimicrobial activity of C-methylolated nitro-hydrocarbons can be considerably increased by incorporation of an activated halogen group in the α-position to the electronegative nitro group (Fig. 21). Such molecules possess two reactive sites (toxophoric groups) which can react with the nucleophilic centres of the microbial cell. As a result they are generally more active than the corresponding compound with only one toxophoric group.

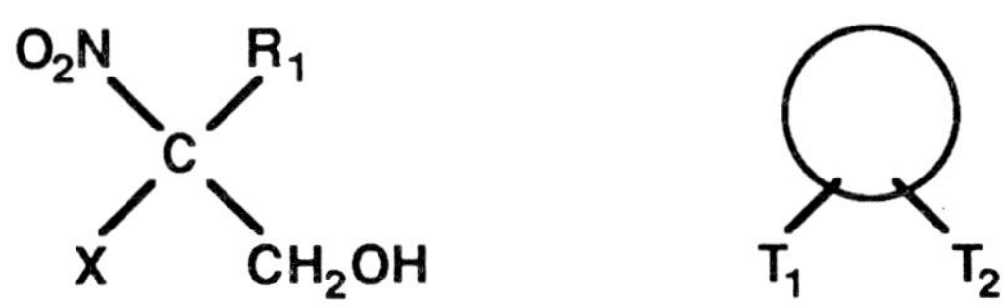

X = halogen T = toxophoric group

Fig. 21. Molecule with two toxophoric groups.

3.2.1 Trimethylol nitromethane — 2-Hydroxymethyl-2-nitro-propane-1,3-diol

```
        NO2
         |
HO-CH2-C-CH2-OH
         |
        CH2-OH
```

$C_4H_9NO_5$

M_r 151·12
CAS-No. 51245-95-5

Chemical and physical properties

Appearance	White, crystalline
CH_2O content	60%
Melting point	175–176°C (decomposition)
Solubility	In water at 20°C 2200 g/litre; very soluble in alcohols, practically insoluble in hydrocarbons such as heptane, kerosene, mineral oil, benzene, toluene
Reaction in water	pH of 0·1 M aqueous solution at 20°C: 5·0
Stability	Very stable when dry and in media with pH values up to 6·5; in solutions buffered to pH 8, 9%, and in those buffered to pH 9, 12% of the formaldehyde content are detectable by the Tannenbaum method (Paulus, 1980).

Toxicity data

LD_{50} oral	1900 mg/kg mouse
LD_{50} intraperitoneal	1850 mg/kg mouse

Trimethylol nitromethane is not a primary irritant and the sensitizing potential is minimal.

Antimicrobial effectiveness/applications

Mould producing fungi are more tolerant of trimethylol nitromethane than bacteria. However, trimethylol nitromethane and other methylolnitro-hydrocarbons cannot be ranked among the very effective formaldehyde releasing compounds, although in alkaline media there is a noticeable increase of the tendency, especially of trimethylol nitromethane, to release formaldehyde accompanied by a corresponding increase in efficacy. A report of Clark *et al.* (1974) surveys the synthesis and antimicrobial activity of aliphatic nitro compounds.

Trimethylol nitromethane as a preservative for industrial fluids in the meantime has been widely substituted by more effective formaldehyde releasing compounds. However, its good skin compatibility and the fact that there is at no

Table 17. Minimum Inhibition Concentrations (MIC) of Trimethylol Nitromethane in Nutrient Agar

Test organism	*MIC (mg/litre)*
Bacillus subtilis	>1000
Escherichia coli	3000
Proteus vulgaris	>1000
Pseudomonas aeruginosa	>1000
Staphylococcus aureus	1000
Candida albicans	≥1000
Aspergillus niger	>1000
Chaetomium globosum	>1000
Penicillium glaucum	>1000

time any detectable odour of formaldehyde is not a negligible advantage especially when the compound is applied for the protection of lubricoolants.

3.2.2 2-Bromo-2-nitro-propane-1,3-diol — Bronopol

```
        NO2
         |
HO-CH2-C-CH2-OH
         |
        Br
```

$C_3H_6BrNO_4$

M_r 199·99
CAS-No. 52-51-7

Chemical and physical properties

Appearance	Colourless to pale yellow crystals	
CH_2O content	30%	
Melting point	Approx. 130°C	
Vapour pressure at 20°C	$1{\cdot}68 \times 10^{-5}$ hPa	
Solubility (g/litre at 22–25°C)	In water	250
	ethanol	500
	isopropanol	250
	propylene glycol	140
	polyethylene glycol 300	110
	diethyl sebacate	100
	isopropyl myristate	< 5
	liquid paraffin	< 5
	olive oil	< 5
	castor oil	< 5
Stability	Very stable when dry, slightly hygroscopic; deactivation, discoloration and corrosion occur to a limited extent when solutions of Bronopol are in contact with iron and aluminium; decomposition in alkaline media; decomposition products: formaldehyde, bromide, nitrite and nitro-alkylols (Fig. 22).	

Toxicity data

LD_{50} oral	Approx. 325 mg/kg rat
LD_{50} intraperitoneal	Approx. 26 mg/kg rat

There is no evidence that Bronopol produces carcinogenic, mutagenic, embryotoxic or teratogenic effects. Bronopol is slightly irritant to mucous membranes and the human skin, but not in concentrations which are normally applied. The sensitization potential of Bronopol is regarded as less than mild.

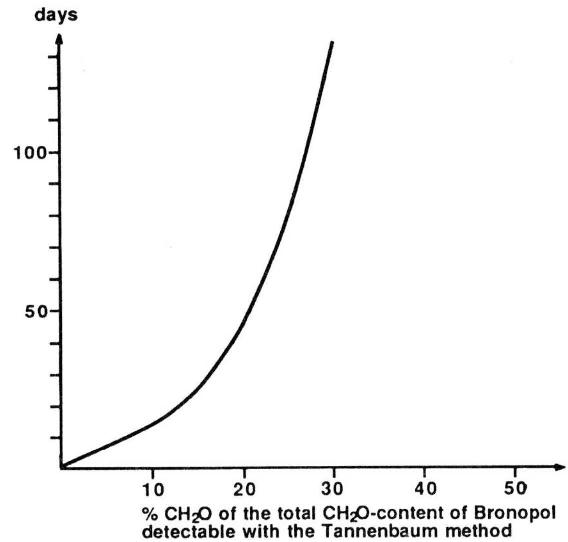

Fig. 22. Release of formaldehyde in a 0·1% solution of Bronopol in water at pH 9 (20°C).

Antimicrobial effectiveness/applications

Table 18. Minimum Inhibition Concentrations (MIC) of Bronopol in Nutrient Agar

Test organism	*MIC (mg/litre)*
Aerobacter aerogenes	500
Aeromonas punctata	200
Bacillus mycoides	500
Bacillus subtilis	150
Escherichia coli	50
Leuconostoc mesenterioides	200
Proteus mirabilis	50
Pseudomonas aeruginosa	50
Pseudomonas fluorescens	50
Staphylococcus aureus	50
Candida albicans	1000
Alternaria alternata	200
Aspergillus niger	2000
Aureobasidium pullulans	750
Chaetomium globosum	750
Cladosporium cladosporioides	1000
Penicillium glaucum	1000
Sclerophoma pityophila	500
Trichoderma viride	>1000

Further data regarding the activity of Bronopol can be found in the reports of Croshaw *et al.* (1964) and Dryce *et al.* (1978).

Comparing the efficacy of Bronopol with that of other formaldehyde releasing compounds and considering that the product is slow to release formaldehyde (see Fig. 22), it is obvious that Bronopol occupies a special position: despite the retarded formaldehyde release, it is superior in efficacy to other formaldehyde releasing compounds over a wide pH range (5–9) and offers the advantage of being particularly effective against Pseudomonades. This results from the combined action of two toxophoric groups (activated halogen and methylol) in one molecule, which may react with the microbe cell's nucleophilic centres, e.g. with SH-group-carrying enzymes. That mechanism leads to a broader range of efficacy, but also to inactivation by the effect of organic matter. Blood and serum at higher concentrations (50%), sulphhydryl compounds such as cysteine, sodium thiosulphate and sodium metabisulphite are markedly antagonistic to the antimicrobial efficacy of Bronopol.

Bronopol, like the majority of electrophilic active ingredients, has a slow microbicidal effect: Bronopol concentrations 2–4 times higher than the MIC in Table 18 take as long as 24 h to display bactericidal activity.

Thanks to its characteristics 'colourless and odourless, easily soluble in water and alcohol, low toxicity, good skin compatibility, broad effective spectrum', Bronopol is being used on a large scale as a preservative for cosmetics and pharmaceuticals (concentrations: 0·01–0·1%). It is listed in the EC list of preservatives allowed for the in-can protection of cosmetics (max. authorized concentration: 0·1%; limitations and requirements: avoid formation of nitrosamines). Since it is in acidic solutions that Bronopol features the highest stability, weakly acidic media are the ideal field of application. In neutral or weakly alkaline formulations there is a risk that Bronopol releases nitrite (see above) which, with defined amines and amides, forms nitrosamines and nitrosamides which are regarded as carcinogens.

This imposes limitations on the applicability of Bronopol as preservative, especially in alkaline reacting industrial functional fluids which, such as for example some metal working fluids, contain amines and amides able to form *N*-nitroso compounds.

Neither anionic nor non-ionic surfactants impair Bronopol's antimicrobial efficacy. For that reason the product is a most suitable preservative for detergent solutions, bath foams, shampoos and hair rinses. Bronopol is also compatible with quaternary ammonium compounds (Section 16.1) and cationic active ingredients such as benzalkonium chloride.

3.2.3 2-Bromo-2-nitro-propan-1-ol (BNP)

Br
|
H_3C-C-CH_2OH
|
NO_2

$C_3H_6BrNO_3$

M_r 183·99
CAS-No. 24403-04-1

Chemical and physical properties

Appearance	Colourless to pale yellow crystals
CH_2O content	16%
Melting point	Approx. 40°C
Solubility	Approx. 130 g/litre water; highly soluble in polar organic solvents
Stability	Very stable when dry; in water decreasing stability as pH and temperature increase; decomposition products: formaldehyde, bromide, nitrite and nitro alcohols; more stable under acid conditions than bronopol (Section 3.2.2)

Toxicity data

According to preliminary data BNP shows a toxicity profile similar to Bronopol (Section 3.2.2).

Antimicrobial effectiveness/applications

BNP possesses a broad spectrum of activity covering bacteria, yeasts, fungi and algae (see Table 19). The efficacy for fungi is not as distinctive as the activity for bacteria, but by far not as limited as the antifungal activity of Bronopol (Section 3.2.2.). Altogether BNP is more effective than Bronopol which is not surprising considering the fact that BNP bearing only one hydroxymethyl group displays better lipoid solubility in comparison to Bronopol bearing two hydroxymethyl groups. The remaining characteristics of BNP are similar to Bronopol; that applies also to the application and fields of application. However BNP's significantly enhanced antifungal activity gives additional benefits in many areas of application.

Table 19. Minimum Inhibition Concentrations (MIC) of BNP in Nutrient Agar (Elsmore & Guthrie, 1991)

Test organism	*MIC (mg/litre)*
Gram-positive baceteria	
Micrococcus flavus	50
Staphylococcus aureus NCIB 9518	50
Staphylococcus aureus	25
Staphylococcus epidermidis NCTC 7291	25
Streptococcus faecalis NCTC 8213	50
Gram-negative bacteria	
Pseudomonas aeruginosa NCIB 11338	25
Pseudomonas aeruginosa NCTC 6750	50
Pseudomonas putida NCIB 9034	25
Pseudomonas stutzeri NCIB 9040	12·5
Pseudomonas cepacia NCIB 9085	25
Pseudomonas fluorescens NCIB 9046	25
Pseudomonas sp.	50
Proteus vulgaris NCTC 4635	25
Proteus morganii NCTC 10041	25
Escherichia coli NCTC 5934	25
Escherichia coli NCTC 9517	25
Klebsiella aerogenes NCTC 418	25
Enterobacter cloacae	25
Salmonella typhimurium NCTC 74	25
Serratia marcescens	25
Yeasts	
Candida albicans ATCC 10231	25–50
Candida tropicalis	25–50
Saccharomyces cerevisiae NCYC 87	50
Spoilage yeast	50
Spoilage fungi	
Stachybotrys atra IMI 82021	12·5–25
Myrothecium verrucaria IMI 45541	50
Amorphotheca resinae IMI 89560	200
Aspergillus niger ATCC 16404	100–200
Aspergillus niger	100–200
Aspergillus sp.	25
Chaetomium globosum IMI 45550	25-50
Cladosporium herbarum	50
Margarinomyces fasiculatis	50
Spoilage mould	100–200
Penicillium funiculosum IMI 87160	25–50
Penicillium sp.	50–100
Spoilage mould	200
Spoilage mould	100
Spoilage mould	100–200
Trichoderma viride	400
Algae	
Phormidium foveolarum CCAP 1462/1	12·5
Nostoc sp. CCAP 1463/4A	9·4
Chlorella emersonii CCAP 2211/8A	6·25

3.2.4 **2-Trifluoromethyl-2-nitro-propane-1,3-diol** (Baasner *et al.*, 1988)

$$\begin{array}{c} NO_2 \\ | \\ HO\text{-}CH_2\text{-}C\text{-}CH_2\text{-}OH \\ | \\ CF_3 \end{array}$$

$C_4H_6F_3NO_4$

M_r 189·10
CAS-No. 3857-06-5

Chemical and physical properties

Appearance	Colourless crystals
CH_2O content	32%
Melting point	128–130°C
Solubility	Sparingly soluble in water, soluble in acetone, DMF

Toxicity data

Not yet available.

Antimicrobial effectiveness/applications

Table 20. Minimum Inhibition Concentrations (MIC) of 2-Trifluoromethyl-2-nitro-propane-1,3-diol in Nutrient Agar

Test organism	*MIC (mg/litre)*
Aerobacter aerogenes	500
Aeromonas punctata	500
Bacillus mycoides	500
Bacillus subtilis	500
Escherichia coli	200
Leuconostoc mesenterioides	500
Proteus mirabilis	200
Pseudomonas aeruginosa	350
Pseudomonas fluorescens	500
Staphylococcus aureus	200
Alternaria alternata	200
Aspergillus niger	500
Aureobasidium pullulans	500
Chaetomium globosum	350
Cladosporium cladosporioides	200
Penicillium glaucum	500
Sclerophoma pityophila	200
Trichoderma viride	500

The spectrum of effectiveness covers bacteria and fungi and is interesting in its equality; it is in that respect superior to that of Bronopol.

2-Trifluoromethyl-2-nitro-propane-1,3-diol may be used as a broad spectrum preservative for industrial fluids, e.g. detergents, cosmetics, polymer emulsions and other water based formulations which can be deteriorated due to microorganisms.

3.2.5 5-Bromo-5-nitro-1,3-dioxane — Bronidox

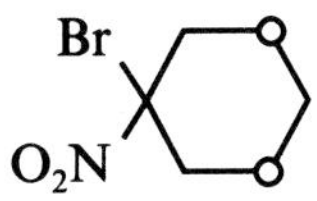

$C_4H_6BrNO_4$ M_r 212
CAS-No. 126958-12-1

Chemical and physical properties

Appearance	White crystals	
CH_2O content	42·3%	
Melting point	58–60°C	
Boiling point at 17 hPa	113–116°C (decomposition)	
Solubility (g/litre at 20°C)	In water	4
	ethanol	250
	isopropanol	100
	propylene glycol	100
	trichloro methane	500

Stability

Bronidox is chemically the formal of 2-bromo-2-nitro-propane-1,3-diol (Bronopol) and therefore more stable than Bronopol. Figure 23 demonstrates that formaldehyde is released from Bronidox in a small amount only and very slowly even at pH 9. However, finally the degradation pathway follows the Bronopol pathway leading to formaldehyde, bromide, nitro-alcohols and nitrite.

Toxicity data

LD_{50} oral	590 mg/kg mouse;
	455 mg/kg rat

Skin irritation at concentrations >0·5%; mucous membrane irritation at concentrations >0·1%.
No skin sensitization according to guinea pig test.
Bronidox is partly resorbed by the skin.

Antimicrobial effectiveness/application

On account of its affinity with Bronopol, Bronidox too is more effective than one would expect considering its stability and its formaldehyde content. Bronidox has a more equilibrated effective spectrum than Bronopol.

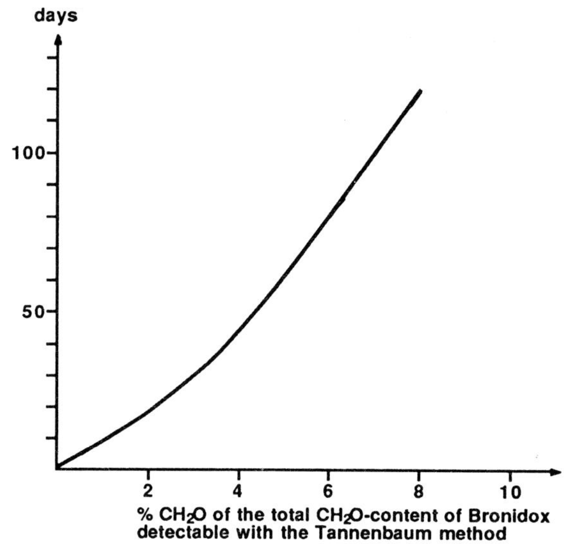

Fig. 23. Release of formaldehyde in a 0·1% solution of Bronidox in water at pH 9 (20°C).

Table 21. Minimum Inhibition Concentrations (MIC) of Bronidox in Nutrient Agar

Test organism	*MIC (mg/litre)*
Escherichia coli	35
Proteus vulgaris	50
Pseudomonas aeruginosa	35
Pseudomonas fluorescens	50
Staphylococcus aureus	50
Streptococcus faecalis	75
Candida albicans	50
Aspergillus niger	50
Chaetomium globosum	10
Penicillium glaucum	10

Bronidox can be used as preservative for functional industrial fluids including cosmetics within a comparatively broad pH range (5–9). The average concentration is 0·1%. In terms of degradation and potential nitrosamine formation Bronidox and Bronopol behave analogously. Bronidox is mentioned in the EC list of preservatives for cosmetics (max. authorized concentration 0·1%; limitations and requirements: rinse-off products only; avoid formation of nitrosamines).

3.2.6 Mixture of *N*-(2-nitrobutyl)morpholine and *N,N'*-(2-ethyl-2-nitrotrimethylene)dimorpholine

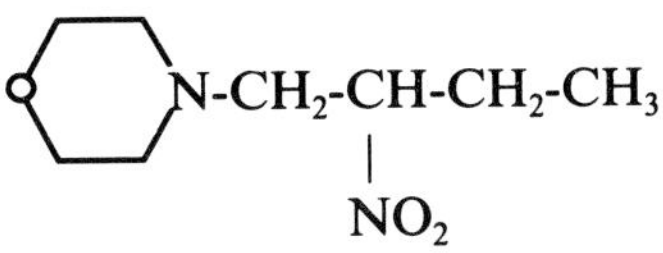

$C_8H_{16}N_2O_3$ — M_r 188·23 — CAS-No. 2224-44-4

Liquid — CH_2O content: 15·9%

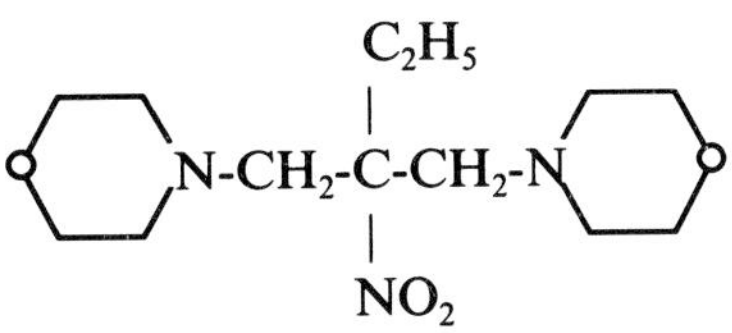

$C_{13}H_{25}N_3O_4$ — M_r 287·36 — CAS-No. 1854-23-5

Crystals — CH_2O content: 20·9%

These compounds are listed under 'C-methylols' because monomethylol nitropropane and dimethylol nitropropane are the intermediates in the synthesis which react with morpholine to form the end products.

Chemical and physical properties

Appearance	Yellow liquid
Density at 20°C	1·098 g/ml
Viscosity at 20°C	40·9 mP s
Vapour pressure at 20°C	< 100 hPa
Refractive index (20°C)	n_D 1·472
Solidification point	Approx. −2°C
Boiling point	> 200°C
Flash point	> 90°C
Solubility (g/litre at 20°C)	In water 11; soluble in organic solvents including non-polar solvents
Reaction in water	Slightly alkaline (pH of the saturated solution in water is 7·5–8).
Stability	Stable and effective between pH 6 and 10

Toxicity data

LD_{50} oral 310–455 mg/kg rat

The concentrated product is corrosive to the skin and mucous membranes; no skin irritation at concentrations less than 1%.

Antimicrobial effectiveness/applications

Table 22. Minimum Inhibition Concentrations (MIC) of the Mixture of *N*-(2-nitrobutyl)morpholine and *N,N'*-(2-ethyl-2-nitro-trimethylene)morpholine in Nutrient Agar

Test organism	*MIC (mg/litre)*
Bacillus subtilis	750
Desulfovibrio desulfuricans	75
Escherichia coli	500
Pseudomonas aeruginosa	500
Staphylococcus aureus	500
Streptococcus faecalis	500
Aspergillus niger	100
Penicillium chrysogenum	100
Candida albicans	100
Chaetomium globosum	1000

The broadness of the spectrum of effectiveness is noteworthy. The a.m. mixture of morpholine derivatives is distinguished by high solubility in non-polar solvents, an advantage which is of value, when the mixture has to be incorporated into oil concentrates (e.g. lubricoolants) as an antimicrobial that is effective as a preservative in the emulsions prepared later out of such oil concentrates in water. It is recommended to establish concentrations between 500 and 1000 ppm of a.i. in such emulsions for a good protection against microbial deterioration.

3.3 Reaction products of amines and formaldehyde

Products obtained by reaction of formaldehyde and amines are always characterized by, among other properties, antimicrobial effectiveness. Here, too, the introduction of the methylol group is the first reaction step; it may be followed by water being split off intra- or intermolecularly or by reaction with other formaldehyde molecules, thus allowing a wide variety of formaldehyde releasing compounds to be obtained (see Fig. 24).

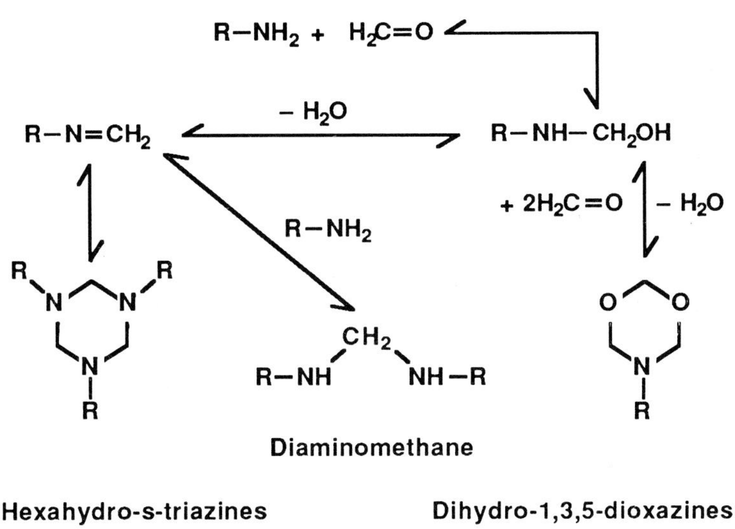

Fig. 24. Reaction of formaldehyde with amines (Paulus, 1980).

Probably the first synthesized formaldehyde releasing compound is hexamethylenetetramine (HTA) which was obtained in 1860 from the condensation of ammonia and formaldehyde according to the reaction pathway illustrated in Fig. 25.

The antimicrobial efficacy of the amine–formaldehyde reaction products essentially corresponds to the formaldehyde content of these compounds. Known exceptions are HTA, hexahydro-oxadiazines and octahydro-s-tetrazines, which derive from ammonia respectively certain alkylolhydrazines (Paulus, 1980); here the detection of formaldehyde by the Tannenbaum methods gives a negative result, which means that these substances have no significant antimicrobial effect at neutral to alkaline pH; they release formaldehyde in acidic media only. This pH dependency is broken off, if, for example, HTA is quaternized (Jacobs *et al.*, 1916). In contrast to HTA the quaternary hexaminium salts release formaldehyde widely independent of pH and therefore may be used as preservatives also for media of neutral to alkaline pH. They are not comparable with the surface active conventional quaternary ammonium compounds (QACs; see Section 16.1)

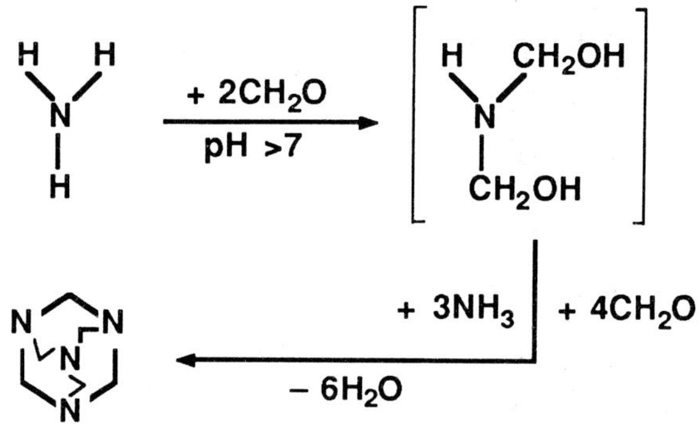

Fig. 25. Condensation of NH_3 and CH_2O to hexamethylenetetramine.

as their antimicrobial effectiveness is based on the release of formaldehyde. The efficacy is however lower than one would expect considering the formaldehyde content of the hexaminium salts (Scott & Wolf, 1962). So it has been observed that in aqueous solutions, especially with pH values above 7, the hexaminium salts' antimicrobial efficacy rapidly decreases to quickly stabilize at a constant inhibition value. This phenomenon is explained by Fig. 26 which illustrates how the hydrolytic cleavage takes place.

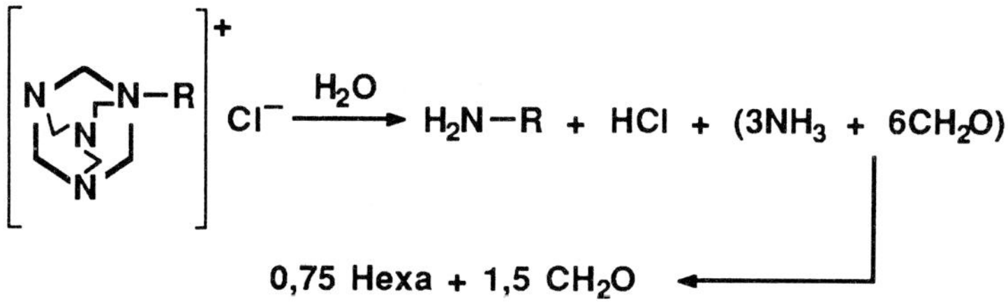

Fig. 26. Hydrolytic cleavage of quaternary hexaminium salts.

The 3 M NH_3 and 6 M CH_2O first liberated during hydrolytic cleavage at pH values >7 will soon become rearranged to form 0·75 M hexamethylenetetramine and 1·5 M CH_2O, the latter remaining available for the antimicrobial efficacy. Examination of MICs of different quaternary hexaminium salts confirms that in actual fact a mere 25% (approx.) of the hexaminium salts' calculated formaldehyde content is available for the antimicrobial efficacy (see Table 23).

Formaldehyde releasing compounds of superior and, above all, more balanced antimicrobial effectiveness are obtained by reacting hydroxy-group-containing formaldehyde releasing compounds and arylisocyanates, provided that the hydroxy group does not form part of a methylol group that may be split off. The microbicides thus formed are formaldehyde releasing *N*-aryl-carbamates (Paulus *et al.*, 1975).

Table 23. MIC of Quaternary Hexaminium Salts for *Bacillus subtilis* Found and Calculated from the CH_2O Content of the Compounds

Compound	*MIC (mg/litre)*		
	Found	*Calculated*	*25% of calculated efficacy*
$[Hexa\text{-}CH_2\text{-}CH{=}CHCl]^+ Cl^-$	100	21	84
$[Hexa\text{-}CH_2\text{-}CO\text{-}NH_2]^+ Cl^-$	100	20	80
$[Hexa\text{-}CH_2\text{-}CO\text{-}NH\text{-}CH_2OH]^+ Cl^-$	50	18	72
Formaldehyde	15	—	—

3.3.1 Hexamethylenetetramine (HTA) — Hexamine — Urotropin

$C_6H_{12}N_4$

M_r 140·19
CAS-No. 67931-01-5

Chemical and physical properties

Appearance	White crystalline, odourless
CH_2O content	6 mol
Melting point	180°C (sublimation)
Flash point	250°C
Solubility (g/litre):	In water 660
	ethanol 28
	very soluble in chloroform, insoluble in ether
Reaction in water	pH of a 0·2 M aqueous solution: 8·4
Stability	Stable at neutral and basic pH, releases formaldehyde in acidic media.

Toxicity data

LD_{50} subcutaneous	200 mg/kg rat
90 days feeding test with rats	0·4 g HTA/day did not cause toxic effects.

No carcinogenic effects when applied orally.

ADI value	0–0·15 mg/kg/day

Antimicrobial effectiveness/applications

The efficacy at HTA is based on its formaldehyde content which, however, is only released in acid media. According to its composition HTA is the formaldehyde releasing compound with the highest percentage of bound formaldehyde. As with formaldehyde, HTA preferably attacks bacteria and not mould producing fungi to such an extent.

HTA is listed in the EC list of preservatives for cosmetics (maximum authorized concentration 0·15%). The application is very limited, as HTA is only effective in acid formulations. As a preservative for acid food preparations or technical functional fluids, HTA today is no longer of importance.

3.3.2 *Cis* isomer of 1-(3-chloroallyl)-3,5,7-triaza-l-azoniaadamentane chloride — *N*-(3-chlorallyl)-hexaminium chloride

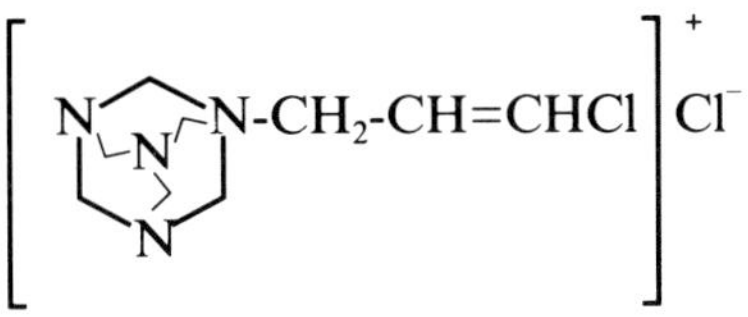

$C_9H_{16}Cl_2N_4$ M_r 251·16
CAS-No. 51229-78-8

Chemical and physical properties

Appearance	White to cream coloured, hygroscopic, odourless powder	
CH_2O content	71·7%	
Decomposition temperature	Starting at 60°C	
Solubility (g/litre at 20°C)	In water	> 2000
	ethanol	44
	isopropanol	9
	methanol	390
	propylene glycol	200
	glycerine	210
	mineral oil	<1
Reaction in water	pH of 1% solution is 5·1	
Stability	Stable in the solid state at storage temperatures up to 40°C; releases formaldehyde in aqueous solution widely independent of pH	

Toxicity data

LD_{50} oral	940–1500 mg/kg rat;
	710 mg/kg guinea pig;
	40–80 mg/kg rabbit
Mutagenicity	Ames test negative

The undiluted material and concentrated solutions are moderately irritant to the skin. 2% solutions are not irritant and do not cause sensitization according to the guinea pig test.

Antimicrobial effectiveness/applications

Table 24. Minimum Inhibition Concentrations (MIC) of *N*-(3-chloroallyl)-hexaminium Chloride in Nutrient Agar and Minimum Lethal Concentrations (MLC) Determined through a Tube Suspension/Dilution Technique (Exposure Time: 48 h)

Test organism	*MIC (mg/litre)*	*MLC (mg/litre)*
Aerobacter aerogenes	100	250
Bacillus subtilis	75	250
Escherichia coli	250	300
Pseudomonas aeruginosa	300	350
Salmonella typhosa	50	200
Salmonella choleraesuis	50	200
Staphylococcus aureus	100	250
Aspergillus terreus	250	
Aureobasidium pullulans	500	
Rhizopus nigricans	500	
Trichophyton mentagrophytes	250	300

The efficacy of *N*-(3-chloroallyl)-hexaminium chloride covers a broad spectrum of bacteria and mould producing fungi, yeasts included. However, the activity against bacteria is more pronounced than that against fungi, characterizing the microbicide as a formaldehyde releasing compound.

As a hexaminium salt the active ingredient has a weak cationic character. Nevertheless it is compatible with anionic compounds. The formaldehyde which is responsible for the antimicrobial efficacy is released independent of the pH of the substrate. Consequently *N*-(3-chloroallyl)-hexaminium chloride is a suitable preservative for many different water based formulations where it may be used without the need of pH adjustment — an important factor when the substrate is pH sensitive. In concentrations from 0·05 to 0·25% the preservative is applied for the in-can or in-tank protection of polymer emulsions, wet paper coatings, starches, inks, detergents, concrete additives and numerous other water based systems. According to its salt character and its high water solubility *N*-(3-chloroallyl)-hexaminium salt does not tend to concentrate in the organic phase of oil and water emulsion systems but remains in the aqueous phase where the vegetation of microbes is likely to arise. The application should, however, be restricted to formulations which do not contain protein (e.g. casein) and to those which tolerate the tendency of the hexaminium salts to cause a little bit of yellowing especially in formulations containing traces of monomers which may react with the amine compounds released from hexaminium salts.

N-(3-chloroallyl)-hexaminium chloride is listed in the EC list of preservatives for cosmetics with a maximum addition rate of 0·2%.

3.3.3 1,1'-(2-Butenylene)-bis-(3,5,7-triaza-1-azoniaadamentane chloride)

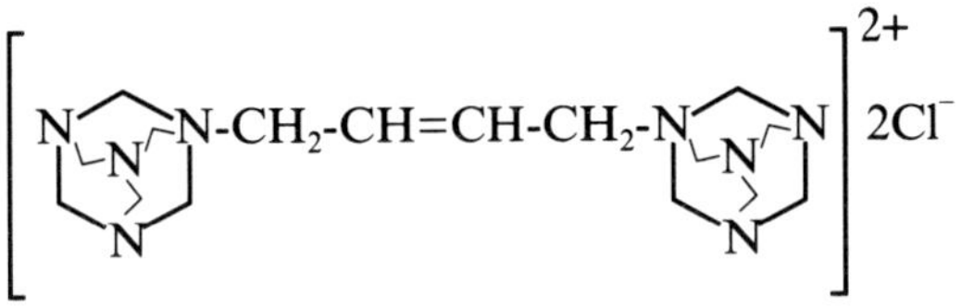

$C_{16}H_{30}Cl_2N_8$ M_r 405·43
CAS-No. 51350-84-6

Chemical and physical properties

Appearance	White, flowable, crystalline, odourless powder (96% a.i.)
CH_2O content	88·8%
Melting point	150°C (decomposition)
Solubility	Highly soluble in water, moderately soluble in polar organic solvents, practically insoluble in non-polar solvents (e.g. oils)
Reaction in water	pH of a 2% solution is 8·5–9
Stability	Stable in the solid state, releases formaldehyde in aqueous solutions widely independent of pH

Toxicity data (Luloff & Eilender, 1975)

LD_{50}	oral	3400 mg/kg
	dermal	8000 mg/kg

Antimicrobial effectiveness/applications

Table 25. Minimum Inhibition Concentrations (MIC) of 1,1'-(2-Butylene)-bis-(3,5,7-triaza-1-azoniaadamentane Chloride) in Nutrient Agar

Test organism	*MIC (mg/litre)*
Aerobacter aerogenes	125
Bacillus subtilis	60
Escherichia coli	250
Pseudomonas aeruginosa	250
Salmonella choleraesuis	60
Staphylococcus aureus	125

The antimicrobial efficacy corresponds with the formaldehyde which is released and disposable from the molecule (see Section 3.3). Although there are two hexaminium moieties in the molecule of the a.i., compared with the *N*-(3-chloroallyl)-hexaminium salt one can expect an increase in efficacy of 20% only due to the considerable ascent of the molecular weight.

The quaternary bis-hexaminium salt described here is particularly useful as a preservative in a wide variety of water based systems, e.g. polymer emulsions, latex paints, pigment and dye slurries, concrete additives, starches, thickener solutions. Addition rates: 0·05–0·2%. For the rest what is mentioned before under Section 3.3 and 3.3.2 is valid.

3.3.4 1-Carboxymethyl-3,5,7-triaza-1-azoniaadamantane chloride

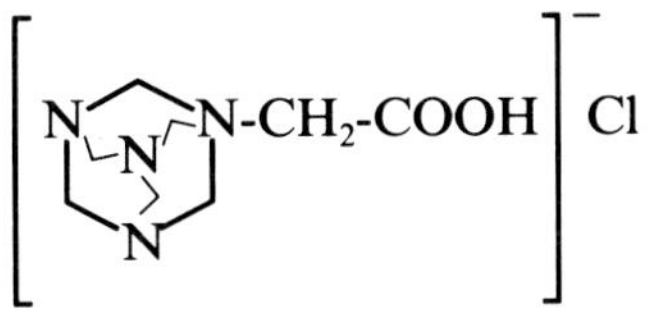

$C_8H_{15}ClN_4O_2$

M_r 234·69
CAS-No. 92623-86-4

CH_2O content 76·7%

Chemical and physical properties of the 40% sodium salt solution

Appearance	Clear brown liquid
Density at 20°C	1·19 g/ml
Solubility	Highly soluble in water, soluble in polar solvents
Reaction in water	pH of a 0·01% solution is between 5·5 and 7·0
Stability	Releases formaldehyde widely independent of the pH of the substrate

Toxicity data

LD_{50}	oral	>5000 mg/kg rat
	dermal	>2000 mg/kg rabbit

No primary skin irritation
Slight irritation of mucous membranes
No sensitization in the guinea pig test
No mutagenic effects according to three different mutagenicity tests.

Antimicrobial effectiveness/applications

Table 26. Minimum Inhibition Concentrations (MIC) of the 40% Aqueous Sodium Salt Solution in Nutrient Agar

Test organism	*MIC (mg/litre)*
Bacillus subtilis	80
Enterobacter aerogenes	100
Escherichia coli	120
Klebsiella pneumoniae	80
Proteus vulgaris	100
Pseudomonas aeruginosa	200
Salmonella choleraesuis	80
Staphylococcus aureus	80

The antimicrobial activity of the a.i. as demonstrated in Table 26 is exhibited over a broad pH range.

The *N*-carboxymethyl-hexaminium salt is a zwitterionic substance that can be used as a preservative for aqueous functional fluids containing nonionic, anionic, or cationic components without deactivation or problems of incompatibility. Addition rates: 0·1–0·3%.

3.3.5 1-Carbamoylmethyl-3,5,7-triaza-1-azoniaadamantane chloride

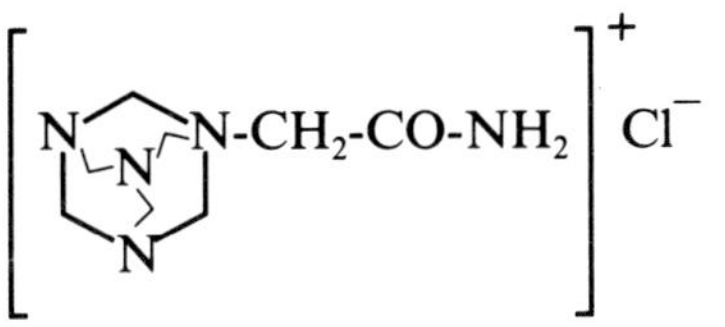

$C_8H_{15}ClN_5O$

M_r 233·70
CAS-No. 92988-93-7

Chemical and physical properties

Appearance	White, crystalline, odourless
CH_2O content	77·0%
Melting point	164°C (decomposition)
Solubility	Highly soluble in water, moderately soluble in polar solvents, practically insoluble in non-polar solvents, e.g. oils
Stability	Stable in the solid state; releases formaldehyde in aqueous solutions widely independent of pH
Toxicity data	Not yet available

Antimicrobial effectiveness/applications

Table 27. Minimum Inhibition Concentrations (MIC) of *N*-carbamoyl-hexaminium Chloride in Nutrient Agar

Test organism	*MIC (mg/litre)*
Aerobacter aerogenes	100
Bacillus subtilis	100
Escherichia coli	120
Pseudomonas aeruginosa	120
Salmonella typhosa	100
Staphylococcus aureus	100
Aspergillus niger	500
Penicillium glaucum	1000
Rhizopus nigricans	250

The antimicrobial activity depends on the release of formaldehyde over a wide pH range as is already described under Sections 3.3–3.3.4.

3.3.6 1-[(*N*-hydroxymethyl-carbamoyl)-methyl]-3,5,7-triaza-1-azonia adamantane chloride — *N*-(*N'*-hydroxymethyl-)carbamoylmethylhexaminium chloride

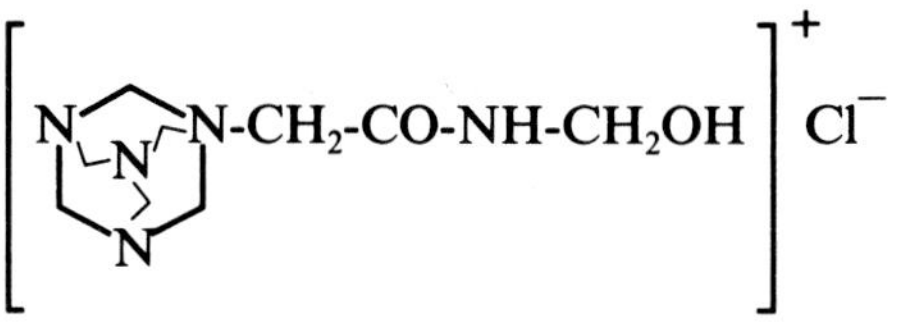

$C_9H_{18}ClN_5O_2$ M_r 263·72
CAS-No. 67508-69-4

Chemical and physical properties

Appearance	White, crystalline, odourless	
CH_2O content	79·6%	
Melting point	155°C (decomposition)	
Solubility (g/litre at 20°C)	In methanol	47
	ethylene glycol	19
	acetone	0·08
	cyclohexane	0·02
	highly soluble in water	
Stability	Stable in the solid state; releases formaldehyde in aqueous solutions widely independent of pH	
Reaction in water	pH of an aqueous solution (5 g/litre) is approx. 4·5	

Toxicity data for a 70% aqueous solution

LD_{50}	oral	>5000 mg/kg rat
	intravenous	>1000 mg/kg rat
	dermal	>1000 mg/kg rat

No primary skin irritation
Slight irritation of mucous membranes

Antimicrobial effectiveness/applications

N-(*N'*-hydroxymethyl-)carbamoylmethyl-hexaminium chloride is compared with other quaternary hexaminium salts of superior efficacy as the starting material for quaternizing hexamethylenetetramine and in this case is a formaldehyde releasing compound, too, namely *N*-hydroxymethyl-chloracetamide (see Section 3.4.1) — Applications as described under Sections 3.3.2–3.3.4 — Addition rates: 0·05–0·2%.

Table 28. Minimum Inhibition Concentrations (MIC) of *N*-(*N*'-hydroxy methyl-)carbamoylmethyl-hexaminium Chloride in Nutrient Agar

Test organism	*MIC (mg/litre)*
Bacillus mycoides	120
Bacillus subtilis	50
Bacterium punctatum	120
Escherichia coli	80
Proteus vulgaris	80
Pseudomonas aeruginosa	120
Pseudomonas fluorescens	120
Staphylococcus aureus	120
Aspergillus flavus	180
Aspergillus niger	400
Aureobasidium pullulans	250
Chaetomium globosum K.	180
Penicillium glaucum	800
Rhizopus nigricans	250

3.3.7 Other quaternary hexaminium salts (Scott & Wolf, 1962)

	CAS-No.
N-methyl-hexaminium chloride	76902-90-4
N-benzyl-hexaminium chloride	5400-93-1
N-tolylmethyl-hexaminium chloride	97159-48-3
N-(3,4-dichloro-)benzyl-hexaminium chloride	96433-36-2
N-(4-chloro-)benzyl-hexaminium chloride	96634-13-8
N-propyl-hexaminium bromide	92987-56-9
N-propylene-hexaminium chloride	35511-29-6
N-propargyl-hexaminium bromide	13496-17-8
N-butyl-hexaminium bromide	94584-23-3

The list is not complete and may be continued.

3.3.8 1,3-Oxazolidines — reaction products of 2-hydroxy-alkylamines and formaldehyde

After the addition of 1 M CH_2O to primary or secondary 2-hydroxy-alkylamines, 1,3-oxazolidines are formed by intramolecular elimination of water (1 M); if at least 1·5 M CH_2O are added to a primary 2-hydroxy-alkylamine dioxazolidinyl methanes can be synthesized by the elimination of 3 mol of water intra- and intermolecularly according to the reaction pattern in Fig. 27.

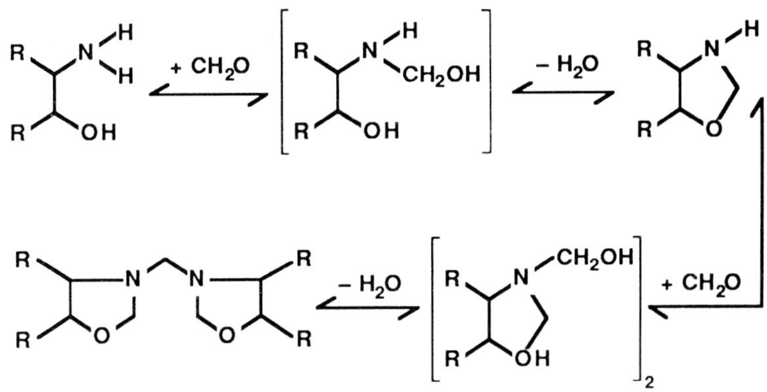

Fig. 27. Reaction scheme for the formation of 1,3-oxazolidines and dioxazolidinyl methanes.

The 1,3-oxazolidines listed here are highly soluble in water and polar solvents, and in oil, too. Solutions of 1,3-oxazolidines in water have a basic pH. The CH_2O releasable is detectable by the Tannenbaum method; in consequence the antimicrobial effectiveness corresponds to the CH_2O content of the compounds.

Chemical and physical properties

Compound	*Structural formula*	*Empirical formula*	M_r *CAS-No.*	*CH_2O content*	*Appearance*
N-methyl-1,3-oxazolidine	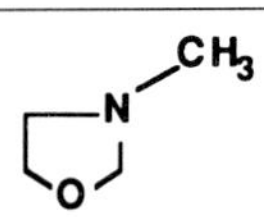	C_4H_9NO	87.12 27970–32–7	34%	liquid bp 101°C
N-ethylol-1,3-oxazolidine	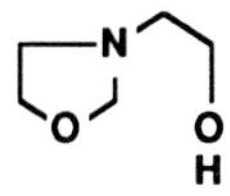	$C_5H_{11}NO_2$	117.15 20073–50–1	26%	liquid bp (0.7) 58-9°C
5-methyl-1,3-oxazolidine	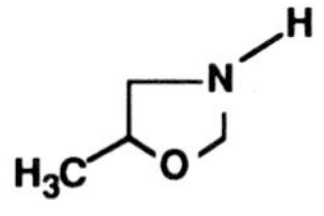	C_4H_9NO	87.12 58328–22–6	34%	liquid bp 124-6°C

continued

Chemical and physical properties — continued

Compound	*Structural formula*	*Empirical formula*	M_r *CAS-No.*	*CH_2O content*	*Appearance*
4,4-dimethyl-1,3-oxazolidine	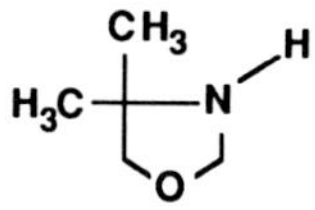	$C_5H_{11}NH$	101.14 51200–87–4	30%	liquid bp (25) 45°C
bis-(5,5'-dimethyl-1,3-oxazolidin-3-yl)-methane	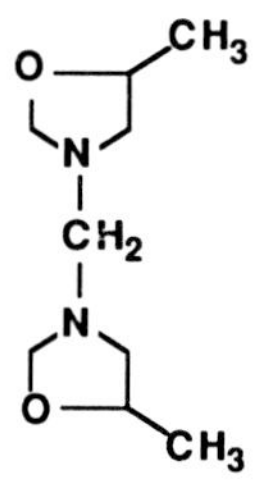	$C_9H_{18}N_2O_2$	186.26 66204–44–2	48%	liquid bp (25) 123°C
bis-(4,4'-tetramethyl-1,3-oxazolidin-3-yl)-methane	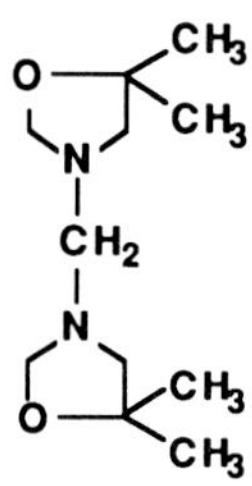	$C_{11}H_{22}N_2O_2$	214.31	42%	liquid bp (10) 120°C
1-aza-3,7-dioxa-5-methylol-(3.3.0)-bicyclooctane	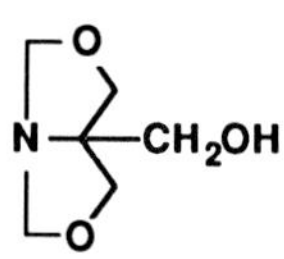	$C_6H_{11}NO_3$	145.16 6542–37–6	41%	solid mp 64°C
1-aza-3,7-dioxa-5-ethyl-(3.3.0)-bicyclooctane	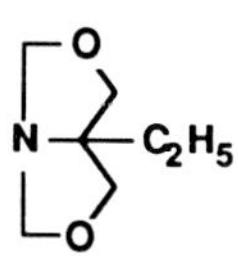	$C_7H_{13}NO_2$	143.18 7747–35–5	42%	liquid bp (15) 71°C

Toxicity data

(a) For 4,4-dimethyl-1,3-oxazolidine

LD_{50} oral	950 mg/kg rat
dermal	1450 mg/kg rabbit
Inhalation LC_{50} for rats	11·7 mg/litre

Mutagenicity: Ames test negative
Sensitization: not observed for the concentrations used (0·05–0·2%) in application

The concentrated compound is irritant to skin and mucous membranes.

LC_{50} after 96 h exposure for

Bluegill sunfish	59 mg/litre
Rainbow trout	93 mg/litre

(b) For bis(5,5'-dimethyl-1,3-oxazolidin-3-yl)-methane

LD_{50} oral Approx. 1000 mg/kg rat

Irritant to skin and mucous membranes.
Acute toxicity for Brachydanio rerio:
LC_{50} approx. 50 mg/litre (exposure: 96 h)

(c) for 1-aza-3,7-dioxa-5-ethyl-(3.3.0)-bicyclooctane

LD_{50} oral	3680–5250 mg/kg rat
dermal	1948 mg/kg rabbit
Inhalation LC_{50} for rats	3·1 mg/litre (Exposure: 4 h)

Irritant to skin and mucosa.
Not a skin sensitizer.
Non-mutagenic.

In subacute studies with rats, oral and dermal, an unobserved effect level of 100 mg/kg/day was determined.

LC_{50} after 96 h exposure for

Bluegill sunfish	240 mg/litre
Rainbow trout	130 mg/litre

Antimicrobial effectiveness/applications

As the 1,3-oxazolidines are compatible with cationic, anionic and non-ionic compounds and are active over a wide pH range (4–11), they are suitable preservatives for many different functional fluids. However, one has to bear in mind that the 1,3-oxazolidines have the same deficiency in their antimicrobial spectrum mentioned before for CH_2O and other formaldehyde releasing compounds, i.e. a relative decreased activity against fungi. Compared with other amine formaldehyde reaction products where water is not eliminated and an equilibrium exists between N-methylols and condensation products, the 1,3-oxazolidines have a much more advantageous partition coefficient favouring them for incorporation into oil concentrates (e.g. lubricoolants) and for the application in fuel to inhibit microbial growth in fuel oil bottom water.

Table 29. Minimum Inhibition Concentrations (MIC) of bis-(5,5'-dimethyl-1,3-oxazolidin-3-yl)-methane in Nutrient Agar

Test organism	*MIC (mg/litre)*
Aeromonas punctata	150
Bacillus subtilis	450
Desulfovibrio desulfuricans	100
Escherichia coli	200
Proteus vulgaris	150
Pseudomonas aeruginosa	200
Pseudomonas fluorescens	200
Staphylococcus aureus	200
Aspergillus flavus	300
Aspergillus niger	600
Aureobasidium pullulans	600
Chaetomium globosum	600
Penicillium glaucum	100
Rhizopus nigricans	300
Candida albicans	600
Candida crusei	300
Torula utilis	300

4,4-Dimethyl-1,3-oxazolidine is listed in the EC list of preservatives allowed for the in-can protection of cosmetics (maximum authorized concentration 0·1%; rinse-off products only; the pH of the finished product shall not be lower than 6). The latter limitation is made, as at pH values lower than 5,1,3-oxazolidines degrade to formaldehyde and the corresponding amines, the skin compatibility of which is limited.

3.3.9 Tetrahydro-1,3-oxazine

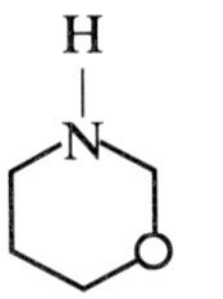

C_4H_9NO M_r 87·12
CAS-No. 85367-17-5

Chemical and physical properties

Appearance	Colourless fluid with a weak pungent odour
CH_2O content	34%
Boiling point	135–6°C
Solubility	Highly soluble in water and polar solvents
Reaction in water	Basic
Stability	Relatively stable at neutral and basic pH, degrades in acidic media quickly to the starting products

Tetrahydro-1,3-oxazines are formed by the addition of CH_2O to primary or secondary 3-hydroxy-alkylamines followed by intramolecular elimination of water, i.e. in analogy to the formation of 1,3-oxazolidines (see Section 3.3.8). Accordingly they behave similar, but do not present advantages. Due to their higher molecular weight they release a lower amount (%) of formaldehyde compared with the corresponding 1,3-oxazolidines. Tetrahydro-1,3-oxazines therefore have not much practical importance as preservatives for functional fluids.

3.3.10 Bis-(tetrahydro-1,3-oxazin-3-yl)methane

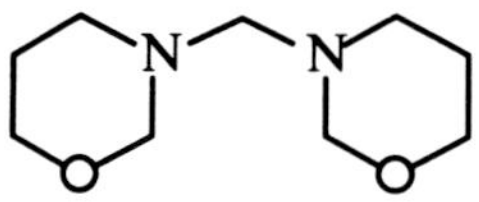

$C_9H_{18}N_2O_2$

M_r 186·25
CAS-No. 63489-63-4

Chemical and physical properties

Appearance	Colourless fluid with a weak odour
CH_2O content	48%
Boiling point at 15 hPa	125°C
Solubility	Highly soluble in water and polar solvents
Reaction in water	Basic
Stability	Stable at basic pH, degrades in acidic media quickly to the starting products

Although bis-(tetrahydro-1,3-oxazin-3-yl) methane is very effective according to its relatively high content of releasable formaldehyde and although it does not emit a pungent odour, it has not yet gained practical importance as a microbicide.

3.3.11 *N*-ethyl-dihydro-1,3,5-dioxazine

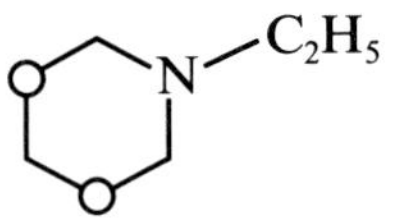

$C_5H_{11}NO_2$ M_r 117·15
CAS-No. 132292-80-9

Chemical and physical properties

Appearance	Colourless fluid with a strong pungent odour
CH_2O content	77%
Boiling point at 55 hPa	62–64°C
Solubility	Soluble in water and polar solvents
Reaction in water	Basic
Stability	Of limited stability only, degrades in water based systems quickly to the starting products

Dihydro-1,3,5-dioxazines are the reaction/condensation products of primary amines with 3 mol of formaldehyde. Although they release a higher quantity of formaldehyde than, for example, 1,3-oxazolidines and are therefore more effective than other formaldehyde releasing compounds they are not of practical importance as microbicides because of their limited stability and very pungent odour.

3.3.12 1,3,5-Substituted-hexahydro-s-triazines

These compounds are built up out of formaldehyde and primary amines by condensation and trimerization in an exothermic reaction (see Fig. 24).

3.3.12.1 Hexahydro-1,3,5-tris(2-hydroxyethyl)-s-triazine

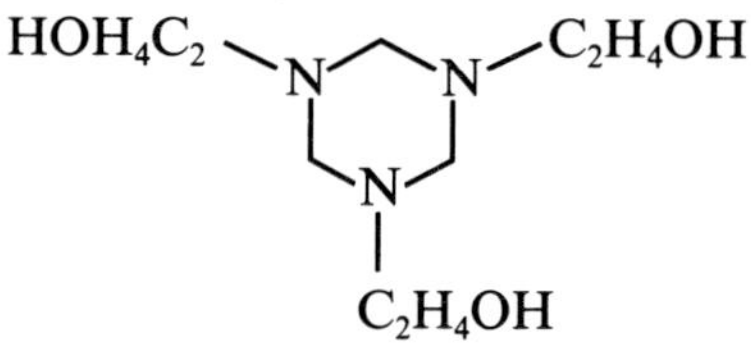

$C_9H_{21}N_3O_3$ M_r 219·29
CAS-No. 4719-04-4

Chemical and physical properties

Appearance	Colourless or yellow viscous liquid with a weak amine odour
CH_2O content	41%
Boiling point at 9 hPa	100°C; the distillate is a fluid of low viscosity which rearranges exothermically to the viscous hexahydro-*s*-triazine
Density at 20°C	1·157 g/ml
Melting point	−20°C
Flash point	>55°C
Ignition temperature	250°C
Solubility	Highly soluble in water and in polar solvents
Reaction in water	Basic; pH of a 0·15% solution in water: approx. 9·5
Stability	Stable at basic pH; degrades in acidic media quickly to the starting products: ethanolamine and formaldehyde

Toxicity data

LD_{50} oral	600–1200 mg/kg rat; 950 mg/kg mouse
LD_{50} dermal	>5 ml/kg rat
Effects upon skin and mucous membranes	Irritation
No sensitization in the guinea pig test	
Mutagenicity	Negative
Fish toxicity	LC_{50} for trouts approx. 70 mg/litre (exposure: 96 h)

Antimicrobial effectiveness/applications

Table 30. Minimum Inhibition Concentrations (MIC) of Hexahydro-1,3,5-tris(2-hydroxyethyl)-*s*-triazine in Nutrient Agar

Test organism	*MIC (mg/litre)*
Escherichia coli	120
Proteus mirabilis	200
Pseudomonas aeruginosa	200
Staphylococcus aureus	200
Formaldehyde resistant bacteria	2500
Candida albicans	750
Aspergillus niger	800
Penicillium glaucum	1500

The efficacy and the spectrum of effectiveness correspond to the amount of formaldehyde which may be released from the hexahydro-*s*-triazine derivative.

The favourable solubility properties combined with low toxicity and good skin compatibility at the dilutions used allow the application of the compound in very different fields of application, mainly as a preservative for functional fluids. Although the hexahydro-1,3,5-tris(2-hydroxyethyl)-*s*-triazine is a formaldehyde releasing compound it is less pungent. Additionally it is cost effective. Therefore it is not surprising that the compound is the main preservative for lubricoolants; estimated market share approximately 50%.

Despite the figures for fungal inhibition problems, fungal growth may occur in practical preservation situations as a result of under-dosing with respect to fungi (Rossmoore & Holtzmann, 1974). Problems with formaldehyde resistant bacteria may occur, too, in consequence of the application of the hexahydro-*s*-triazine derivative alone (Paulus, 1976). These problems are hardly overcome by increasing the dosage as concentrations higher than 0·2% of a.i. are no longer cost effective and include the risk that formaldehyde is released in doses which are not tolerated at work places.

In the EC list of preservatives for cosmetic products the compound is listed among the preservatives provisionally allowed (maximum authorized concentration 0·2%; only for products rinsed off; label: contains hexahydro-1,3,5-tris(2-hydroxyethyl)-*s*-triazine).

3.3.12.2 Hexahydro-1,3,5-triethyl-*s*-triazine (HTT)

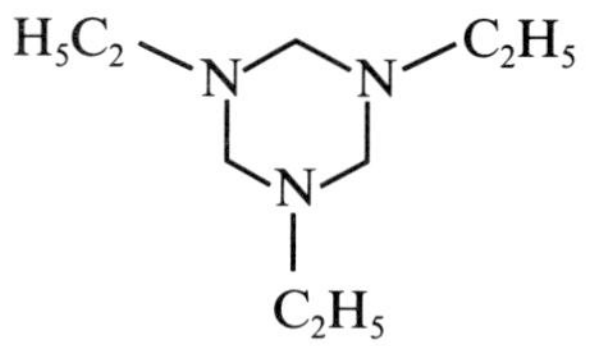

$C_9H_{21}N_3$ M_r 171·29
CAS-No. 7779-27-3

Chemical and physical properties

Appearance	Colourless liquid with a strong to pungent amine odour
CH_2O content	53%
Refractive index (25°C)	n_D 1·4588
Boiling point	196–8°C
Solubility	Soluble in water and in polar and non-polar solvents, e.g. in petroleum distillate fractions (naphtha), kerosene, gas oil, mineral oils, propane, butane, toluene, xylenes, halogenated hydrocarbons
Reaction in water	Basic; pH of a 0·05% solution: 10·4
Stability	Stable at basic pH; degrades in acidic media quickly to the starting products: ethyl amine and formaldehyde

Toxicity data

Similar to those listed under Section 3.3.12.1.

Antimicrobial effectiveness/applications

The spectrum of efficacy corresponds to that of the tris(2-hydroxyethyl)-derivative listed under Section 3.3.12.1 characterized by reduced fungicidal activity and a gap for 'formaldehyde resistant bacteria'. The latter are apparently identical with those which were identified as *Pseudomonas putida* by Hall & Eagon (1985) and given the strain designation 3-T-15². In the experiments of Barnes & Eagon (1986), neither HTT nor ethylamine nor CH_2O could be used by *P. putida* 3-T-15² as a carbon source for growth. However, according to Eagon & Barnes (1986) *P. putida* 3-T-15² has an active formaldehyde dehydrogenase. These findings are in line with those of Paulus (1976), who found that in a nutrient solution containing CH_2O, the latter is no longer detectable after incubation with 'formaldehyde resistant bacteria'.

Among the microbicides based on formaldehyde releasing compounds the hexahydro-*s*-triazines in Sections 3.3.12.1 and 3.3.12.2 have the bulk of the market because of their cost effectiveness, compatibility and relative lack of toxicity. Main application field: metal working fluids (lubricoolants). Compared with the 2-hydroxyethyl compound in Section 3.3.12.1 the ethyl compound has a much more advantageous partition coefficient making it the triazine of choice for oil concentrates. Additionally its solubility in kerosene and related hydrocarbons make it a candidate for overcoming microbiological problems in jet fuels, heating fuel oils, etc., where there is a demand for microbicides combustible without any ash remaining.

Further applications: as a preservative in industrial functional fluids, as a sanitizing agent in dry cleaning processes and in oil based mop saturants.

3.3.12.3 Hexahydro-1,3,5-tris[(tetrahydro-2-furanyl)-methyl]-s-triazine

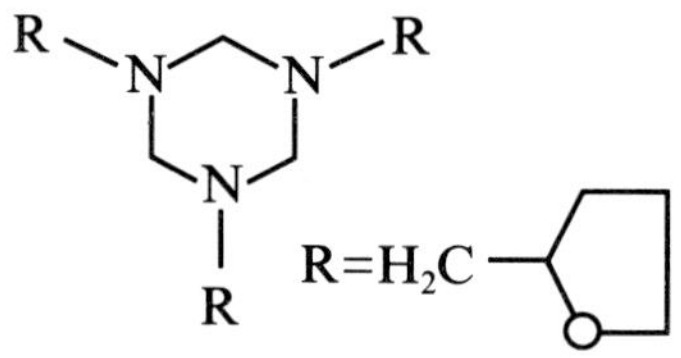

$C_{18}H_{33}N_3O_3$

M_r 339·48
CAS-No. 69141-51-1

Chemical and physical properties

Appearance	Yellow oil
CH_2O content	27%
Boiling point at 1hPa	>50°C
Solubility	Miscible with water, alkanols, lipid-like solvents, oils, lubricoolant concentrates
Reaction in water	Basic
Stability	Stable at basic pH; degradation in acidic media

Toxicity data (Grier *et al.*, 1980)

LD_{50} oral	1060 mg/kg mouse
subcutaneous	>500 mg/kg mouse
intravenous	142 mg/kg mouse
Ames test	Negative
Skin compatibility	A 5% w/v solution was non-irritant on the skin of rabbits (exposure: 24 h); a 1% solution in the eye was only very slightly irritant, a 0·1% solution was non-irritant

Antimicrobial effectiveness/applications

The minimum inhibition concentrations (MIC) for a broad spectrum of fungi associated with metal working fluid deterioration or of economic importance in industrial problems were determined by Grier *et al.* (1980) in comparison to the corresponding MIC of hexahydro-1,3,5-tris(2-hydroxyethyl)-*s*-triazine (Table 31). Apparently there is no significant difference between the two hexahydro-*s*-triazine derivatives in their activity against fungi.

Main application for the triazine I: preservation of industrial functional fluids, especially lubricoolants.

There is a significantly prolonged antifungal action obtained in lubricoolants when compared to failure with the tris(2-hydroxyethyl)-derivative (Section 3.3.12.1), although the MIC of the two hexahydro-*s*-triazines do not differ very much.

However, the fact that the (tetrahydro-2-furanyl)-methyl derivative is fivefold more lipoidal than the 2-hydroxyethyl analogue may be responsible for the effect (Grier *et al.*, 1980).

Table 31. Antifungal Spectrum of Hexahydro-*s*-triazines. (I = (tetrahydro-2-furanyl)-methyl-derivative; II = 2-hydroxyethyl-derivative)

Test fungi	*MIC (mg/litre)*		
	MF[a]	*Triazine I*	*Triazine II*
Aspergillus niger	442	300	200
Alternaria solani	3550	400	200
Aureobasidium pullulans	4341	200	100
Botrytis alli	3587	100	100
Cephalosporium sp.	4641	400	>400
Ceratocystis pilifera	4339	100	50
Ceratocystis ulm	4042	200	200
Cladosporium fulvum	35	100	100
Cochliobolus miyabeanus	4626	50	100
Cochliobolus miyabeanus	4630	50	100
Fusarium oxysporum	4014	400	400
Fusarium sp.	4642	>400	>400
Helminthosporium biforme	3640	400	300
Helminthosporium cymodontis	3642	100	100
Penicillium digitatum	4591	400	100
Phoma sp.	4332	100	100
Scopulariopsis brevicaulis	3769	100	100
Pithomyces chartarum	4395	50	100
Trichoderma lignorum	3560	300	300
Trichoderma viride	4064	200	50
Ustilago zeae	1996	100	50
Verticillium albo-atrium	3793	200	100
Verticillium serrae	3794	100	100

[a] Merck fungi.

3.3.13 *N*-methylene-cyclohexylamine

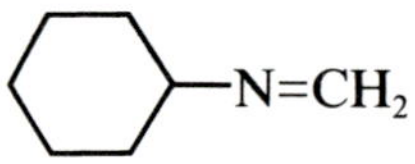

$C_7H_{13}N$ M_r 111·19
CAS-No. 4705-14-0

Chemical and physical properties

Appearance	White, crystalline, weak amine odour
CH_2O content	27%
Melting point	75°C
Solubility	Sparingly soluble in water, soluble in polar and non-polar solvents
Reaction in water	Basic
Stability	Stable at basic pH, degradation in acidic media

The substance is synthesized according to the reaction scheme in Fig. 24. The intermediate *N*-methylol-cyclohexylamine does not last, it is converted immediately without any additional effort to the stable *N*-methylene compound by the elimination of water.

Toxicity data

Not available.

Antimicrobial effectiveness/applications

The compound does not exhibit either special activity or cost effectiveness. It proved well in trials when used as a sanitizing agent in the dry cleaning process because of its good solubility in chlorinated hydrocarbons.

3.3.14 Diaminomethanes — Aminals

R-N-CH_2-N-R
  | |
  R   R

The reaction/condensation of 1 mol CH_2O with 2 mol secondary amine leads to aminals as is demonstrated in Fig. 24

3.3.14.1 Dimorpholino-methane

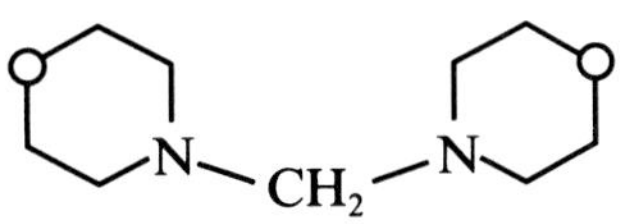

$C_9H_{18}N_2O_2$

M_r 186·26
CAS-No. 5625-90-1

Chemical and physical properties

Appearance	Clear, colourless fluid with a weak amine odour
CH_2O content	16%
Boiling point at 0·2 hPa	74°C
Solubility	Very soluble in water and organic solvents
Reaction in water	Basic; pH of a 0·5% solution: 8·5
Stability	Stable at basic pH, degradation in acidic media

Toxicity data

Not available

Antimicrobial effectiveness/applications

When compared with other amine based formaldehyde releasing compounds the substance does not offer an advantage with respect to its antimicrobial activity. It is not cost effective and therefore has not gained much importance in practical application, although it is distinguished by favourable solubility properties and a weak odour.

3.3.15 1.4.6.9-Tetraaza-tricyclododecane(4.4.1.14.9)

$C_8H_{16}N_4$ M_r 168·24
CAS-No. 54159-21-6

Chemical and physical properties

Appearance	White, odourless crystals
CH_2O content	71%
Melting point	199°C
Solubility	Very soluble in water, polar and non-polar solvents, e.g. in chlorinated hydrocarbons
Reaction in water	Basic

Toxicity data

Not available.

Antimicrobial effectiveness/applications

Minimum inhibition concentrations (mg/litre) for	
Escherichia coli	100
Pseudomonas aeruginosa	100
Aspergillus niger	1500
Penicillium glaucum	1000
Rhizopus nigricans	500

Due to its high formaldehyde content the reaction product of ethylene diamine and CH_2O is one of the most effective formaldehyde-amine condensates. It may be produced as a 50% solution in water which is useful as a preservative for aqueous functional fluids. Additionally the compound can serve as a sanitizing/disinfecting agent in dry cleaning fluids. Addition rates 0·05–0·15%.

3.3.16 3,5-Dimethyl-tetrahydro-1,3,5-2H-thiadiazine-2-thiono — Dazomet

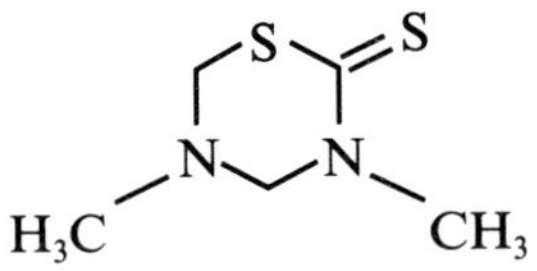

$C_5H_{10}N_2S_2$ M_r 162·28
CAS-No. 533-74-4

Chemical and physical properties

Appearance	Colourless crystals (powder) with a weak characteristic odour	
CH_2O content	37%	
Melting point	104–105°C (decomposition)	
Vapour pressure at 20°C	0·4 hPa	
Flash point	>360°C	
Solubility (g/kg at 20°C)	In water	3
	ethanol	15
	acetone	173
	chloroform	391
	benzene	51
	cyclohexane	0·4
Stability	Stable at temperatures up to 35°C. Sensitive to moisture; hydrolysis: in acidic media to carbon disulphide, methylamine and formaldehyde; in neutral and alkaline media to H_2S, methylamine and formaldehyde	

The compound is listed here under the reaction products of formaldehyde and amines, because it is synthesized by condensation of formaldehyde with methylamine and carbon disulphide.

Toxicity data

LD_{50} oral	520 mg/kg rat; 320–620 mg/kg rabbit; 420 mg/kg mouse
No-effect level (Cramer & Ford, 1978):	<0·5 mg/kg body weight

Eye and skin irritant.

Fish toxicity	LC_{80} 2 mg/litre for *Lebistes reticulatus* (exposure: 96 h)

Antimicrobial effectiveness/applications

Table 32. Minimum Inhibition Concentrations (MIC) of Dazomet in Nutrient Agar

Test organism	*MIC (mg/litre)*
Aeromonas punctata	150
Bacillus mycoides	100
Bacillus subtilis	200
Escherichia coli	200
Proteus vulgaris	200
Pseudomonas aeruginosa	150
Pseudomonas fluorescens	100
Staphylococcus aureus	200
Formaldehyde resistant bacteria	200
Alternaria alternata	35
Aspergillus flavus	100
Aspergillus niger	100
Chaetomium globosum	100
Penicillium glaucum	100
Rhizopus nigricans	100
Trichoderma viride	100
Candida krusei	100
Torula utilis	20

As the MIC in Table 32 prove, Dazomet has an extraordinary broad spectrum of high activity which covers bacteria including 'formaldehyde resistant bacteria', fungi and yeasts, indicating that the substance is a very special formaldehyde releasing compound. One finds an explanation by looking at the pattern of synthesis which is reversible; that means an intermediate of neutral to alkaline hydrolysis is *N*-methyl-dithiocarbamate (see Section 9.9), a known antifungal slimicide. A pH between 4 and 9 is the optimum for Dazomet.

The spectrum of activity is attractive for application of the compound in a number of industrial systems, e.g. as a slimicide in water circuits, paper machine systems, as a broad spectrum microbicide which prevents fungal blooms in metal working fluid systems. However, there are limitations: poor water solubility, instability, release of H_2S and coloration by reaction with heavy metal salts.

3.3.17 5-Amino-1,3-bis(2-ethylhexyl)-5-methyl-hexahydropyrimidine — Hexetidine

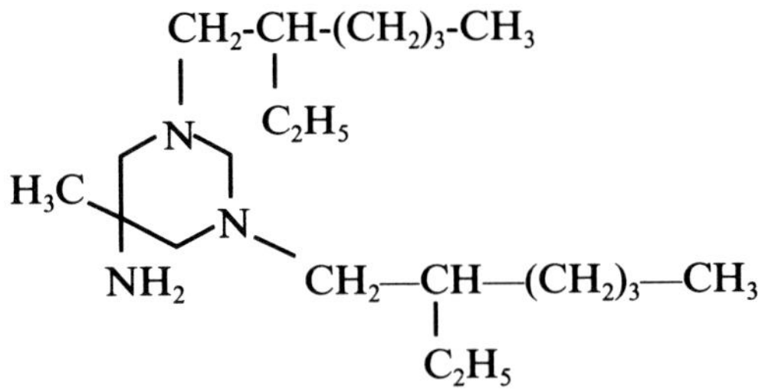

$C_{21}H_{45}N_3$ M_r 339·61
CAS-No. 141-94-6

Chemical and physical properties

Appearance	Colourless, viscous fluid with a slight amine odour
CH_2O content	27%
Boiling point at 0·5 hPa	160°C
Density at 20°C	0·89 g/ml
Refractive index at 20°C	n_D 1·4640
Solubility	Sparingly soluble in water, soluble in ethanol, methanol and acetone
Stability	Heat resistant; does not separate formaldehyde under conditions of use

Toxicity data

LD_{50}	oral	1·43 g/kg rat;
	intraperitoneal	0·03–0·085 g/kg mouse;
	dermal	1·86 ml/kg rat
		>4·0 ml/kg rabbit

Irritant to skin and mucous membranes; 0·5% solutions were not irritant.
No sensitization.
Ames test: negative.

Antimicrobial effectiveness/applications

From the spectrum of effectiveness one can conclude that the activity of Hexetidine is not the result of released formaldehyde. This is in line with the fact that the formaldehyde contained in Hexetidine is not traceable with the Tannenbaum method or other more severe methods, although the synthesis pathway for Hexetidine

Table 33. Minimum Inhibition Concentrations (MIC) of Hexetidine in Nutrient Agar

Test organism	*MIC (mg/litre)*
Staphylococcus aureus	5–10
Escherichia coli	1250
Pseudomonas aeruginosa	>5000
Candida albicans	5000
Aspergillus niger	800
Chaetomium globosum	75
Penicillium brevicaule	400

starts with the reaction of 2-ethylhexylamine and nitroethane with formaldehyde. The intermediate (substituted 5-nitro-hexahydro-pyrimidine) is reduced to Hexetidine. Apparently Hexetidine is effective as a membrane active substance which is inactivated by acids and soaps. Non-ionic detergents can support the activity of Hexetidine. One has to pay attention to the deep gaps in the spectrum of effectiveness with regard to Pseudomonades. The extraordinary high activity of Hexetidine against fungi is remarkable.

Hexetidine is used as a microbicide in cosmetics and in pharmaceutical products. In the EC positive list of preservatives which cosmetics may contain it is mentioned with a maximum authorized concentration of 0·1% (limitation: for rinse-off products only). The gaps in the spectrum of effectiveness of Hexetidine have to be closed by combination with other suitable microbicides.

3.3.18 3,5-Dimethyl-1-hydroxymethyl-pyrazole

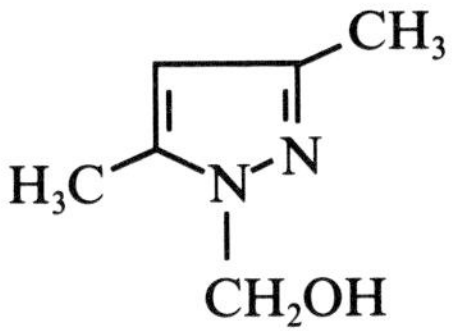

$C_6H_{10}N_2O$ M_r 126·16
CAS-No. 85264-33-1

Chemical and physical properties

Appearance	White crystalline powder	
CH_2O content	23·8%	
Melting point	110–113°C	
Bulk density at 25°C	493 g/litre	
Solubility (g/litre)	In water	42
	In polypropylene glycol	200
pH values in water	4·3% solution:	6·1
	0·01% solution:	7·2
Stability	Stable to 100°C; releases formaldehyde in water-based solutions	
Vapour pressure	$3{\cdot}4 \times 10^{-3}$ hPa	

Toxicity data

LD_{50}	oral	2600 mg/kg rat
	dermal	>2000 mg/kg rat

Irritant to skin and eyes.
Not a mutagen via Ames (Salmonella); not a teratogen.

Antimicrobial effectiveness/applications

Table 34. Minimum Inhibition Concentrations (MIC) of 3,5-dimethyl-2-hydroxymethyl-pyrazole in Nutrient Agar

Test organism	*MIC (mg/litre)*
Aspergillus niger	>800
Chaetomium globosum	400
Scopulariopsis brevicaulis	>800
Lentinus tigrinus	800
Pseudomonas aeruginosa	800
Staphylococcus aureus	400

As a *N*-methylol-pyrazole derivative the substance releases formaldehyde which is quantitatively detectable by the Tannenbaum method; in consequence the antimicrobial effectiveness of the compound corresponds to its formaldehyde content. It may be used for the in-can/in-tank protection of water based functional fluids, such as latex paints, adhesives, lubricoolants, polymer emulsions, detergents, cosmetics; it is effective over a broad pH range (3–11·5).

3.4 Reaction products of amides and formaldehyde

When heated under neutral to weakly alkaline conditions amides react quantitatively with formaldehyde to give *N*-hydroxymethyl amides (R-CO-NH-CH_2OH). These have a higher degree of stability than *N*-hydroxymethyl amines or their condensation products. The formaldehyde releasable from *N*-hydroxymethyl amides is not detectable using the Tannenbaum method. As a result *N*-hydromethyl amides generally fail to produce substantial antimicrobial activity. That is the price for the higher degree of stability

However, an exception must be made for those *N*-hydroxymethyl amides that derive from antimicrobial effective compounds having an amide configuration in the molecule, e.g. α-halogen-amides (introduction of a second toxophoric group into the molecule).

At first sight the effectiveness of these amides does not seem to be improved, but rather reduced by the introduction of the hydroxymethyl group. But in practice, on using the substances as preservatives in neutral to alkaline, aqueous functional fluids, a partial splitting into formaldehyde and the starting compound will be observed (see Fig. 28), the antimicrobial effects of the breakdown products complementing each other in such a way that the effective spectrum is broadened. Finally, this is the effect of substances bearing two toxophoric groups or structural elements in one molecule.

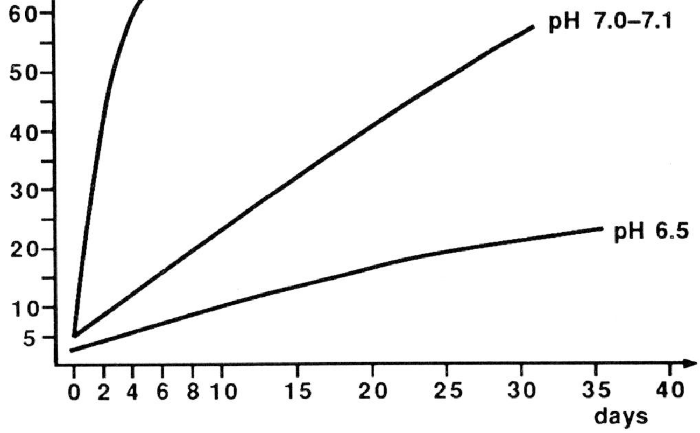

Fig. 28. Separation of CH_2O in 0·5% solutions of *N*-hydroxymethyl-chloracetamide in water at 20°C with different pH values. CH_2O released determined by the method of Tannenbaum.

3.4.1 *N*-hydroxymethyl-chloracetamide — *N*-methylol-chloracetamide (MCA)

Cl-CH_2-CO-NH-CH_2OH	$C_3H_6ClNO_2$	M_r 123·54 CAS-No. 2832-19-1

Chemical and physical properties

Appearance	White, crystalline, odourless	
CH_2O content	24%; not detectable using the Tannenbaum method.	
Melting point	102°C	
Solubility (g/litre at 20°C)	In water	260
	methanol	200
	Sparingly soluble in non-polar solvents	
Reaction in water	Neutral	
Stability	Releases CH_2O in neutral to alkaline media and chloride ions at pH values >9	

Toxicity data

LD_{50}	oral	340 mg/kg rat
	dermal	>500 mg/kg rat (exposure: 7 days)

Slightly irritant to skin and mucous membranes.

Antimicrobial effectiveness/applications

MCA's efficacy increases with pH; the optimum activity is achieved between pH 7·5 and 9. This is due to the fact that the amount of formaldehyde liberated from MCA increases with pH (see Fig. 28).

The MIC in Table 35 allow the following conclusions: In slightly acidic media CA (see Section 8.1) in comparison to MCA is the more active microbicide. However, at pH 8, where MCA releases approx. 75% of its formaldehyde, MCA is more effective, as the breakdown products (CA and CH_2O) now complement each other.

MCA is an ideal preservative for the protection of technical functional fluids having a slightly alkaline pH, e.g. polymer emulsions, water based paints, adhesives. For all that the good water solubility and compatibility with many different aqueous formulations is an important advantage. Besides that a favourable partition coefficient guarantees that MCA remains in the water phase of two-phase systems, thus attacking the microbes in the phase where they are vegetating. Colourlessness and odourlessness complete the advantageous properties of MCA as a preservative.

Table 35. Minimum Inhibition Concentrations (MIC) of MCA and Chloroacetamide (CA) at Different pH Values in Nutrient Agar

Test organism	*MIC (mg/litre)*			
	MCA		*CA*	
	pH 5	*pH 8*	*pH 5*	*pH 8*
Escherichia coli	>2500	1000	>2500	>2500
Pseudomonas aeruginosa	1500	1200	1000	1200
Staphylococcus aureus	>2500	600	800	2500
Aspergillus niger	4000	1500	1000	1000
Chaetomium globosum	1000	750	500	500
Penicillium glaucum	4000	1000	500	750
Rhizopus nigricans	4000	750	500	500

Generally the activity of formaldehyde releasing compounds is reduced in alkaline media containing ammonia, because of the formation of inactive hexamethylene tetramine (Section 3.3.1). In the case of MCA, however, the hexamethylene tetramine formed reacts with MCA or CA to give quaternary hexaminium salts (Sections 3.3.5 and 3.3.6) which are effective, as they release formaldehyde widely independent of pH (see Section 3.3).

3.4.2 2,2,3-Trichloro-*N*-hydroxymethyl-propionamide

Cl-CH_2-C(Cl_2)-CO-NH-CH_2OH $C_4H_6Cl_3NO_2$ M_r 206·46
CAS-No. 7321-41-7

Chemical and physical properties

Appearance	White, crystalline, odourless
CH_2O content	15%; not detectable using the Tannenbaum method
Melting point	98°C
Solubility	Soluble in water, polar and non-polar solvents
Reaction in water	Neutral
Stability	Liberates CH_2O in neutral to alkaline media and chloride ions at pH values >9

Toxicity data

Not available.

Antimicrobial effectiveness/applications

Compared with MCA (Section 3.4.1) 2,2,3-trichloro-*N*-hydroxymethyl-propionamide does not exhibit special activity. Because of the excellent solubility properties — soluble not only in water but in non-polar solvents, too — the active ingredient is very much appropriated as a preservative for oil based lubricoolants (Paulus *et al.*, 1967).

3.4.3 *N*-hydroxymethyl-ureas

These urea derivatives are listed among the reaction products of amides and formaldehydes because urea is regarded as the diamide of carbonic acid.

Compound	*Structural formula*	*Empirical formula*	M_r	*CH_2O content*	*Appearance*	*CAS-No.*
N-hydroxymethyl-urea	HN-CO-NH_2 | CH_2OH	$C_2H_6N_2O_2$	90·08	33%	Solid, m.p. 111°C	1000-82-4
N,N'-bis-hydroxy-methyl-urea	HN-CO–NH | | CH_2OH CH_2OH	$C_3H_8N_2O_3$	120·11	50%	Solid, m.p. 126°C	140-95-4

The hydroxymethyl ureas are colourless, odourless and distinguished by good water solubility; they release formaldehyde in water based solutions, especially in alkaline solutions, and therefore have been used to some extent as preservatives in alkaline functional fluids, e.g. lubricoolants. In more neutral solutions the hydroxymethyl ureas release a small amount only of their formaldehyde content. Using hydroxymethyl ureas as preservatives one has to bear in mind that a nutrient for microbes is left, when the formaldehyde is separated and used up by irreversible reactions with other compounds or microorganisms.

3.4.4 *N,N'*-bis(hydroxymethyl)thiourea — *N*-hydroxymethyl-*S*-hydroxymethyl-thiourea — Dimethylol thiourea

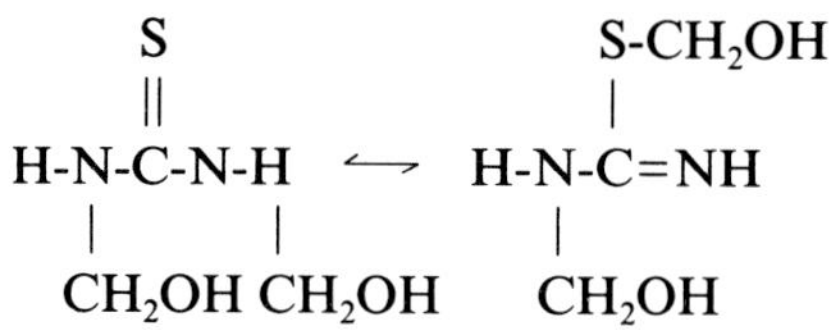

$C_3H_8N_2O_2S$ M_r 136·18
CAS-No. 3084-25-1

Formaldehyde can react with thioureas at the *N* atoms and at the *S* atoms. The *S*-methylol isomers are formed as the principal products under alkaline conditions; however it is not believed that the materials obtained are 100% isomerically pure (Walker, 1975).

Chemical and physical properties

Appearance	White crystalline solid
CH_2O content	44%
Melting point	Approx. 86°C
Solubility	Soluble in water and alcohols
Stability	Tolerates heating to 80°C; releases formaldehyde in water based media

Antimicrobial effectiveness/applications

The efficacy is in line with the formaldehyde separated and covers bacteria, yeasts and fungi accordingly. The compound is only occasionally used as a preservative in functional fluids.

3.4.5 *N*-hydroxymethyl-*N*'-methyl-thiourea

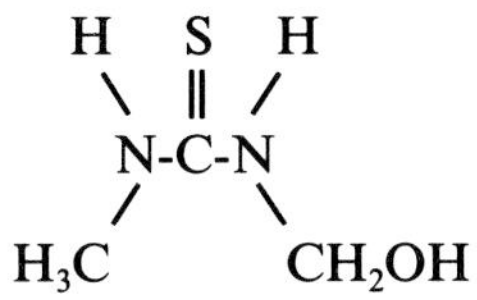

$C_3H_8N_2OS$

M_r 120·18
CAS-No. 15599-39-0

Chemical and physical properties

Appearance	White, crystalline powder
CH_2O content	25%
Melting point	84–86°C
Solubility (g/litre) at 20°C	In water 100; soluble in ethanol
Stability	Heat resistant to 80°C; sensitive to light, separates formaldehyde slowly in water based solutions (in a 0·5% solution 0·035% CH_2O are detectable after 4 days by the Tannenbaum method)

Toxicity data

LD_{50}	oral	>3 g/kg mouse;
	subcutaneous	approx. 1·6 g/kg mouse

Antimicrobial effectiveness/applications

Especially mentioned is the efficacy against Pseudomonades. However, up to now the compound is only occasionally used as a microbicide.

3.4.6 *N*-(hydroxymethyl)-*N*-[1,3-bis(hydroxymethyl)-2,5-dioxo-imidazolidin-4-yl]-*N*'-hydroxymethyl-urea — Diazolidinyl urea

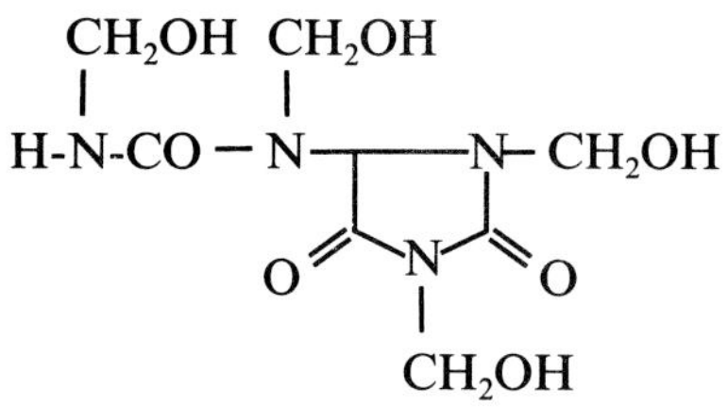

$C_8H_{14}N_4O_7$

M_r 278·23
CAS-No. 78491-02-8

Chemical and physical properties

Appearance	White, crystalline powder with a slight odour
CH_2O content	43%
Melting point	Approx. 120°C (decomposition)
Solubility	Very soluble in water
Reaction in water	Neutral
Stability	Very stable; the release of formaldehyde in water based formulations occurs very slowly

Toxicity data

LD_{50}	oral	2·57 g/kg rat;
	dermal	>2 g/kg rabbit

1% and 5% solutions are not irritant to skin and mucous membranes.
A slight sensitization is observed in the guinea pig test.

Antimicrobial effectiveness/applications

The efficacy of diazolidinyl urea is mainly a bacteriostatic one because of the slow release of formaldehyde. The addition rates which are generally applied (0·1–0·3%) are not effective against mould producing fungi and yeasts. However, because of its favourable toxicological data diazolidinyl urea is often used as a preservative for cosmetics. Accordingly it is listed in the corresponding EC positive list (maximum authorized concentration: 0·5%). The spectrum of effectiveness may be completed by using diazolidinyl urea in combination with *p*-hydroxy-benzoates (see Section 6.1.10); such combinations exhibit a marked synergism (Berke & Rosen 1982*a*).

3.4.7 Bis-[N'-hydroxymethyl-2,5-dioxo-imidazolidin-4-yl)-ureido]-methane — Imidazolidinyl urea

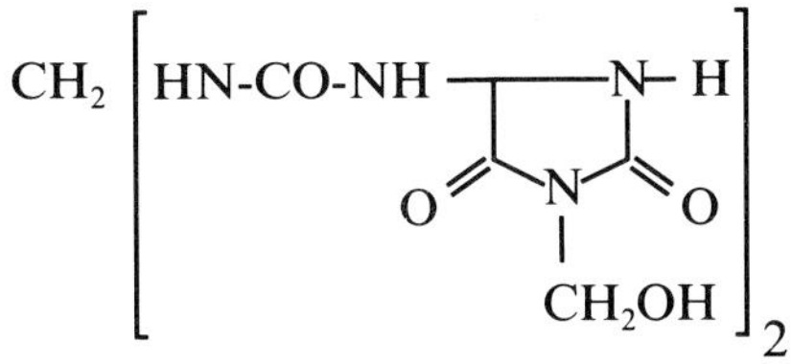

$C_{11}H_{16}N_8O_8$

M_r 388·3
CAS-No. 118215-45-5

Chemical and physical properties

Appearance	White, crystalline, odourless powder	
CH_2O content	23%	
Melting point	Decomposition starting at 140°C	
Solubility (g/litre) at 25°C	In water	2000
	ethylene glycol	1500
	propylene glycol	1200
	glycerine	1000
	methanol	0·5
	ethanol	<0·5
	isopropanol	<0·5
Reaction in water	Neutral	
Stability	Stable; releases formaldehyde in water-based media; however, only approx. 60% of the formaldehyde content of the molecule is detectable by the Tannenbaum method	

Toxicity data

LD_{50} oral	5·2 g/kg rat	
	7·2 g/kg mouse	
dermal	> 8 g/kg rabbit	

The compound is not irritant to skin and mucous membranes; no sensitization was observed in patch tests with 200 persons.
No teratogenic effects were observed with mouse (oral applicated maximum dosage: 300 mg/kg).

Antimicrobial effectiveness/applications

The bacteriostatic activity of imidazolidinyl urea is not very distinguished; the efficacy against fungi is not worth a mention. Nevertheless the compound is an important preservative for cosmetics and pharmaceutical products, as it may be applied without difficulty at relatively high concentrations because of its favourable physical and chemical properties and toxicity data. In the EC list of preservatives permitted for the in-can protection of cosmetics imidazolidinyl urea is mentioned with a maximum authorized concentration of 0·6%. Combinations of imidazolidinyl urea with *p*-hydroxybenzoates (see Section 6.1.10), e.g. 0·3–0·5% imidazolidinyl urea with 0·2% *p*-hydroxy methylbenzoate and 0·1% *p*-hydroxy propylbenzoate, have proved a success in killing the most resistant strains of Pseudomonades.

The ability of imidazolidinyl urea to be active over a wide pH range is as remarkable as its compatibility with almost all cosmetic ingredients.

Table 36. Minimum Inhibition Concentrations (MIC) of Imidazolidinyl Urea in Nutrient Agar

Test organism	*MIC (mg/litre)*
Escherichia coli	2000
Klebsiella pneumoniae	2000
Pseudomonas aeruginosa	2000
Staphylococcus aureus	1000
Candida albicans	8000
Aspergillus niger	8000

3.4.8 1-(Hydroxymethyl)-5,5-dimethyl-2,4-dioxo-imidazolidine — Monomethylol-dimethyl-hydantoin

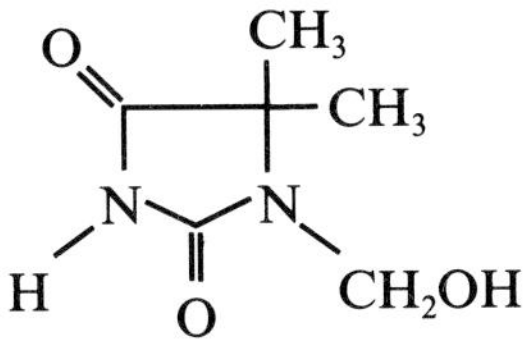

$C_6H_{10}N_2O_3$ M_r 158·16
CAS-No. 28453-33-0

2,4-Dioxo-imidazolidines, also called hydantoins, also include the carbonamide structure, and therefore the corresponding *N*-hydroxymethyl compounds are listed among the reaction products of amides and formaldehyde.

Chemical and physical properties

Appearance	White, crystalline, odourless
CH_2O content	19%
Melting point	Approx. 100°C
Solubility	Very soluble in water, ethanol, methanol, acetone, less soluble in ethylacetate, practically insoluble in ether, trichloroethylene, tetrachloromethane
Reaction in water	Neutral
Stability	Stable up to 85°C; releases formaldehyde in solution at pH 6. A part only (approx. 30%) of the formaldehyde content is detectable by the Tannenbaum method.

Toxicity data

The maximum working concentration is fixed at 10 mg/m^3.

Antimicrobial effectiveness/applications

Monomethylol-dimethyl-hydantoin produces a spectrum of activity which is characteristic of formaldehyde releasing compounds. The compound is proposed for the preservation of cosmetics and pharmaceutical products; addition rate: 0·2%.

3.4.9 1,3-Bis(hydroxymethyl)-5,5-dimethyl-2,4-dioxo-imidazolidine — Dimethylol-dimethyl-hydantoin

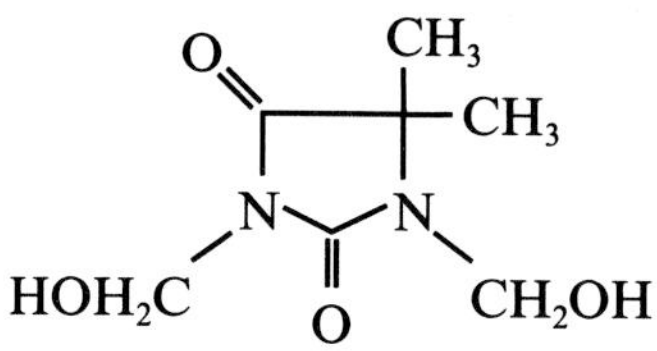

$C_7H_{12}N_2O_4$ M_r 188·19
CAS-No. 6440-58-0

Chemical and physical properties (55% solution in water)

Appearance	Colourless fluid with slight odour of formaldehyde
CH_2O content	17·5%; only approx. 2% are detectable by the Tannenbaum method
Solidification point	−7·5°C
pH value at 25°C	6·5–7·5
Solubility	Highly soluble in water
Stability	Stable between pH 6 and 8; at higher pH values release of formaldehyde; tolerates heating to 90°C

Toxicity data

LD_{50} oral	3·0–4·32 g/kg mouse, 3·8 g/kg femal rat;
intraperitoneal	1·32–2·0 g/kg mouse;
dermal	>20 g/kg rabbit
LC_{50} for the inhalation	>204 mg/litre

The compound is not irritant to the skin, but irritant to mucous membranes — no sensitization was observed — mutagenicity tests were negative.

Antimicrobial effectiveness/applications

The effectiveness of dimethylol-dimethyl-hydantoin is based on the release of formaldehyde and is not negatively influenced by anionic, cationic or non-ionic compounds. The active ingredient therefore is proposed as a suitable preservative of very low toxicity for cosmetics and pharmaceutical products; addition rates: 0·1–0·2%. Combinations with *p*-hydroxybenzoates (Section 6.1.10) have proved successful. Isothiazolinones (Section 13.3) may also be used in mixtures for increasing the efficacy against fungal growth.

Table 37. Minimum Inhibition Concentrations (MIC) of Dimethylol-dimethyl-hydantoin in Nutrient Agar

Test organism	*MIC (mg/litre)*
Escherichia coli	500
Proteus mirabilis	800
Pseudomonas aeruginosa	600–700
Pseudomonas fluorescens	800
Staphylococcus aureus	300–400
Candida albicans	1000
Aspergillus niger	1500
Penicillium notatum	1500

3.4.10 *N*-hydroxymethyl derivatives of substances containing a carbamate (>*N*-CO-O-) or a thiocarbamate (>*N*-CS-O-) or a dithiocarbamate (>*N*-CS-S-) structure, thus displaying two toxophoric structural elements in one molecule, e.g. the antifungally active dithiocarbamate structure and the *N*-hydroxymethyl group, starting point for the release of bactericidal formaldehyde

3.4.10.1 Sodium *N*-hydroxymethyl-*N*-methyldithiocarbamate

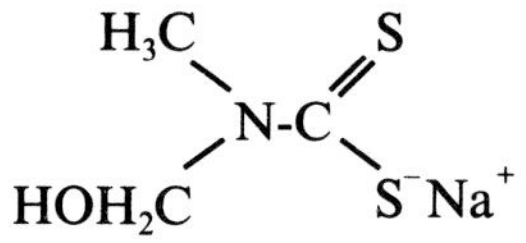

$C_3H_6NOS_2Na$ M_r 159·21
CAS-No. 60487-28-7

Chemical and physical properties (40% water based solution)

Appearance	Yellow to brown coloured liquid
CH_2O content	Approx 7·5%
Density at 25°C	1·23 g/ml
pH of a dilution in H_2O	8–10 (approx. 0·1 g/litre)
Solubility	Completely soluble in water and mixtures of alcohol and water
Stability	Degradation in acid media under liberation of CS_2, methylamine and CH_2O; coloration with traces of heavy metal

Toxicity data

Moderately toxic by ingestion in single doses.
Irritating to skin and eyes.

Antimicrobial effectiveness/applications

According to its chemical composition the formulation exhibits antibacterial and antifungal activity. Concentrations of 2–10 mg/litre are used for slime control in pulp and paper mill systems. Pulp that may be held in storage for 8 h to 1 week may be protected by the addition of 0·01–0·03% of the 40% a.i. solution. The hydroxymethylated dithiocarbamate may also be used as a preservative for papermaking chemicals, such as glues, starch and clay slurries and coating formulations.

3.4.10.2 1-Hydroxymethyl-2-thiono-1:2-dihydro-benzothiazol — *N*-hydroxymethyl-benzothiazolin-2-thione (MBTT)

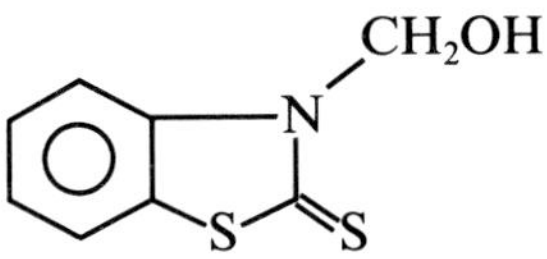

$C_8H_7NOS_2$ M_r 197·28
CAS-No. 3161-57-7

Chemical and physical properties

Appearance	Yellow, odourless crystals	
CH_2O content	15%; detectable by the Tannenbaum method	
Melting point	130°C	
Solubility (g/litre at 20°C)	In acetone	180
	dioxane	220
	dimethylformamide	550
	soluble in NaOH solution; practically insoluble in H_2O	
Stability	Decomposition in alkaline solutions to CH_2O and the corresponding salt of 2-mercapto benzthiazol (MBT) (see Section 13.9).	

N-hydroxymethyl-benzothiazolin-2-thione is listed under the reaction products of amides and formaldehyde because of its thiocarbonamide structure (Paulus, 1980).

Toxicity data

Not available.

Antimicrobial effectiveness/applications

As the formaldehyde contained in MBTT can be determined by quantitative analysis using the Tannenbaum method, it is not surprising that the range of activity of MBTT is such that gaps observed in the effective spectrum of MBT are successfully filled. The regularity of the activity spectrum of MBTT is demonstrated in Table 38, the efficacy of MBTT against Pseudomonades is 15-fold higher than that of MBT.

In alkaline functional fluids MBTT can serve as a broad spectrum preservative which additionally inhibits corrosion of non-ferrous metals as does MBT. Because of its solubility properties MBTT may be incorporated into lubricoolants.

Table 38. Minimum Inhibition Concentrations (MIC) of MBT and MBTT in Nutrient Agar

Test organism	*MIC (mg/litre)*	
	MBT	*MBTT*
Aspergillus flavus	500	500
Aspergillus niger	300	300
Chaetomium globosum	100	100
Penicillium glaucum	150	200
Aureobasidium pullulans	150	200
Rhizopus nigricans	500	500
Bacillus mycoides	500	150
Bacillus punctatus	450	100
Bacillus subtilis	700	200
Bacterium vulgare	500	100
Escherichia coli	200	150
Pseudomonas aeruginosa	8000	500
Pseudomonas fluorescens	8000	750
Staphylococcus aureus	450	150

3.4.10.3 3-Hydroxymethyl-5,6-dichloro-benzoxazolinone

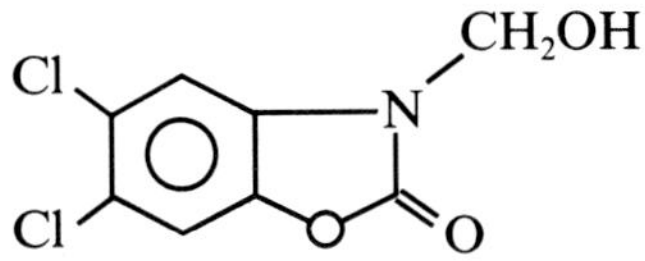

$C_8H_5Cl_2NO_3$

M_r 234·04
CAS-No. 19986-60-8

Chemical and physical properties

Appearance	Brownish, odourless powder
CH_2O content	13% detectable by the Tannenbaum method
Melting point	138–40°C
Solubility	Practically insoluble in H_2O, soluble in polar solvents and in NaOH solutions
Stability	In alkaline solutions transformation to the corresponding salt under elimination of CH_2O

The starting product, the 5,6-dichlorobenzoxazolinone (see Section 9.6) contains the carbonamide or rather the carbamate structure and is a known fungicide especially for the protection of textile material. The introduction of the hydroxymethyl group leads to a compound with an effective spectrum more balanced than that of the starting product (see Table 39).

Table 39. Minimum Inhibition Concentrations (MIC) of *N*-hydroxymethyl-5,6-dichlorobenzoxazolinone (A) and *N*-hydroxymethyl-5-chlorobenzoxazolinthione (B) (see Section 3.4.10.4) in Nutrient Agar

Test organism	*MIC (mg/litre)*	
	A	*B*
Escherichia coli	250	250
Pseudomonas aeruginosa	1500	800
Aspergillus niger	120	1500
Penicillium glaucum	60	500
Rhizopus nigricans	120	250

3.4.10.4 3-Hydroxymethyl-5-chloro-benzoxazoline-2-thione

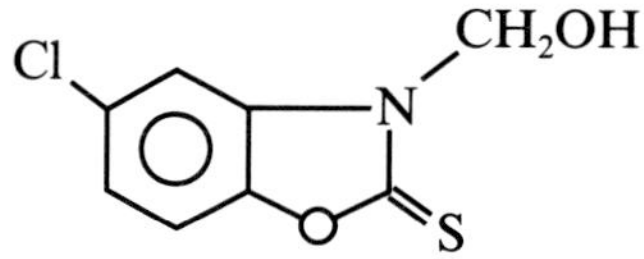

$C_8H_6ClNO_2S$ M_r 215·66
CAS-No. 3998-53-6

Chemical and physical properties

Appearance	White, odourless needles
CH_2O content	14%; detectable by the Tannenbaum method
Melting point	161°C
Solubility	Practically insoluble in H_2O, soluble in polar solvents and in alkaline solutions
Stability	In alkaline solutions transformation to the corresponding salts with elimination of CH_2O

Antimicrobial effectiveness (see Table 39)

Up to now compounds A and B have not gained importance in practical applications such as microbicides for the protection of materials, probably because of not being cost effective.

3.5 Reaction products of amino acids with formaldehyde

3.5.1 Sodium 2-hydroxymethylaminoacetate — Sodium hydroxymethylglycinate

$HO\text{-}CH_2\text{-}NH\text{-}CH_2\text{-}COO^-Na^+$ $C_3H_6NO_3Na$ M_r 127·08
CAS No. 70161-44-3

Chemical and physical properties (50% aqueous solution)

Such a solution is prepared by reacting a mixture of glycine and sodium hydroxide in water with formaldehyde. The addition of formaldehyde to the amino group of glycine leads to a molecule which may react further; it could dehydrate to a Schiff base, react with another molecule of glycine or polymerize (Berke & Rosen, 1982*b*). It is therefore more accurate to characterize the a.m. solution as an equilibrium mixture of compounds produced by the reaction of the sodium salt of glycine with formaldehyde.

Appearance	Clear, colourless to pale yellow liquid with a mild pungent odour
CH_2O content	Approx. 12%
Density at 20°C	1·28–1·30 g/ml
pH	10–12
Solubility	Highly soluble in water and alcohols
Stability	Releases formaldehyde, increasingly with decreasing pH

Toxicity data

LD_{50} oral	Approx. 1200 mg/kg rat;
dermal	>2000 mg/kg rat

Slightly to moderately irritant to skin and eyes.
No sensitization was observed in the guinea pig test according to Magnusson-Kligman — Bacterial mutagenicity tests were negative.

Antimicrobial effectiveness/applications

Sodium hydroxymethylglycinate is a formaldehyde releasing compound; its formaldehyde content is detectable with the Tannenbaum method. The antimicrobial activity corresponds to its formaldehyde content. The compound inhibits preferably the growth of bacteria, but at higher concentrations also the growth of fungi and yeasts. High water solubility accompanied by a favourable partition coefficient make sodium hydroxymethylglycinate an appropriate microbicide for the in-can/in-tank protection of aqueous functional fluids. It is mainly recommended for the preservation of cosmetics, especially hair care products, to meet a current need for a safe but economical microbicide.

3.5.2 Bis(1,1-dioxo-perhydro-1,2,4-thiadiazinyl-4) methane — Taurolin

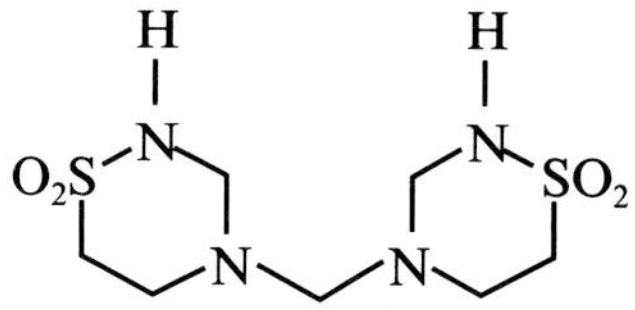

$C_7H_{16}N_4O_2S_2$

M_r 284·36
CAS-No. 19388-87-5

Chemical and physical properties

CH_2O content	32%
Solubility	Very soluble in water and polar organic solvents
Stability	Releases formaldehyde in aqueous media

Antimicrobial effectiveness/application

The starting material for the synthesis of Taurolin is 2-aminoethanolsulphonic acid (Taurin) which by the addition of 2 M NH_3 and 3 M CH_2O in a condensation reaction is converted to Taurolin. The spectrum of effectiveness corresponds to that of formaldehyde, however, according to findings of Myers *et al.* (1980) and Allwood & Myers (1981) the activity of Taurolin is considerably greater than that of formaldehyde; apparently the 4-hydroxymethyl-1,1-dioxoperhydro-1,2,4-thiadiazine which is released in the first step on hydrolysis of Taurolin plays an important role in the mechanism of Taurolin's antimicrobial activity.

Taurolin is not an important microbicide for the protection of aqueous functional fluids and is mentioned here more for the sake of completeness.

3.6. Tetra-(hydroxymethyl)-phosphonium sulphate (THPS)

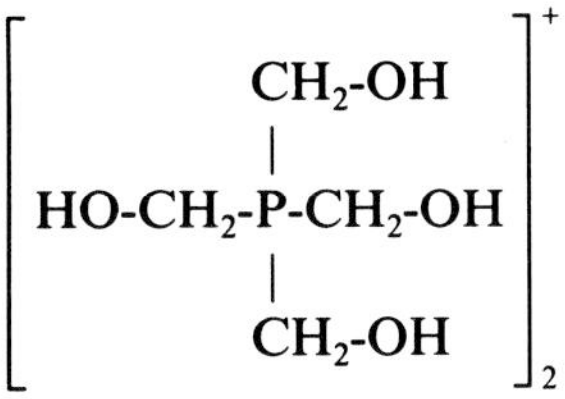

SO_4^{2-} $C_8H_{24}O_{12}P_2S$ M_r 406·29
CAS-No. 55566-30-8

Chemical and physical properties (33% (approx.) formulation)

Appearance	Clear, straw coloured liquid of pungent odour
CH_2O content	59% for 100 a.i.; detectable by the Tannenbaum method
Solubility	Miscible with water in all proportions
Density at 25°C	1·3 g/ml
pH	Approx. 5
Stability	Releases formaldehyde; sensitive to oxidizing agents and strong alkalis

Toxicity

Strongly irritant to the skin, eyes, and mucous membranes — not mutagenic in bacterial test systems — no evidence of carcinogenicity after a 2-year gavage study in rats and mice.

Ecotoxicity

Acute toxicity to *Salmo gaidneri* (Rainbow trout): 99 mg/litre (exposure: 96 h). THPS is inherently biodegradable.

Antimicrobial effectiveness/applications

THPS is a quaternary phosphonium salt which releases formaldehyde. It is claimed to be effective against algae, fungi and bacteria and recommended for use in badly fouled cooling systems and injection water for secondary oil recovery as a slimicide which is especially active against sulphate reducing bacteria at addition rates of 50–500 mg/litre.

Table 40. Minimum Inhibition Concentrations (MIC) of the THPS Formulation in Nutrient Agar

Test organism	*MIC (mg/litre)*
Aerobacter aerogenes	400
Aeromonas punctatum	400
Bacillus mycoides	400
Bacillus subtilis	400
Escherichia coli	800
Leuconostoc mesenteroides	400
Proteus mirabilis	800
Pseudomonas aeruginosa	≥800
Pseudomonas fluorescens	≥800
Staphylococcus aureus	50
Candida albicans	>800
Candida krusei	600
Rhodotorula mucilaginosa	400
Torula rubra	>800
Torula utilis	800
Aspergillus niger	>800
Chaetomium globosum	600
Penicillium brevicaule	>800

4 6-ACETOXY-2,4-DIMETHYL-1,3-DIOXANE — DIMETHOXANE

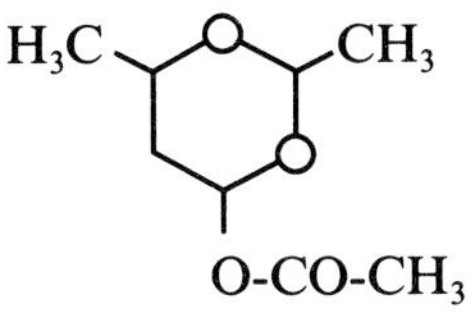

$C_8H_{14}O_4$ M_r 172·2
CAS-No. 828-00-2

Chemical and physical properties

Appearance	Colourless to yellow fluid with a typical odour
Acetaldehyde content	77% (3 M)
Boiling point at 4 hPa	66–68°C
Solidification point	<25°C
Density (at 25°C)	1·055–1·070 g/ml
Refractive index (20°C)	n_D 1·4270–1·4320
Solubility	Miscible with water; very soluble in organic solvents and oils
Stability	Sensitive to acids because of its acetal nature; hydrolyses slowly in aqueous solutions producing acetaldehyde; tolerates 120°C for a short while

Toxicity data

LD_{50} oral 1·9 ml/kg rat

1% solution is not irritant neither for skin nor mucous membranes; no sensitization.
MWC value for acetaldehyde: 50 ml/m^3 (90 mg/m^3).

Antimicrobial effectiveness/applications

Apparently Dimethoxane is effective against bacteria, yeasts and fungi.

As a preservative for functional fluids, such as detergent solutions, emulsions, water based paints, metal working fluids, Dimethoxane profits from its insensitiveness to non-ionic, anionic and cationic emulsifiers and detergents. Addition rates: 0·03–0·1%.

Table 41. Minimum Inhibition Concentrations (MIC) of Dimethoxane in Nutrient Agar

Test organism	*MIC (mg/litre)*
Aerobacter aerogenes	625
Bacillus subtilis	625
Escherichia coli	625
Pseudomonas aeruginosa	625
Pseudomonas fluorescens	625
Salmonella choleraesuis	300
Staphylococcus aureus	1250
Candida albicans	1250
Aspergillus niger	1250

5 PHENOLICS

Phenol, also termed carbolic acid, and the phenol derivatives could also have been discussed in Section 6 'Acids' because of their acidity and the resulting capacity to form stable salts. However, since they represent a large, important substance class, they are described in a separate section.

Coal tars, which form during the destructive distillation of coal, are one source of phenol and phenol derivatives. Today, however, the production of microbicidally effective phenol derivatives for the protection of materials and the formulation of disinfectants from coal tars has for the most part been replaced by efficient, industrial-scale, target-specific syntheses which supply very pure active substances. The so-called pressure phenol synthesis which uses chlorobenzene produces phenol as well as 2-phenylphenol and 4-phenylphenol as by-products. Today a more significant synthesis for the production of 2-phenylphenol is one starting with cyclohexane via cyclohexanone. This is first condensed to 2-cyclohexenylcyclohexanone and its catalytic dehydration provides a good yield of very pure 2-phenylphenol. On heating benzyl chloride in excess phenol, 2- and 4-benzyl phenol form. These in turn can be used to produce the corresponding chlorobenzyl phenols by adding SO_2Cl_2, e.g. 2-benzyl-4-chlorophenol = Chlorophen. The most effective compounds form on chlorination of phenol and phenol derivatives. On the other hand, chlorophenols are the starting products for manufacture of the microbicides Dichlorophen and Hexachlorophen which form during irreversible reaction of formaldehyde with 4-chlorophenol or 2,4,5-trichlorophenol.

However, the development of microbicidally active phenol derivatives started from phenol itself, from carbolic acid, the antiseptic properties of which were detected in 1860 and first used by Lister in 1867 to kill bacteria on medical instruments, surgical dressings and wounds. The development of chemical disinfection revolutionized progress in hospitals, particularly in surgery. It was possible to successfully employ and develop surgical techniques which in the past had been feasible but unpractical because of the associated unavoidable massive and generally fatal infections.

The substance class of microbicidally active phenol derivatives, in short phenolics, was also developed. Hundreds of different derivatives were isolated, synthesized and investigated with the aim of finding phenol derivatives which were more effective and at the same time less toxic and less irritating to the skin than the parent chemical carbolic acid. Another objective was to find microbicidal phenol derivatives with chemico-physical properties which make them suitable for the protection of materials. This development, which can now be considered as completed, provided knowledge on the relationship between structure and effectiveness and on the mechanism of action of phenol derivatives. This is summarized in the following.

The free hydroxyl group together with the aromatic system of the phenyl ring constitutes the reactive centre of the phenol molecule. Its reactivity can be influenced in various ways by introducing different substituents into the phenyl nucleus.

Alkylated phenols are less soluble in water and less acidic than phenol. Furthermore, the acidity decreases in the direction *m*–>*p*–>*o*-substitution. Nevertheless, the alkyl phenols which are of interest as microbicides are still capable of forming alkaline salts which dissolve easily in water. With reduced water-solubility, a property which is important for the use of phenol derivatives as microbicides also changes, namely the ratio of the distribution between the aqueous and non-aqueous phases, including bacterial phases. However, not only the distribution factor but also the capability of reducing surface tension changes; in fact it changes increasingly with the length of the alkyl chain. As to be expected for membrane-active substances, a consequence of these property changes is increasing antimicrobial effectiveness with increasing alkyl chain length. This reaches a maximum with the unbranched C_6 chain in the *p*-position. Branched chains with the same number of C atoms do not increase the effectiveness of the corresponding phenol derivatives to such an extent. The introduction of alkyl chains with more than 6 C atoms does not lead to a further rise in effectiveness. This is due to the decreasing water-solubility.

Halogenation of phenol also leads to phenol derivatives which are much more effective than the starting substance. At the same time, the dissociation constant increases with an increase in the number of halogen atoms, i.e. the acidic character of the phenol derivatives becomes more distinctive. The combination of alkylation and halogenation (the latter with preference in the *p*-position) has led to microbicides which have attained great practical significance as substances for the protection of materials and disinfection, e.g. *p*-chloro-*o*-benzyl phenol, *p*-chloro-*m*-cresol, *p*-chloro-*m*-xylenol.

For the sake of completeness, the nitration of phenol and cresol must be mentioned. It particularly strengthens the bactericidal effect. Furthermore, nitrophenols have specific biological properties since they are able to interfere with oxidative phosphorylation. However, nitrophenols are no longer of practical importance as microbicides.

As already mentioned, phenol derivatives are membrane-active microbicides. They adsorptively coat the surface of the microbe cell then, at a higher concentration, they are dissolved more or less rapidly and well by lipoids depending on their chemico-physical properties (see above). They attack the cell wall and penetrate into the cell. There are reactions with the protoplasm and the cellular protein; enzymes are also inhibited as a result; the oxidoreductases and the enzymes of carbohydrate and protein metabolism react particulary sensitively. Whether the phenol derivatives act microbistatically or microbicidally is purely a question of the application concentration. At low concentrations in ambient medium, there is only reversible adsorption of the phenolic active substance at the cytoplasmic membrane and the related inhibiting effect. As stated above the cell wall is only penetrated and destroyed and the microbe cell killed at higher concentrations.

Phenolics act optimally in acidic and neutral media, i.e. in their undissociated state. In this respect, the pK_a values of phenol derivatives are interesting. The values for some important phenol microbicides are listed in Table 42.

For practical use of microbicidal phenol derivatives, it is very frequently necessary to improve their solubility, especially their solubility in water and hence to shift the distribution ratio towards the aqueous phase. If phenol derivatives

Table 42. pK_a Values of Phenolic Microbicides

Phenol derivative	*pK_a value*
3-Methyl-6-isopropyl-phenol (Thymol)	10·6
o-Phenyl-phenol	11·6
Benzyl-phenol	11·6
o-Benzyl-*p*-chloro-phenol	9·7
p-Chloro-*m*-cresol	9·6
5,5'-Dichloro-2,2'-dihydroxy-diphenyl methane	8·7/12·6
2,4,6-Trichloro-phenol	8·5
Pentachlorophenol	5·26

are converted into sodium or potassium salts which dissolve easily in water for this purpose, it is important to remember that the dissociated phenolate anion is not nearly as effective as the undissociated phenol. As solubilization is an urgent requirement for effectiveness, one has to find a compromise when using alkalis or amines as solubilizing agents. Sometimes half of the alkali quantity required for salt formation is sufficient to attain adequate solubility in water; in other cases it is necessary to use excess alkali. It is particularly advantageous if one has and can use phenol derivatives which show considerable water-solubility in their undissociated condition, however this is seldom the case. One example is *p*-chloro-*m*-cresol with a water-solubility of 0·4%. For many applications, alkaline phenolate solutions are used as transport media to deposit the active ingredients in the form of sparingly soluble and non-leachable precipitates on acidic surfaces such as wood, leather, cardboard or textiles. Frequently the influence of the carbon dioxide in the atmosphere is enough to release and fix the free phenol on material impregnated with phenolate solution. In aqueous stock-piled phenolate solutions with slight excess alkalinity, the introduction of CO_2 and the related reduction of the pH value can cause precipitation of the free ingredients. Since such precipitations are undesirable and restrict applicability of the phenolate solutions, counteractions must be taken: storage in closed reservoirs or provision of distinct excess alkalinity.

Corresponding to the mechanism of action, the microbicidal phenol derivatives are effective over a wide spectrum including bacteria, yeasts and fungi. However, there are different effectiveness peaks depending on the type and number of the substituents. Therefore the combined application of various phenol derivatives sometimes allows the application concentrations to be reduced; this is to be aimed at for many reasons. Phenol derivatives are not effective against resistant bacterial spores, at least not at room temperature. However, they are very effective against lipophilic viruses but deficient in their activity against viruses with hydrophilic properties.

Table 43 presents an overview of the activity spectrum of phenol derivatives which have gained especial importance as microbicides.

In detergent solutions, primarily in solutions of non-ionic detergents but also in solutions of anionic detergents, phenol derivatives frequently only display the antimicrobial effect expected of them at unusually high concentrations. This also

Table 43. Minimum Inhibition Concentrations (MIC) of Phenol Derivatives in Nutrient Agar (for abbreviations see Fig. 29)

Test organism	*MIC (mg/litre)*						
	OPP	*BP*	*CBP*	*PCMC*	*PCMX*	*DC*	*PCP*
Aeromonas punctata	200	100	10	200	100	50	10
Bacillus subtilis	100	100	10	150	75	100	10
Escherichia coli	200	500	3500	250	200	100	500
Leuconostoc mesenteroides	100	100	10	200	100	5	35
Proteus vulgaris	200	200	100	200	200	50	100
Pseudomonas aeruginosa	1500	5000	5000	800	1000	>5000	500
Pseudomonas fluorescens	1500	5000	>5000	800	500	3500	500
Staphylococcus aureus	100	100	20	200	100	5	10
Desulfovibrio desulfuricans	50	100	50	35	50	20	35
Formaldehyde resistant bacteria	750	4000	>5000	250	300	300	250
Candida albicans	100	100	50	200	75	50	35
Torula rubra	100	100	50	50	100	50	100
Alternaria tenuis	100	75	20	200	75	50	1
Aspergillus flavus	85	200	75	100	100	50	100
Aspergillus niger	75	100	100	100	100	100	50
Aureobasidium pullulans	35	100	20	30	50	35	20
Chaetomium globosum	60	50	20	80	50	20	20
Cladosporium herbarum	60	200	100	200	100	200	50
Coniophora puteana	50	35	5	100	35	2	35
Lentinus tigrinus	100	75	20	3500	75	5	10
Paecilomyces variotii	100	100	50	200	100	50	50
Penicillium citrinum	35	100	75	100	50	50	50
Penicillium glaucum	80	100	50	100	35	50	50
Polyporus versicolor	65	100	50	5000	75	50	20
Rhizopus nigricans	50	100	50	100	100	35	15
Sclerophoma pityophila	100	100	20	100	75	20	10
Stachybotrys atra corda	50	35	20	100	35	15	15
Trichoderma viride	75	200	100	140	100	50	200
Trichophyton pedis	20	20	10	100	50	10	10

applies to the esters of *p*-hydroxy-benzoic acid (Section 6.1.10) which can also be considered as phenol derivatives. The cause is inclusion of the phenolic ingredients in the micelles which detergents form as soon as their concentration exceeds the critical micelle concentration. Below this concentration which is characteristic for different detergents, the micelles dissolve again, releasing the active phenol derivatives.

It is frequently claimed that phenolics, particularly chlorinated phenolics, possess high oral toxicity, are percutaneously toxic and are generally difficult to degrade. Such blanket condemnations of the phenolics are based on invalid generalizations not confirmed by the facts. It is therefore in no way justified to disqualify the whole class of microbicidal phenolic compounds out of hand simply because some of them are described, correctly, as highly toxic, persistent, ecotoxic or dioxin-containing. The phenol derivatives enumerated in the following have many applications as microbicides, not only because they have broad spectra of effectiveness or because of the special nature of their chemical and physical properties, but also because they include derivatives whose toxicological properties and effects on ecosystems comply with the demands which should be made on microbicides for material protection by a civilization with an awakened sense of responsibility for the environment. There is nothing inappropriate or irresponsible in the continued large-scale use of those phenol derivatives as microbicides for the protection of materials and as active ingredients for disinfectants. There would appear to be little point in listing the phenols in order of preference according to toxicity data, since the relevance of these data depends on the intended application. Thus *o*-phenyl-phenol will continue to be the only phenol derivative which is used to preserve citrus fruits because it has the most favourable toxicity data, which is particularly important in this application. It should be pointed out here that there is no foundation for allegations to the effect that phenolics are incapable of undergoing biodegradation. As can be seen from the curves in Fig. 29 there are phenolics which are degraded easily and quickly, whereas others, e.g. pentachlorophenol, are degraded slowly and only at low concentrations. But microbial degradation of phenolics does not occur in biological sewage treatment plants only; also the aquatic environment possess a considerable biodegradation potential against phenolics, even against pentachlorophenol (Liu, 1989).

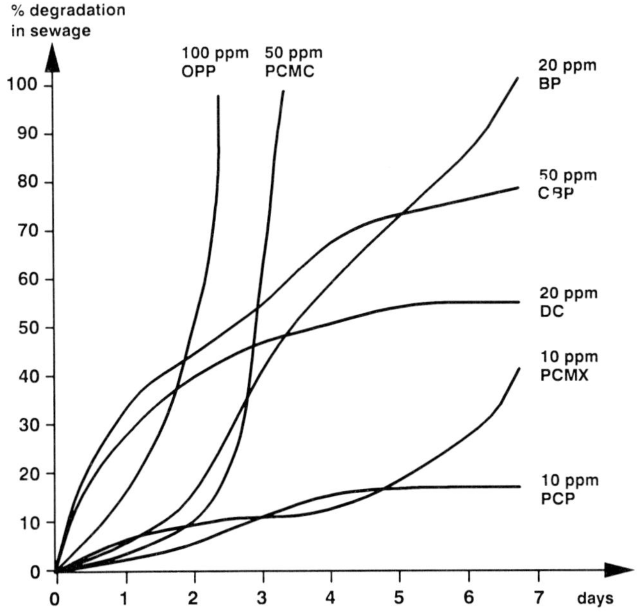

Fig. 29. Decrease in concentration of phenolics in activated sludge (Paulus & Genth, 1983). (OPP, *o*-phenyl-phenol; PCMC, *p*-chloro-*m*-cresol; BP, benzyl-phenol; CBP, *p*-chloro-*o*-benzyl-phenol; DC, (2,2'-dihydroxy-5,5'dichloro-)diphenyl methane; PCMX, *p*-chloro-*m*-xylenol; PCP, pentachlorophenol.)

From the facts there is no foundation for a blanket rejection of phenol derivatives. On the contrary it must be assumed, for many reasons, that it will not be possible to dispense with special phenolics as microbicides for material protection. After all, some of them, such as *o*-phenyl-phenol and *p*-chloro-*m*-cresol, meet the requirement now expected of such microbicides very satisfactorily.

5.1 Phenol — Carbolic acid

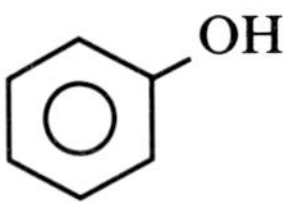

C_6H_6O M_r 94·11
CAS-No. 108-95-2

Chemical and physical properties

Appearance	Colourless crystals with a characteristic odour
Density at 20°C	1·071 g/ml
Melting point	41–43°C
Boiling point	182°C
Flash point	79°C
Solubility	Soluble in water (6·7% at 16°C) and organic solvents
Dissociation constant	$1{\cdot}05 \times 10^{-10}$
Stability	Volatile with water vapour

Toxicity data

LD_{50} oral	414 mg/kg rat
dermal	250 mg/kg rat
subcutaneous	344 mg/kg rat

Phenol is corrosive for mucous membranes and the skin and exhibits its toxic effect also percutaneously.

MWC value 5 ml/m^3 (19 mg/m^3)

Antimicrobial effectiveness/applications

The minimum inhibition concentrations of phenol for bacteria, yeasts and fungi are significantly higher (>1000 mg/litre in nutrient agar) than those of other phenol derivatives. Moreover phenol is highly toxic. Therefore it is not important any longer as a preservative or as an active ingredient in disinfectants.

5.2 Alkylphenols

5.2.1 3,5-Dimethylphenol — Xylenol

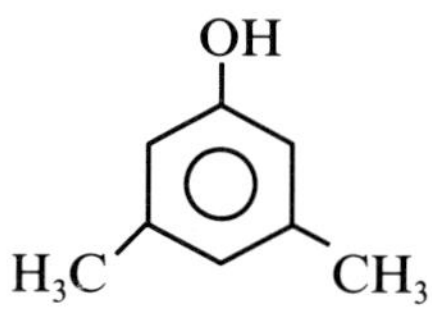

$C_8H_{10}O$ M_r 122·17
CAS-No. 108-68-9

Chemical and physical properties

Appearance	Colourless crystals with phenolic odour
Melting point	65–66°C
Boiling point	222°C
Solubility	Moderately soluble in water, very soluble in alcohols, acetone and other organic solvents
Stability	Volatile with water vapour

Toxicity data

Toxic and corrosive.

Antimicrobial effectiveness/applications

Table 44. Minimum Inhibition Concentrations (MIC) of Xylenol in Nutrient Agar

Test organism	*MIC (mg/litre)*
Escherichia coli	500
Staphylococcus aureus	1000
Aspergillus niger	500
Chaetomium globosum	500
Penicillium glaucum	500

Xylenol may be used as an active ingredient in preservatives and disinfectants, but is, however, no longer of practical importance in these applications.

5.2.2 Isopropyl-methyl-phenols

$C_{10}H_{14}O$ M_r 150·22

5.2.2.1 2-Isopropyl-5-methyl-phenol (Thymol)

CAS-No. 89-83-8

5.2.2.2 4-Isopropyl-3-methyl-phenol (*o*-Cymenol)

CAS-No. 3228-03-3

5.2.2.3 5-Isopropyl-2-methyl-phenol (Carvacrol)

CAS-No. 499-75-2

Chemical and physical properties

Structural formula (*Thymol*, o-*Cymenol*, *Carvacrol*)

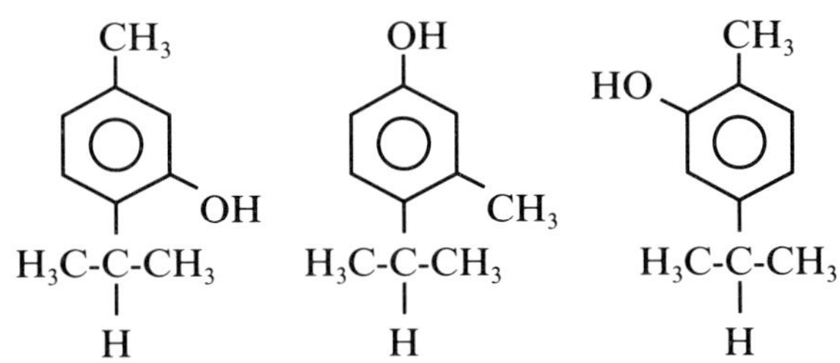

	Thymol	o-*Cymenol*	*Carvacrol*
Appearance	Crystals	Crystals	Fluid
	with a characteristic, distinctive odour		
Melting point	49–51°C	111–114°C	3·0–3·5°C
Boiling point	232°C	245°C	236°C
Density (g/ml at 20°C)	0·965		0·976
Refractive index (at 20°C)	1·5227		1·5233
Solubility (g/litre) in water	0·8	Poor	Poor
in organic solvents	high	high	high
pK_a value	10·62	10·31	10·32
Stability	Volatile with water vapour		

Toxicity data

	Thymol	*Carvacrol*
LD_{50} oral, mouse	1800 mg/kg	5000 mg/kg

Antimicrobial effectiveness/applications

Table 45. Minimum Inhibition Concentrations (MIC) of Isopropyl-methyl-phenols in Nutrient Agar

Test organism	*MIC (mg/litre)*		
	Thymol	o-*Cymenol*	*Carvacrol*
Bacillus subtilis	200	200	
Escherichia coli	500	750	200
Pseudomonas aeruginosa	1000	1000	
Pseudomonas fluorescens	1000	300	
Staphylococcus aureus	200	200	200
Aspergillus niger	200	200	200
Alternaria tenuis	200	200	
Aureobasidium pullulans	200	200	
Chaetomium globosum	100	100	100
Coniophora puteana	50	50	
Lentinus tigrinus	150	100	
Penicillium glaucum	350	500	200
Polyporus versicolor	200	200	
Sclerophoma pityophila	200	500	
Trichoderma viride	350	500	

The broad spectrum of effectiveness and the high antimicrobial activity of isopropyl-methyl-phenols is enlightened by the MIC in Table 45. In spite of that the ispropyl-methyl-phenols have not gained much importance as preservatives for material protection or as active ingredients in disinfectants. The reason for this is the distinctive odour of these phenol derivatives, their poor water solubility and their unfavourable distribution between water and organic phases. Last but not least they are more expensive than other phenol derivatives with more favourable properties. *o*-Cymenol is listed in the EC list of preservatives for cosmetics with a maximum allowed concentration of 0·1%.

5.2.3 4-(2-Methyl-2-butyl)-phenol — *p*-Tert.pentylphenol — *p*-Tert.amylphenol

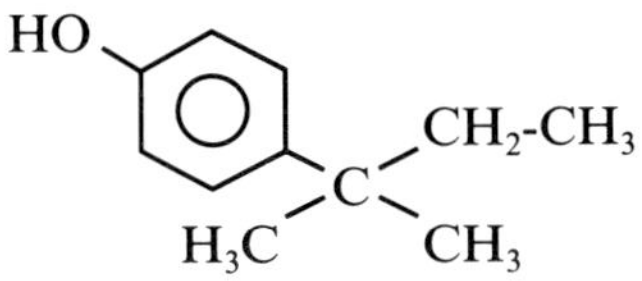

$C_{11}H_{16}O$ M_r 164·25
CAS-No. 80-46-6

Chemical and physical properties

Appearance	Colourless crystals with a characteristic odour
Melting point	93–94°C
Boiling point	265–67°C
Solubility	Sparingly soluble in water, highly soluble in ethanol and other organic solvents

Toxicity data

LD_{50} oral 5·4 ± 1·2 g/kg rabbit (Block, 1983).

Corrosive for mucous membranes and skin.

Antimicrobial effectiveness/applications

Table 46. Microbicidal Concentrations within 10 min according to AOAC Use-Dilution Method (Block, 1983)

Test organism	*Microbicidal conc. (g/litre)*
Pseudomonas aeruginosa	>1·0
Salmonella choleraesuis	0·4
Staphylococcus aureus	0·5

p.Tert.amylphenol is used as an active ingredient in disinfectants together with other microbicidal phenolic compounds which close the gap for Pseudomonades in the activity spectrum of *p*-tert.amylphenol.

5.2.4 Cyclohexylphenols

$C_{12}H_{16}O$ M_r 176·26

5.2.4.1 2-Cyclohexylphenol

CAS-No. 119-42-6

5.2.4.2. 4-Cyclohexylphenol

CAS-No. 1131-60-8

Chemical and physical properties

	2-Cyclohexylphenol	4-Cyclohexylphenol
Structural formula	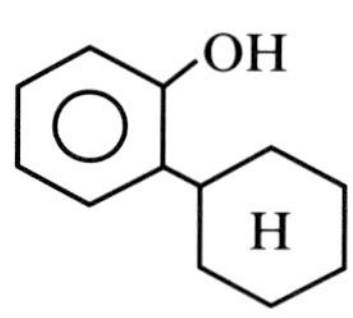	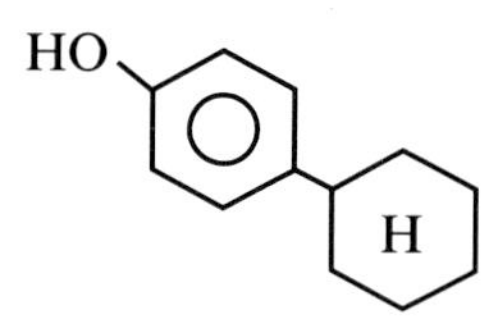
Appearance	Colourless crystals	Colourless crystals
Melting point	55·7°C	128°C
Boiling point	238°C	295°C
pK_a value	10·59	10·30
Solubility	Sparingly soluble in water, highly soluble in organic solvents	

Antimicrobial effectiveness/applications

The cyclohexylphenols are more active against moulds than against bacteria; but in total their activity spectrum is very much unequalized. This in connection with poor water solubility and unfavourable partition coefficients have up to now inhibited the practical application of the cyclohexylphenols as microbicides.

5.2.5 Benzylphenols — Monohydroxydiphenylmethanes

$C_{13}H_{12}O$ M_r 184·24

5.2.5.1 2-Benzylphenol

CAS-No. 28944-41-4

5.2.5.2 4-Benzylphenol

CAS-No. 101-53-1

Chemical and physical properties

Structural formula	*2-Benzylphenol*	*4-Benzylphenol*
	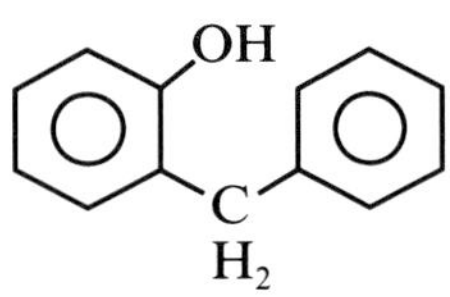	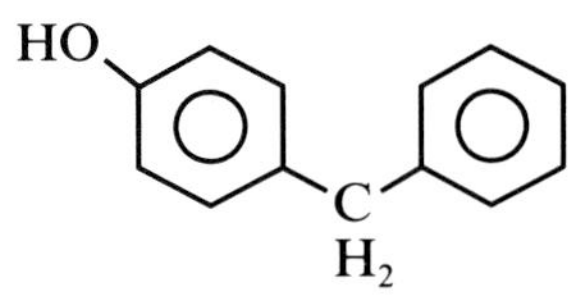
Appearance	Colourless crystals	Colourless crystals
Melting point	51·5–53°C	84°C
Boiling point	312°C	325–30°C
Flash point	>110°C	
Solubility	Sparingly soluble in water, highly soluble in organic solvents	

Toxicity data (40/60 mixture of 2-benzyl- and 4-benzylphenol)

LD_{50} oral 3·36 g/kg rat
dermal 1·78 ml of a 80% solution in C_2H_5OH/kg rat

Irritant and corrosive to skin and mucous membranes.

Antimicrobial effectiveness/applications (efficacy data for a 40/60-mixture of 2-benzyl- and 4-benzylphenol)

The a.m. mixture of benzylphenols is used as an active ingredient in disinfectants, generally in combination with other active phenolics able to close the gap for Pseudomonades. The fact that the MICs are higher than the microbicidal concentrations demonstrates that the benzylphenols are sensitive to organic matter. As a preservative for material protection the mixture has not gained importance because of unfavourable solubility properties.

Table 47. Minimum Inhibition Concentrations (MIC) in Nutrient Agar

Test organism	*MIC (mg/litre)*
Bacillus punctatus	100
Bacillus subtilis	100
Escherichia coli	500
Leuconostoc mesenteroides	100
Proteus vulgaris	200
Pseudomonas aeruginosa	5000
Pseudomonas fluorescens	5000
Staphylococcus aureus	100
Alternaria tenuis	75
Aspergillus flavus	200
Aspergillus niger	100
Aspergillus terreus	200
Aureobasidium pullulans	100
Candida albicans	100
Chaetomium globosum	50
Cladosporium herbarum	200
Coniophora puteana	35
Lentinus tigrinus	75
Paecilomyces varioti	100
Penicillium citrinum	100
Penicillium glaucum	100
Polyporus versicolor	100
Sclerophoma pityophila	100
Stachybotrys atra corda	35
Trichoderma viride	200
Trichophyton pedis	20

Table 48. Microbicidal Activity within 10 min — Use-Dilution Test

Test organism	*Microbicidal conc. (g/litre)*
Escherichia coli	0·3
Pseudomonas aeruginosa	0·8
Staphylococcus aureus	0·25

5.3 Halogenated alkylphenols

As a rule it has to be stated that halogenated alkylphenols are more active and broader in effectiveness than the alkylphenols. It is therefore in no way astonishing that some of the most important phenol derivatives in practical application are found in this class of phenolics. Among others Klarmann *et al.* (1933) have carried out systematic examinations of the relationship between chemical structure and antimicrobial activity with regard to halogenated alkylphenols. However, one should not overestimate the value of the data obtained when decisions and selections have to be made for practical application, as other properties of the microbicidal compounds, such as water-solubility, partition coefficient, activity in the presence of interfering factors encountered in practice, and toxicity, are of the same or of even more importance. Although there are available a lot of data and experience, often the optimum compound and formulation must be determined by experiment.

5.3.1 4-Chloro-3-methyl-phenol — *p*-Chloro-*m*-cresol (PCMC)

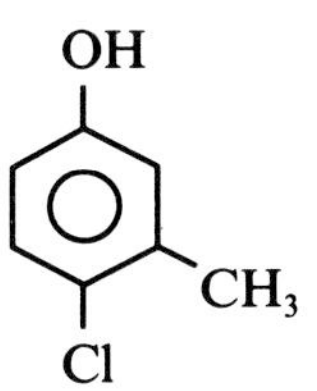

C_7H_7ClO M_r 142·59
CAS-No. 059-50-7

Chemical and physical properties

Appearance	Colourless crystals with phenolic odour
Density at 20°C	1·37 g/ml
Bulk density	approx. 800 g/litre
Vapour pressure (hPa)	0·08 at 20°C; 7 at 100°C
Melting point	63–65°C
Boiling point	Approx. 239°C
Flash point	Approx. 118°C
Ignition temperature	Approx. 590°C
Solubility (g/litre at 20°C)	In water 4
	10% NaOH 320
	ethanol 500
	highly soluble in other organic solvents
pK_a value	9·6
Reaction in water	pH of a 0·1% solution in water: 6·5
Stability	Volatile with water vapour

The PCMC sodium salt (C_7H_7ClNaO) is available in white to slightly yellow flakes or as a powder; it contains generally 3·5 mol crystal water. The following data have been published:

Molar mass	227·63 incl. 3·5 M crystal water	
PCMC content	Approx. 63%	
Density at 20°C	1·36 g/ml	
Vapour pressure at 100°C	0·001 hPa	
Ignition temperature	> 250°C	
Solubility (g/litre at 20°C)	In water	580
	ethanol	2000
	isopropanol	450
Reaction in water	pH of a 1% solution in water: 10·5–11·5	

Toxicity data

PCMC is a microbicide well known for many years, with a long history of human exposure without any adverse effects. This fact is an advantage the importance of which one cannot overestimate; it is in line with the toxicity data which in the meantime have been very thoroughly investigated (Paulus & Genth, 1983). Of the data available the most important ones are listed below:

LD_{50} oral	5129 mg/kg male rat
	3636 mg/kg female rat
LD_{50} percutaneous	>500 mg/kg rat (7 days exposure)

Inhalation hazard assessment with rats:

exposure to air saturated with PCMC at room temperature did not cause any symptoms;

exposure to air enriched with PCMC by dynamic vaporization at a concentration of 704 mg/m^3 air caused transitory irritating effects on the mucosa of the eyes and the inhalatory tract.

PCMC is irritant and corrosive to mucous membranes and the skin.

The sensitization potential of PCMC is not significant.

A subchronic dermal study in rabbits (max. dose 160 mg PCMC/kg/day) resulted only in dermal or local effects, but no treatment related systemic effects.

A pharmacokinetic investigation showed that PCMC administered orally to rats (300 mg/kg) is eliminated rapidly through the kidneys and that the elimination accounts for a high proportion of the administered dose.

A biokinetic investigation — 13 weeks feeding test in rats using dosages of PCMC up to 1500 ppm — accordingly gave no indication of an accumulation of the substance in liver and fat.

No mutagenic effects were observed neither in the Salmonella/microsomes test nor in the micronucleus test on mice.

Ecotoxicity

Toxicity for fish (*Leuciscus idus*)	LC_0	0·5 mg/litre
	LC_{50}	1·2 mg/litre
	LC_{100}	2·0 mg/litre

Phytotoxicity — PCMC was found to be non-phytotoxic when applied at the rate of 2 kg PCMC-Na/ha.

PCMC is not adsorbed in soil because of its relatively high water solubility; it is volatile out of water, e.g. flat rivers or lakes; it is quickly and completely biodegraded (see Fig. 29) additionally it is photochemically degradable. Due to these facts PCMC does not accumulate in the environment.

Antimicrobial effectiveness/applications

As is demonstrated in Table 43. PCMC presents a broad and equalized spectrum of antimicrobial efficacy; it covers bacteria (including myobacteria and Pseudomonades), yeasts and mould producing fungi (exception: some wood destroying basidiomycetes). In connection with the application of PCMC as an active ingredient in disinfectants the fact that PCMC is effective against hepatitis B virus, even in the presence of heavy organic contamination, is of special importance.

PCMC is most effective between pH 4 and 8, where, in fact, contrary to other phenol derivatives only PCMC is sufficiently water soluble. By comparing the microbicidal concentrations of PCMC with those of PCMC-Na, solutions of which have pH values higher than 8, one notices a distinctive decrease in the activity of PCMC in alkaline media (Table 49).

Table 49. Microbicidal Activity within 5 and 10 min of PCMC and PCMC-Na — Use-Dilution Test

Test organism	*Microbicidal concentration (g/litre)*			
	PCMC		*PCMC-Na*	
	5 min	*10 min*	*5 min*	*10 min*
Candida albicans	1·0	1·0	6·0	6·0
Escherichia coli	1·5	1·0	8·0	4·0
Pseudomonas aeruginosa	1·5	0·7	6·0	6·0
Staphylococcus aureus	1·0	1·0	10·0	10·0

Starting from a concentrate containing:

10% PCMC
25% alkylsulphonate
20% isopropanol
45% demineralized water

Table 50. Microbicidal Concentration of PCMC within 5 min — Use-Dilution Test

Test organism	*Microbicidal concentrations (g/litre)*	
	In hard water	*In hard water + 'serum'*[a]
Escherichia coli	0·75	2·5
Pseudomonas aeruginosa	1·0	2·5
Staphylococcus aureus	0·75	2·5

[a] The serum load consists of 10% albumin + 1% yeast extract.

PCMC exhibits within 5 min a microbicidal effect in water, contaminated with 10^7 viable cells/ml, at the concentrations listed in Table 50.

Due to their broad activity spectrum and their favourable properties PCMC and its sodium salt are proven preservatives for the in-can/in-tank protection of very different functional fluids, e.g. solutions of thickeners and adhesives, pigment and filler slurries, concrete additives, textile and leather auxiliaries, paper coatings and lubricoolants. Especially for the preservation of protein containing products PCMC is the first choice because of its compatibility with protein and its unbroken activity in the presence of protein. For the in-tank protection of lubricoolants PCMC-Na is a very important supplement to formaldehyde releasing compounds as it is also highly effective against formaldehyde resistant bacteria. Problems in so-called bio-resistant lubricoolants, which exhibit bacteriostatic efficacy, are caused by uninhibited fungal growth leading to clogging of pipes and filters; such problems may be easily overcome by the application of PCMC or PCMC-Na, which are very active against a great variety of different species of fungi.

In the EC positive list of preservatives allowed for the in-can protection of cosmetics PCMC is mentioned with a maximum authorized concentration of 0·2% and the limitation: prohibited in products intended to come into contact with mucous membranes.

Because of its distinctive and reliable effectiveness, uninfluenced by detergents and heavy contamination, PCMC is a very important active ingredient in disinfectants for hospitals, public buildings, households and animal stables.

The application of PCMC as a preservative is widespread in the leather industry, e.g. for the protection of wet blues.

5.3.2 4-Chloro-3,5-dimethyl-phenol — *p*-Chloro-*m*-xylenol (PCMX)

Structural formula

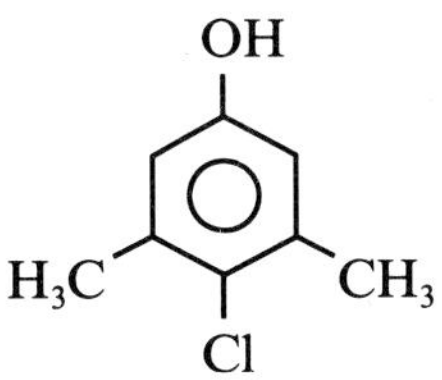

C_8H_9ClO M_r 156·61
CAS-No. 88-04-0

Chemical and physical properties

Appearance	Crystals with phenolic odour
Vapour pressure (at 100°C)	3 hPa
Melting point	114–116°C
Boiling point	246°C
Solubility	0·33 g/litre H_2O at 20°C highly soluble in alkaline solutions and organic solvents; fats and oils
pK_a	9·7
Stability	Volatile with water vapour

Toxicity data

LD_{50} oral 3830 mg/kg rat

Not mutagenic (Ames test).
Not teratogenic according to a test with rats.
Moderately irritant to skin and mucosa.
Low sensitization potential.

Ecotoxicity

LC_0 for *Leuciscus idus*	1 mg/litre
LC_{50} for *Poecilia reticulata*	3·4 mg/litre
EC_{50} for activated sludge bacteria	36 mg/litre

In activated sludge PCMX (15 mg/litre) is degraded slowly (approx. 40% within 7 days); see Fig. 29.

Antimicrobial effectiveness/applications

The MICs in Table 43 show the efficiency of PCMX and the broadness of its activity spectrum.

Starting from a pine oil based PCMX concentrate/disinfectant (pH 9) the PCMX concentrations listed in Table 51 proved microbicidal within 15 min.

Due to its strong microbicidal activity PCMX is used as an active ingredient in disinfectants; however, one has to bear in mind that it is relatively sensitive to organic matter. Moreover PCMX has gained some importance as a preservative for the protection of functional fluids, especially those containing protein, although the poor-water solubility of PCMX does not particularly favour this application. In the EC PCMX is listed among the preservatives allowed for the application in cosmetics with a maximum authorized concentration of 0·5%.

Table 51. Microbicidal PCMX Concentration — Use-Dilution Test

Test organism	*Microbicidal PCMX concentration (g/litre)*	
	In water	*In water 'loaded with serum'*[a]
Escherichia coli	0·2	0·5
Proteus mirabilis	0·2	0·5
Pseudomonas aeruginosa	0·1	2·0
Staphylococcus aureus	0·3	>2·0

[a] 20% bovine albumin.

5.3.3 2,4-Dichloro-3,5-dimethyl-phenol — Dichloro-*m*-xylenol (DCMX)

Structural formula

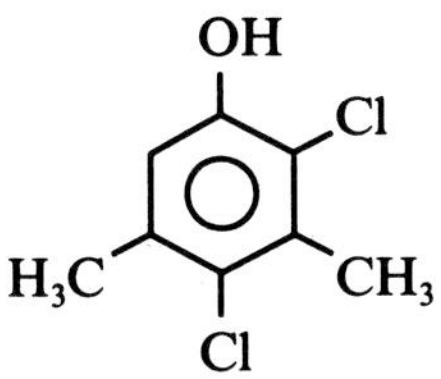

$C_8H_8Cl_2O$ M_r 191·06
CAS-No. 133-53-9

Chemical and physical properties

Appearance	Crystals with distinctive phenolic odour	
Melting point	95–96°C (sublimation)	
Boiling point	250°C	
Flash point	134°C	
Solubility (g/litre at 20°C)	In water	0·2
	acetone	> 700
	diethylketon	> 700
	toluene	180;
	soluble in alkaline solutions	
Stability	Volatile with water vapour	

Toxicity data

LD_{50} oral 2·81–4·12 g/kg rat

Antimicrobial effectiveness/applications

Table 52. Minimum Inhibition Concentrations (MIC) of DCMX in Nutrient Agar

Test organism	*MIC (mg/litre)*
Escherichia coli	2000
Pseudomonas aeruginosa	1000
Staphylococcus aureus	50
Aspergillus niger	200
Chaetomium globosum	200
Penicillium glaucum	100

DCMX behaves similarly to PCMX in activity and properties, but is of inferior importance in application as an active ingredient in pine oil based disinfectants or as a preservative for functional fluids, as it is even less soluble in water than PCMX and has a more distinctive phenolic odour.

5.3.4 4-Chloro-3-methyl-6-isopropyl-phenol — 4-Chloro-thymol

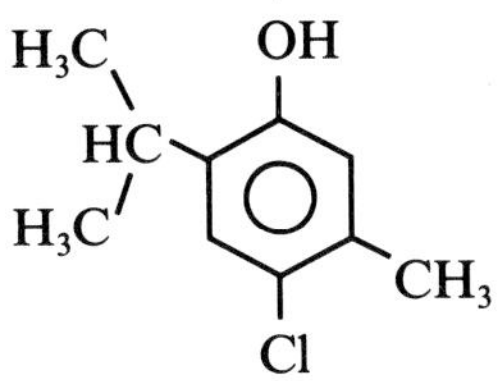

$C_{10}H_{13}ClO$ M_r 184·67
CAS-No. 89-68-9

Chemical and physical properties

Appearance	Crystalline, with a characteristic phenolic odour
Melting point	64°C
Boiling pint	259–263°C
Solubility	0·3 g/litre H_2O at 20°C; soluble in alkaline solutions, highly soluble in organic solvents
Stability	Volatile with water vapour

Toxicity data

LD_{50} oral	2·46 g/kg mouse

Antimicrobial effectiveness/applications

Apparently 4-chloro-thymol is in particular effective against mould producing fungi, but less active against bacteria and insufficient in activity against Pseudomonades. 4-Chloro-thymol has up to now not gained importance as a preservative or an active ingredient in disinfectants.

Table 53. Minimum Inhibition Concentrations (MIC) of 4-Chloro-thymol in Nutrient Agar

Test organism	*MIC (mg/litre)*
Escherichia coli	200
Pseudomonas aeruginosa	4000
Aspergillus niger	100
Chaetomium globosum	20–50
Penicillium glaucum	50
Rhizopus nigricans	20–50

Table 54. Microbicidal Concentrations of 4-Chloro-thymol — Use-Dilution Test

Test organism	*Microbicidal concentration (g/litre) after 10 min exposure*
Escherichia coli	0·2
Pseudomonas aeruginosa	>20
Staphylococcus aureus	0·075

5.3.5 2-Benzyl-4-chloro-phenol — Chlorophen

Structural formula

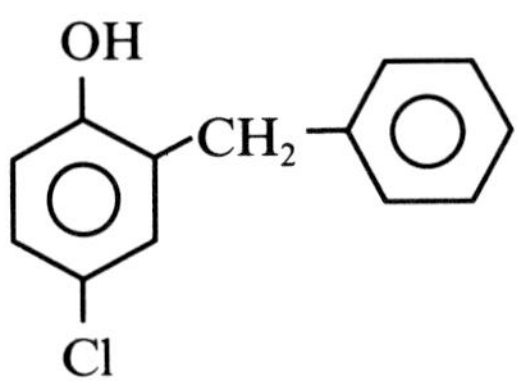

$C_{13}H_{11}ClO$ M_r 218·69
CAS-No. 120-32-1

Chemical and physical properties

Appearance	Colourless crystals which tend to become pink; phenolic odour	
Density at 20°C	1·22 g/ml	
Bulk density	Approx. 650 g/litre	
Vapour pressure at 100°C	Approx. 0·1 hPa	
Melting point	48–49°C	
Solidification point	Approx. 44°C	
Boiling point	327°C; 190–196°C at 15 hPa	
Flash point	Approx. 188°C	
Ignition temperature	Approx. 490°C	
Solubility (g/litre)	In water	0·5 at 20°C
		0·8 at 50°C
	ethanol	>3000
	toluene	2000
	10% NaOH	1000
pK_a value	9·7	
Stability	Volatile with water vapour	

Toxicity data

LD_{50} oral	4·15 g/kg rat
dermal	> 2·5 g/kg rat
LC_{50} inhalation (4 h)	2·5 mg/litre for rats

Strongly irritant and corrosive to mucous membranes and the skin (500 mg/kg rabbit/day over 5 days), but not percutaneously toxic. Non-mutagenic, non-teratogenic.

Ecotoxicity

LC_0 for *Leuciscus idus*	0·5 mg/litre (48 h exposure)
LC_{50} for Bluegill Sunfish	0·3 mg/litre (96 h exposure)
LC_{50} for Rainbow trout	0·7 mg/litre (96 h exposure)
LC_{50} for water flea	0·6 mg/litre (48 h exposure)

Activated sludge organisms tolerate 10–50 mg/litre of Chlorophen and degrade it (75–100% elimination within 7 days; see Fig. 29).

Antimicrobial effectiveness/applications

Minimum inhibition concentrations are listed in Table 43.

Starting from a concentrate containing:

10% Chlorophen
25% alkylsulphonate
20% isopropanol
45% demineralized water

Chlorophen exhibits microbicidal effects in water at concentrations listed in Table 55.

Table 55. Microbicidal Concentrations of Chlorophen within 5 min — Use-Dilution Test

Test organism (approx. 10^7/ml)	*Microbicidal concentration (g/litre)*	
	In demineralized water	*Plus 'serum'*[a]
Escherichia coli	0·1	0·1
Pseudomonas aeruginosa	10	25
Staphylococcus aureus	0·25	1·0

[a] The serum load consists of 10% albumin + 10% yeast extract.

Apparently the activity spectrum of Chlorophen against fungi and yeasts is more equalized than that against bacteria, with deep gaps for Pseudomonades. In disinfectants therefore Chlorophen should be used in combination with other active phenolic compounds which close the gaps. That way it is possible to reduce the concentration of Chlorophen in disinfectants and to minimize the risk of skin irritation through Chlorophen. The unequalized activity spectrum had inhibited Chlorophen from gaining importance as a preservative for material protection. Nevertheless it is listed in the EC list of preservatives allowed for cosmetics (maximum authorized conc. 0·2%).

5.3.6 4-Chloro-2-cyclopentyl-phenol

Structural formula

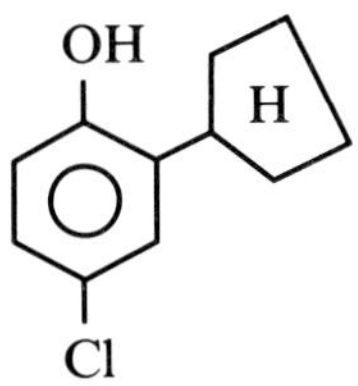

$C_{11}H_{13}ClO$ M_r 196·68
CAS-No. 13347-42-7

Chemical and physical properties

Appearance	Brownish fluid with a characteristic odour
Solidification point	23·7°C
Boiling point	303°C
Solubility	In water 0·24 g/litre H_2O at 20°C; highly soluble in organic solvents
Stability	Volatile with water vapour

Toxicity data

LD_{50} oral	2·46 g/kg rat
	0·42 g/kg rabbit
dermal	0·85 g/kg rabbit

5% solutions are still irritant to the skin and mucosa.
No sensitization was observed.

Antimicrobial effectiveness/applications

4-Chloro-2-cyclopentyl-phenol is very effective against fungi (MIC 10–50 mg/litre) and Staphylococci and Salmonella (microbicidal conc. in 10 min 400 mg/litre) but insufficiently active against Pseudomonades. However, due to its toxicological properties it is no longer used as an active ingredient in disinfectants.

5.3.7 2-Methyl-3,4,5,6-tetrabromo-phenol — 3,4,5,6-Tetrabromo-*o*-cresol

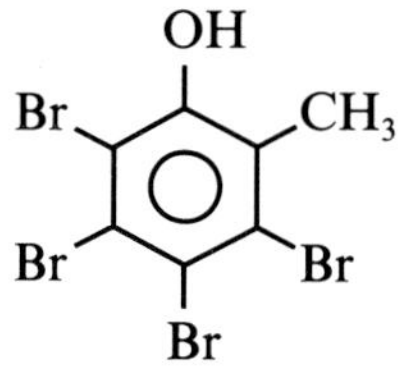

$C_7H_4Br_4O$ M_r 423·75
CAS-No. 576-55-6

Chemical and physical properties

Appearance	Slightly yellowish, odourless, crystalline powder	
Melting point	205–208°C	
Solubility (g/litre at 20°C)	In water practically insoluble	
	ethanol	85
	isopropanol	80
	ethylene glycol	9
	1,2-propylene glycol	8
	vaseline oil	4
Stability	Coloration under light and in contact with traces of iron	

Toxicity data

LD_{50} oral > 6·4 g/kg rat

Very compatible with skin and eye mucosa.

Antimicrobial effectiveness/applications

The compound is very effective against Staphylococci (MIC: approx. 2·5 mg/litre), but very much less active against Pseudomonades (MIC: 2000 mg/litre). Remarkable is the efficacy of tetrabromo-*o*-cresol against algae (MIC for Chlorella species: 1 mg/litre).

Table 56. Minimum Inhibition Concentrations (MIC) of Tetrabromo-*o*-cresol for Fungi in Nutrient Agar

Test organism	*MIC (mg/litre)*
Aspergillus niger	3000
Chaetomium globosum	60
Penicillium glaucum	2500
Rhizopus nigricans	1200

It has been proposed to use the compound as a preservative for cosmetics or as an active ingredient in deodorants or for the antimycotic treatment of textiles. However, the significance of tetrabromo-*o*-cresol has remained very limited (probably due to poor water-solubility, coloration risks and costs).

5.4 Arylphenols

5.4.1 2-Phenyl-phenol — *o*-Phenyl-phenol (OPP)

Structural formula

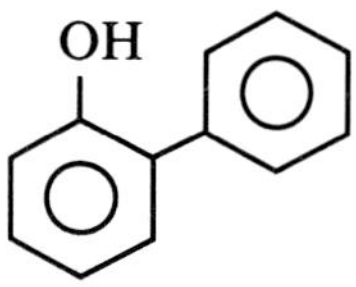

$C_{12}H_{10}O$ M_r 170·21
CAS-No. 90-43-7

Chemical and physical properties

Appearance	White crystals	
Density at 20°C	1·26 g/ml	
Bulk density	650 g/litre	
Vapour pressure (hPa)	7×10^{-4} at 20°C; 1 at 100°C	
Melting point	59°C	
Boiling point	285°C; 152–54°C at 20 hPa	
Flash point	138°C	
Ignition temperature	>250°C	
Solubility (g/litre at 20°C)	In water	0·2
	10% NaOH	500
	ethanol	5900
	isopropanol	3300
	ethylene glycol	2750
	1,2-propylene glycol	3000
	acetone	6600
	white spirit	200
	pine oil	2000
pK_a value	11·6	
Stability	Volatile with water vapour	

The OPP sodium salt ($C_{12}H_9NaO$) is available in white to slightly yellow flakes or as a powder; it contains generally 4 mol crystal water. The following data have been published:

CAS-No.	6152-33-6	
Molar mass	264·28 incl. 4 mol crystal water	
OPP content	Approx. 64%	
Density at 20°C	1·36 g/ml	
Solubility (g/litre at 20°C)	In water	1200
	ethanol	2000
	isopropanol	1500
Reaction in water	pH of a 2% sol. in water: 11·1–11·8	

Toxicity data

Among the non-halogenated phenol derivatives OPP is the one which has been most intensively investigated with regard to its toxicity. The results are in general very favourable. Besides that OPP is a microbicide well known and widely used for many years; that means it presents a long history of human exposure without any adverse effects. The most important toxicity data are indicated in the following:

LD_{50} oral	3000 mg/kg rat

Inhalation hazard assessment with rats: exposure (7 h) to air saturated with OPP at room temperature caused no symptoms.

OPP is only moderately irritant to the skin, but irritant and corrosive to mucous membranes.

The sensitization potential of OPP is negligible.
Non-mutagenic, non-teratogenic.
WHO has fixed an ADI value for humans: 0·02 mg/kg body weight.

Ecotoxicity

OPP is completely and quickly biodegradable. Activated sludge organisms tolerate 100 mg OPP/ml and degrade it completely within 2 days (see Fig. 29).

LC_0 for *Leuciscus idus*	20 mg/litre/96 h
EC_0 for *Daphnia magna*	0·38 mg/litre/48 h

Antimicrobial effectiveness/applications

In Table 57, MIC are tabulated in completion of the MIC listed in Table 43.

Starting from a concentrate containing:

10% OPP
25% alkylsulphonate
20% isopropanol
45% demineralized water

OPP exhibits microbicidal effects in water at concentrations listed in Table 58. There is no gap in the activity spectrum of OPP, which covers bacteria, yeasts and fungi, although the efficacy of OPP against Pseudomonades is not that distinctive.

Due to the fact that OPP is really a broad spectrum microbicide and because of its favourable toxicity and ecotoxicity data OPP has become the most important phenolic microbicide used in numerous applications.

Table 57. Minimum Inhibition Concentrations (MIC) of OPP in Nutrient Agar

Test organism	*MIC (mg/litre)*
Aerobacter aerogenes	100
Bacillus mycoides	100
Candida crusei	200
Rhodotorula mucilaginosa	100
Saccharomyces bailii	100
Saccharomyces cerevisiae	200
Torula utilis	100
Aspergillus terreus	200
Aspergillus ustus	150
Cephaloascus fragrans	200
Ceratocystis pilifera	100
Cladosporium sphaerospermum	40
Gliocladium virens	200
Mucor racemosus	200
Penicillium digitatum	35
Penicillium funicolosum	75
Penicillium italicum	85
Phanerochaete sanguinea	100
Phialophora fastigiata	200
Rhizopus stolonifer	200
Stereum sanguinolentum	100

Table 58. Microbicidal Concentrations of OPP within 5 min — Use-Dilution Test

Test organism (approx. 10^7/ml)	*Microbicidal concentration (g/litre)*	
	In hard water	*In hard water + 'serum'*[a]
Escherichia coli	0·5	1·0
Pseudomonas aeruginosa	2·5	5·0
Staphylococcus aureus	0·75	2·5

[a] The serum load consists of 10% albumin + 10% yeast extract.

OPP is an important active ingredient in disinfectants for hospitals, public buildings, households and animal stables. Very often it is formulated in combination with other phenolic microbicides, e.g. Chlorophen (Section 5.3.5) and/or PCMC (Section 5.3.1).

As a preservative for functional fluids OPP is used in lubricoolants, adhesives, thickeners, textile, leather and paper auxiliaries, ceramic glazes, etc. Textile material, paper, leather and even fruits, e.g. citrus fruits, are protected against microbial deterioration through the application of OPP or OPP-Na. OPP is gaining more and more importance as a substitute for PCP for the temporary protection of freshly cut wood and sawn timber.

In the EC positive list of preservatives for cosmetics OPP is listed with a maximum authorized concentration of 0·2%. Limits are fixed in the USA and the EC for OPP on fruits and vegetables, e.g. 10 mg OPP/kg citrus fruit.

5.4.2 4-Phenyl-phenol — *p*-phenyl-phenol (PPP)

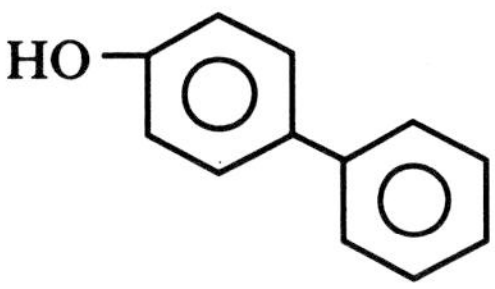

$C_{12}H_{10}O$ M_r 170·21
CAS-No. 92-69-3

Chemical and physical properties

Appearance	White crystals
Melting point	165°C
Boiling point	312°C
Density at 20°C	1·219 g/ml
Solubility at 20°C	In water <0·05 g/litre
	10% NaOH approx. 0·1 g/litre
	highly soluble in organic solvents

Toxicity data

Not available.

Antimicrobial effectiveness/applications

Table 59. Minimum Inhibition Concentrations (MIC) of PPP in Nutrient Agar

Test organism	*MIC (mg/litre)*
Aerobacter aerogenes	150
Bacillus mycoides	75
Bacillus punctatum	75
Bacillus subtilis	100
Escherichia coli	>1000
Leuconostoc mesenteroides	50
Pseudomonas aeruginosa	>1000
Pseudomonas fluorescens	>1000
Proteus mirabilis	>1000
Staphylococcus aureus	50
Aspergillus niger	100
Aureobasidium pullulans	150
Chaetomium globosum	50
Cladosporium cladosporioides	50
Lentinus tigrinus	200
Penicillium glaucum	50–100
Sclerophoma pityophila	200
Trichoderma viride	200

The pressure hydrolysis of chlorobenzene under alkaline conditions to phenol gives OPP and PPP as by-products (5–6%). The ratio of OPP/PPP is 2/1. Separation by fractional distillation is possible. Although the efficacy and activity spectrum of PPP are similar to those of OPP (Section 5.4.1), PPP has no importance as a microbicide for material protection or as an active ingredient in disinfectants. The reasons for this are the unfavourable solubility properties of PPP and the difficulty of providing PPP in larger quantities at reasonable costs.

5.4.3 4-Chloro-2-phenyl-phenol — Monochloro-*o*-phenyl-phenol (MCOPP)

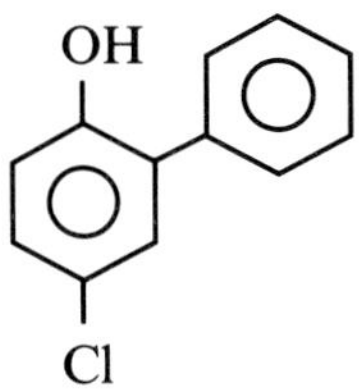

$C_{12}H_9ClO$ M_r 204·66
CAS-No. 1331-46-0

Chemical and physical properties

Appearance	White crystals; the technical product available is a pale straw coloured liquid containing 80% MCOPP, 15% of the 6-chloro-isomer and 5% other chlorinated OPP derivatives
Melting point	36–37°C
Boiling point	250–300°C; 178°C at 15 hPa
Solubility	Sparingly soluble in water, highly soluble in organic solvents.

Toxicity data

LD_{50} oral 3·5 g/kg rat

Irritant and corrosive to skin and mucosa, may cause photo contact dermatitis. Percutaneously toxic.

Antimicrobial effectiveness/applications

Table 60. Microbicidal Concentrations of MCOPP in Water — Use-Dilution Test

Test organism	*Microbicidal conc. (g/litre)*
Escherichia coli	0·05
Pseudomonas aeruginosa	2·0
Staphylococcus aureus	0·05

MCOPP is much more effective than Chlorophen (Section 5.3.5) and was used in pine oil/soap based disinfectants as an active ingredient. However, the unfavourable toxicological properties of MCOPP have led to this application being discontinued.

5.5 Phenoxyphenols

5.5.1 2,4,4'-Trichloro-2'-hydroxy-diphenylether — Triclosan

Structural formula

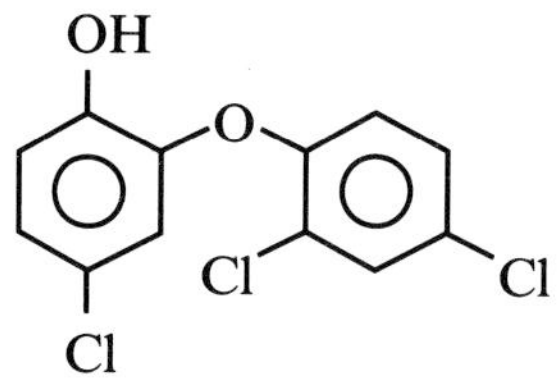

$C_{12}H_7Cl_3O_2$

M_r 289·55
CAS-No. 3380-34-5

Chemical and physical properties

Appearance	White, crystalline powder with weak aromatic odour	
Vapour pressure (hPa)	5×10^{-6} at 20°C; $2{\cdot}6 \times 10^{-2}$ at 100°C	
Melting point	60–61°C	
Decomposition temperature	280–290°C	
Solubility (g/litre)	In water	0·010 at 20°C
		0·039 at 50°C
	0·1 N NaOH	23
pK_a value	7·9	
Stability	Unstable under light; coloration in contact with traces of heavy metals; under certain conditions (alkalinity, heat) conversion into chlorinated dibenzo-*p*-dioxines is possible	

Toxicity data

LD_{50} oral	4530 – >5000 mg/kg rat	
	>5000 mg/kg dog	
intravenous	29 mg/kg rat	
intraperitoneal	198 mg/kg rat	

Non-mutagenic, non-teratogenic.
Good local skin compatibility; rarely sensitization.

Ecotoxicity

LC_0 for fish Approx. 0·6 mg/litres (exposure 48 h)

Concentrations up to 10 mg/litre do not inhibit the activity of activated sludge organisms which degrade Triclosan slowly.

Antimicrobial effectiveness/applications

Table 61. Microbicidal Concentration of Triclosan in Water within 10 min — Use-Dilution Test

Test organism	*Microbicidal concentration (g/litre)*
Escherichia coli	0·05
Pseudomonas aeruginosa	>10·0
Staphylococcus aureus	0·01

The MIC for Staphylococci are in the range 0·03–0·1 mg/litre, whereas Pseudomonades require 100–1000 mg/litre and more for growth inhibition. The growth of several species of mould forming fungi on nutrient agar is inhibited by 5–50 mg Triclosan/litre. Apparently Triclosan is highly active against Staphylococci and has a gap in its activity spectrum for Pseudomonades. However, its antimicrobial activity is widely used in deodorants (sticks, sprays, soaps). The optimum pH for efficacy of Triclosan is between 4 and 8. Triclosan is normally not used in disinfecting cleansing agents or as a preservative for functional fluids or for the protection of materials, as there are available other active phenol derivatives which are cheaper and offer a broader spectrum of effectiveness and better solubility properties.

In the EC positive list of preservatives for cosmetics Triclosan is listed with a maximum authorized concentration of 0·3%.

5.6 Halogenated phenols

The halogenation of phenol, preferably its chlorination, produces halogen phenols that possess a far higher antimicrobial effect and a far higher acidity than the basic product. But halogenating phenol means not only obtaining microbicides of better efficacy, but also of higher persistence, the degree of persistence increasing with the degree of chlorination. Among the microbicides for the protection of materials, chlorinated phenol derivatives, i.e. tri-, tetra- and above all pentachlorophenol, therefore did play a leading role in the past because they largely came up to the (hardly differentiated) demands that were made on microbicides until a few years ago, viz.

— powerful effect and broad effective spectrum,
— highest possible stability = persistence,
— economical formulation.

In the meantime such requirements are out of date or have become more differentiated. The demand is no longer for persistence at any price, but for limited persistence graduated according to the application, for stability or instability as appropriate. Nor is there necessarily a demand for microbicides with so wide a range of effectiveness as pentachlorophenol — fungicidal, bactericidal, algicidal, insecticidal, molluscicidal, herbicidal ... (which therefore should better be referred to as 'biocides'). What is required are such microbicides that display the desired — for instance fungicidal — activity, and no more, at very low concentrations. The toxicity of polychlorinated phenol derivatives increases with the degree of chlorination. On account of their toxicity and ecotoxicity, polychlorinated phenol derivatives are very liable to be substituted by other products, their use for the protection of materials being more and more on the decrease. On account of their lessening importance only the main representatives of polychlorinated phenols will be the subject of the following brief description.

5.6.1 2,4,6-Trichloro-phenol

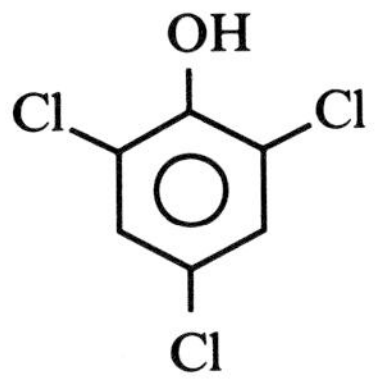

$C_6H_3Cl_3O$ M_r 197·45
CAS-No. 88-06-2

Chemical and physical properties

Appearance	Brownish crystals with a distinctive phenolic odour
Melting point	64–66°C
Density at 75°C	1·4901 g/ml
Boiling point	246°C
Vapour pressure at 20°C	$7{\cdot}3 \times 10^{-3}$ hPa
Solubility (g/litre)	In water 0.9 at 25°C 2·4 at 96°C; soluble in alkaline solutions, highly soluble in organic solvents
pK_a value	8·5 at 25°C
Stability	Volatile with water vapour

Toxicity data

LD_{50} oral	820 mg/kg rat
intraperitoneal	276 mg/kg rat
Skin and eye irritation tests	moderate skin irritation severe eye irritation

National Cancer Institute (NCI) carcinogenesis bioassay (feeding tests); clear evidence: mouse, rat (NCI Technical Report, Bethesda, MD 20014).

Ecotoxicity

Highly toxic for fish (LC_0 < 0·5 mg/litre) and activated sludge organisms.

Antimicrobial effectiveness/applications

Minimum inhibition concentrations of 2,4,6-trichloro-phenol in nutrient media (mg/litre):

For bacteria	approx. 500–100
For fungi	approx. 20
For algae	approx. 20–30

The microbicidal activity is considerably reduced in the presence of organic matter. Much more than for other phenol derivatives the efficacy of trichlorophenols depends on pH variations. 2,4,6-trichlorophenol also exhibits insecticidal effectiveness. Nowadays it is still used as an active ingredient in wood preservatives especially for the anti-sapstain treatment of freshly cut and sawn timber, moreover in preservatives for the leather industry and for functional technical fluids, e.g. protein containing formulations, concrete additives. But altogether the product as a microbicide is on the decline because of its toxicity/ecotoxicity and its intensive adherent phenolic odour.

5.6.2 2,4,5-Trichloro-phenol

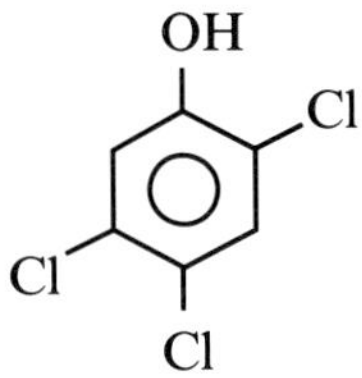

$C_6H_3Cl_3O$ M_r 197·45
CAS-No. 95-95-4

Chemical and physical properties

Appearance	White to brownish crystals with an extremely intensive and adherent phenolic odour
Melting point	67–69°C
Boiling point at 985 hPa	248°C
Solubility	Sparingly soluble in cold water; highly soluble in organic solvents; soluble in alkaline solutions
Stability	Volatile with water vapour

Toxicity data

LD_{50} oral	1000 mg/kg guinea pig
	600 mg/kg mouse
	820 mg/kg rat
LD_{50} subcutaneous	2260 mg/kg rat
intraperitoneal	355 mg/kg rat
intravenous	56 mg/kg mouse

With regard to carcinogenicity, evidence is inadequate according to IARC monographs on the evaluation of the carcinogenic risk of chemicals to man (WHO International Agency for Research on Cancer).

Irritant to skin and mucosa.

Ecotoxicity

Highly toxic for fish (LC_0 < 0·5 mg/litre) and activated sludge organisms.

Antimicrobial effectiveness/applications

Minimum inhibition concentrations in nutrient media (mg/litre):

For bacteria	50–250
For fungi	10

Apparently 2,4,5-trichlorophenol is, compared with 2,4,6-trichlorophenol, the more potent microbicide. It has found application in agents for the preservation of wood and leather (wet blues) and in preservatives for functional technical fluids, especially in protein containing products, e.g. animal glues. In the meantime production and use as a microbicide has been stopped almost completely, mainly because of the risk of forming 2,3,7,8-tetrachloro-dibenzo-*p*-dioxin (TCDD) during production and burning of 2,4,5-tetrachloro-phenol. TCDD, the 'Seveso poison' is an extremely toxic compound whose LD_{50} to the guinea pig is 600 ng/kg.

5.6.3 Pentachlorophenol (PCP)

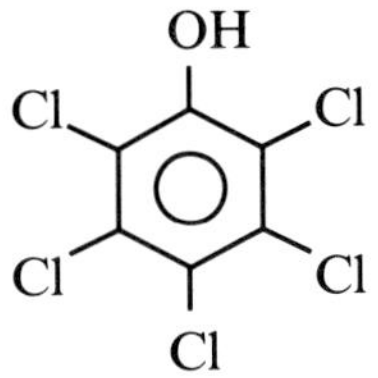

C_6HCl_5O M_r 266·34
CAS-No. 87-86-5

Chemical and physical properties

Appearance	White to grey crystalline powder or flakes with phenolic odour
Melting point	191°C
Boiling point	310°C (decomposition)
Density at 191°C	1·978 g/ml
Vapour pressure (hPa)	$5{\cdot}1 \times 10^{-5}$ at 20°C 0·16 at 100°C
Solubility	0·02 g/litre H_2O; soluble in organic solvents and alkaline media
pK_a value	5·26
Stability	Volatile with water vapour

PCP forms a stable sodium salt as a hydrate (powder, flakes or granules), in general containing approx. 89% PCP-Na. The sodium salt is highly soluble in water, but stable clear solutions only form in the presence of caustic soda in excess.

For pentachlorophenol laurate, see Section 7.8.

Toxicity data

LD_{50} oral	117 mg/kg mouse
	150 mg/kg rat
	168 mg/kg hamster
subcutaneous	100 mg/kg rat
percutaneous	96 mg/kg rat
	100 mg/kg guinea pig
intraperitoneal	56 mg/kg rat
	58 mg/kg mouse
Inhalation LC_{50} for mouse	225 mg/m^3
for rat	355 mg/m^3

Irritant to skin and mucosa. No sensitization.

Ecotoxicity

LC_0 for fish   0·2 mg/ml

Activated sludge organisms tolerate 15 mg PCP/ml.
Biodegradation of PCP occurs extremely slowly (see Fig. 29).

Antimicrobial effectiveness/applications

The minimum inhibition concentrations of PCP listed in Table 43 demonstrate its extraordinary broad and equalized activity spectrum. Moreover PCP exhibits strong algicidal effectiveness in concentrations of 2·5–5 mg/litre. It is indeed more than a simple microbicide, namely a biocide that is also highly toxic for plants, fish, molluscs, insects and mammals.

The antimicrobial effectiveness of PCP coupled with persistence were appreciated and opened the chemical to a wide variety of applications in the field of material protection. Moreover it was used as an algicide and a slimicide in circulating process water systems. For the protection of wood, leather, paper, cardboard and textiles, PCP or PCP-Na is deposited on the materials out of solutions in solvents or water by surface application or pressure impregnation (wood) assuring deep penetration of the active ingredient. PCP deposits possess a reasonable lifetime as the chemical is neither significantly volatile nor leachable. This in particular is valid for pentachlorophenol laurate (see Section 7.8). PCP/PCP-Na are suitable preservatives for glues and adhesives. In water based paints PCP/PCP-Na can provoke discoloration. Moreover one needs relatively high concentrations of active ingredient for the in-can protection of paints because of the unfavourable partition of PCP between water and polymer binder in the paint.

PCP, in the past one of the most important microbicides for the protection of materials, has been for many years under discussion because of its toxicity, especially its dangerous percutaneous effects, its ecotoxicity and its persistence. Today the pressure for substitution becomes effective and the application of PCP is really on the decline. Considerable quantities are still used for wood protection only, in particular for the protection of freshly cut and sawn timber against coloration through mould growth.

5.6.4 2,4,6-Tribromo-phenol

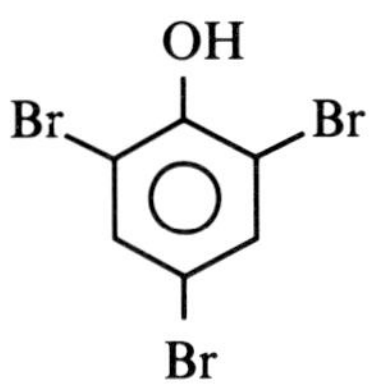

$C_6H_3Br_3O$ M_r 330·80
CAS-No. 118-79-6

Chemical and physical properties

Appearance	Pinkish to white flakes with mild phenolic odour
Melting point	92–94°C
Boiling point at 995 hPa	282–290°C; sublimes
Solubility (g/kg)	In water 0·07 at 15°C
	benzene 720 at 20°C
	heptane 80 at 20°C
	soluble in alkaline solutions
Stability	Decomposes upon excessive heating

Toxicity data

LD_{50} oral 1400 mg/kg rat

Irritant to skin and mucosa.

Mutagenicity Up to 100 μg/plate Ames test negative, above that, toxic; negative in Salmonella assays

Antimicrobial effectiveness/applications

Table 62. Minimum Inhibition Concentrations (MIC) of 2,4,6-Tribromo-phenol Sodium Salt in Nutrient Agar

Test organism	*MIC (mg/litre)*
Aspergillus niger	25
Chaetomium globosum	50
Penicillium citrinum	50
Polyporus versicolor	35
Polystictus versicolor	70
Rhizopus nigricans	500
Trichoderma viride	250
Tyromyces palustris	100

Tribromophenol is comparable in activity to chlorinated phenols and exhibits considerable insecticidal effectiveness. It is worthwhile mentioning that it is more active against *Cephaloascus fragrans* than chlorinated phenols. Because of these properties and its relatively favourable toxicity data (as far as they have been established) tribromophenol and its sodium salt are used in wood preservatives to control fungi and insects. Standard application methods of pressure and vacuum impregnation, dipping, soaking, brushing and spraying are suitable.

5.7 Bisphenols

Bisphenols are composed of two phenolic groups which are not separated or separated by various linkages. Some of them exhibit considerable antimicrobial activity, namely those with phenolic groups separated by $-CH_2-$ or -S- or -O-; bisphenols with the OH groups at the 2,2'-position to the linkage are the most effective ones. Increasing efficacy is observed with an increasing degree of halogenation which is accompanied by an increase in toxicity.

5.7.1 Bis-(4-hydroxyphenyl)-methane — 4,4'-Dihydroxy-diphenylmethane

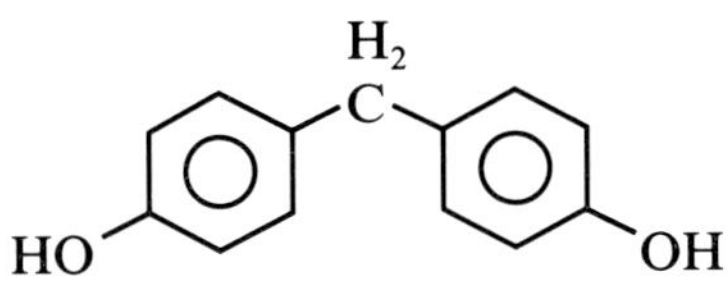

$C_{13}H_{12}O_2$ M_r 200·24
CAS-No. 620-92-8

Chemical and physical properties

Appearance	White powder with phenolic odour
Melting point	162–164°C
Solubility	Practically insoluble in water, soluble in organic solvents

Antimicrobial effectiveness/applications

Table 63. Minimum Inhibition Concentrations (MIC) of Bis-(4-hydroxyphenyl)-methane in Nutrient Agar

Test organism	*MIC (mg/litre)*
Escherichia coli	500
Staphylococcus aureus	500
Aspergillus niger	>1000
Chaetomium globosum	500
Penicillium glaucum	750

Due to its moderate antimicrobial activity the compound has not gained importance as a microbicide in practice.

5.7.2 2,2-Bis-(4-hydroxyphenyl)-propane — Bisphenol A

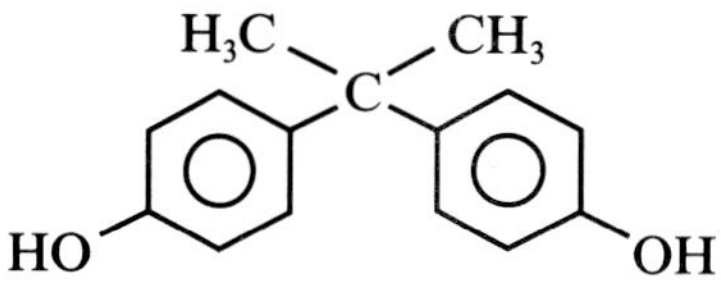

$C_{15}H_{16}O_2$

M_r 228·29
CAS-No. 80-05-7

Chemical and physical properties

Appearance	White crystalline powder with phenolic odour
Melting point	156–157°C
Boiling point at 1·3 hPa	190°C
Solubility	Insoluble in water, soluble in ethanol, acetone, solutions of caustic soda in water

Antimicrobial effectiveness/applications

Bisphenol A as a microbicide has found little interest only, mainly because of its fungicidal activity which, however, is not very distinctive. The antibacterial efficacy of Bisphenol A is not worth a mention.

5.7.3 2,2'-Dihydroxy-5,5'-dichloro-diphenyl methane — Dichlorophen

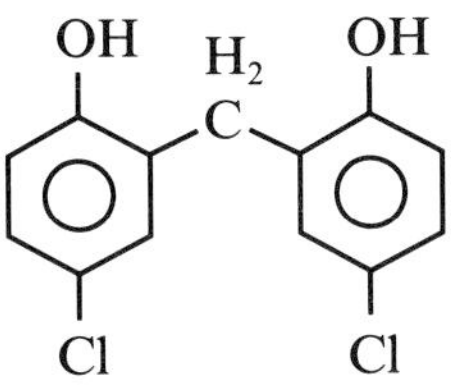

$C_{13}H_{10}Cl_2O_2$ M_r 269·14
CAS-No. 97-23-4

Chemical and physical properties

Appearance	White crystalline powder, sometimes with phenolic odour caused by traces of *p*-chloro-phenol	
Melting point	176°C	
Density at 20°C	1·5 g/ml	
Bulk density	250 g/litre	
Vapour pressure at 100°C	<0·01 hPa	
Solubility (g/litre)	In water	0·07 at 20°C; 0·2 at 50°C
	10% NaOH	500
	butanol	400
	toluene	12
Stability	Not volatile with water vapour	
pK_a values	8·7/12·6	

Toxicity data

LD_{50} oral	3310 mg/kg rat	
cutaneous	>200 mg/kg rat (no systemic effects were observed)	

Not irritant to the skin, moderate irritant to mucous membranes.
No significant sensitization potential.
Subchronical study: 90 day feeding test with dogs: no-effect level: 2000 ppm.
Not mutagenic (Salmonella/microsomes test).
Reproduction study with rats (two generations feeding test): no-effect level: 1000 ppm = 50 mg/kg/day.

Ecotoxicity

LC_0 for *Leuciscus idus*	0·5 mg/litre (exposure: 3 days)
LC_{100}	1·0 mg/litre

For biodegradation see Fig. 29.

Concentrations up to 20 mg/litre have no adverse effects on the degradation activity of sewage plants.

Antimicrobial effectiveness/applications

The minimum inhibition concentrations for Dichlorophen in Table 43 enlighten the activity spectrum of the compound; obviously there is a gap for Pseudomonades. The efficacy against yeasts and fungi is, however, strong and equalized. The optimal pH for bactericidal activity is 5–6. At that pH *Staphylococcus aureus* is killed totally at 20°C within 10 min through 250 mg Dichlorophen/litre. However, in the presence of serum the bactericidal concentration of Dichlorophen is much higher (>1000 mg/litre). There is a big discrepancy between bactericidal and bacteriostatic activity; the latter is especially distinctive at pH 8.

With regard to its application as a microbicide for the protection of materials Dichlorophen offers a lot of advantageous properties: stable, not volatile, not leachable, soluble in oil and alkaline solutions, good skin compatibility. It is used as an active ingredient in slimicides, as a preservative for lubricoolants or the fungicidal treatment of textiles (see also Section 7.9), paper, cardboard, adhesives. Dichlorophen of high purity (practically odourless) may also be used in cosmetics, e.g. deodorants, antiseptic soaps. But there is a risk of coloration for Dichlorophen-containing soaps at exposure to light, especially at pH >7.

5.7.4 3,3'-Dibromo-5,5'-dichloro-2,2'-dihydroxydiphenyl methane — Bromochlorophen

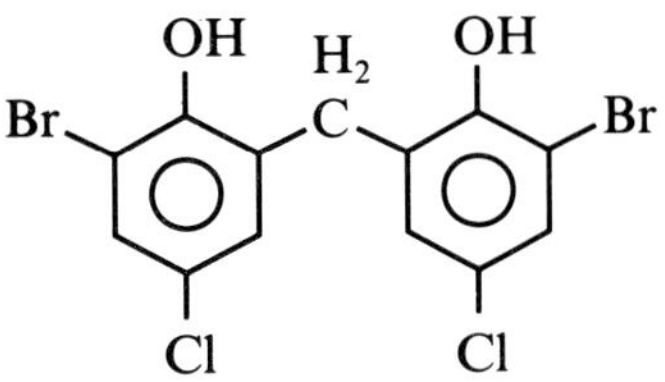

$C_{13}H_8Br_2Cl_2O_2$ M_r 426·92
CAS-No. 15435-29-7

Chemical and physical properties

Appearance	White, crystalline powder with a slight phenolic odour		
Melting point	188–191°C		
Solubility (g/litre)	Sparingly soluble in water		
	In ethanol	approx.	95
	n-propanol	approx.	70
	isopropanol	approx.	40
	1,2-propyleneglycol	approx.	25
	paraffin oil	approx.	5
	glycerine		<1
Stability	Not stable under light		

Toxicity data

LD_{50} oral	3700–8153 mg/kg rat
dermal	>10 000 mg/kg rat
intraperitoneal	580 mg/kg rat
LC_{50} on inhalation	800 mg/m^3 for rats (exposure: 5 h)

Good skin compatibility, only moderately irritant to mucous membranes, no sensitization (guinea pig test).
Photosensitization not observed.
Not mutagenic, not teratogenic (Fiedler, 1989).

Antimicrobial effectiveness/applications

The minimum inhibition concentrations range between 10 mg/litre against *Staphylococcus aureus* and 1000 mg/litre against *Escherichia coli* and *Pseudomonas aeruginosa*. The efficacy against fungi is not very distinctive. Mainly because of its good skin compatibility Bromochlorophen is used as a preservative in cosmetics and an additive in deodorants. The EC positive list of preservatives for cosmetics mentions a maximum authorized concentration of 0·1%.

5.7.5 Bis-(3,5,6-trichloro-2-hydroxyphenyl)-methane — Hexachlorophen

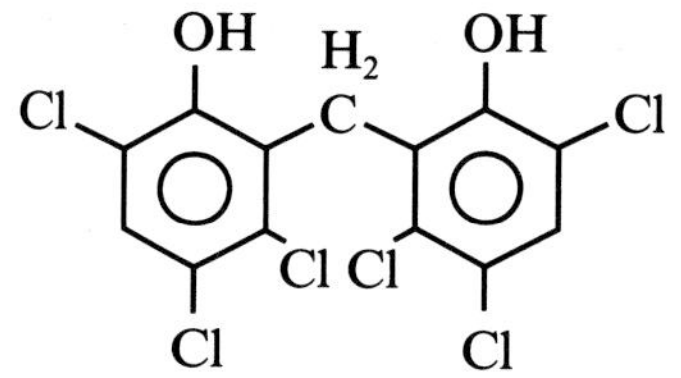

$C_{13}H_6Cl_6O_2$ M_r 406·92
CAS-No. 70-30-4

Chemical and physical properties

Appearance	White crystalline powder	
Melting point	164–165°C	
Solubility (g/litre)	In acetone	1010
	ethanol	500
	toluene	56
	practically insoluble in water, soluble in alkaline solutions	
pK_a values	5·4; 10·9	

Toxicity data

LD_{50} oral	59 mg/kg rat
	80 mg/kg mouse

Not mutagenic, not teratogenic; neurotoxic.
Irritant to the skin and mucous membranes — absorption through the skin.
No-effect level (Cramer & Ford, 1978): 3·7 mg/kg body weight.

Antimicrobial effectiveness/applications

Hexachlorophen is highly effective against bacteria, especially against Staphylococci, not as effective against fungi. Minimum inhibition concentrations for

Staphylococcus aureus	Approx. 1 mg/litre
Pseudomonas aeruginosa	Approx. 500 mg/litre
Fungi	500–1000 mg/litre

Optimum pH for bacteriostatic activity is 8. A maximum of bactericidal action is achieved at pH 5–6. Under these conditions *Staphylococcus aureus* is killed within 5 min by 1000 mg Hexachlorophen/litre.

Hexachlorophen has been used as a microbicide in cosmetics, medical soaps and detergents, and in textiles with antimicrobial effectiveness. However, because of its toxicity, especially its neurotoxicity and the potential absorption through the skin, the application has been banned worldwide.

5.7.6 Bis-(2-hydroxy-5-chlorophenyl)-sulphide — 2,2'-Dihydroxy-5-5'-dichloro-diphenylsulphide — Fentichlor

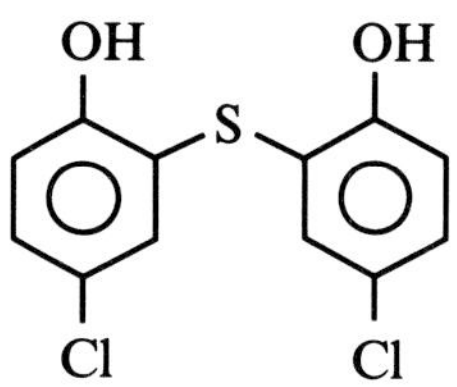

$C_{12}H_8Cl_2O_2S$ M_r 287·18
CAS-No. 97-24-5

Chemical and physical properties

Appearance	White powder
Melting point	174°C
Solubility	30 mg/litre H_2O; very soluble in organic solvents and alkaline solutions; the sodium salt is available as a 50% solution
Stability	Solutions exposed to light change colour (brown)

Toxicity data

LD_{50} oral	3·25 g/kg rat

Can cause photosensitization.

Antimicrobial effectiveness/applications

The similarity of Fentichlor to other bisphenols is demonstrated by the lack in activity against Pseudomonades. Remarkable is the broad and equalized spectrum of effectiveness against fungi and yeasts and even algae.

Low water solubility, tendency to coloration, lack in effectiveness against Pseudomonades have limited the applications of Fentichlor as a microbicide for material protection. It can be used as a preservative in lubricoolants to overcome problems with fungal growth, especially in so-called 'bioresistant' lubricoolants which inhibit the growth of bacteria but not that of fungi. The 50% sodium salt solution of Fentichlor finds application as a slimicide and algicide.

Table 64. Minimum Inhibition Concentrations (MIC) of Fentichlor in Nutrient Agar

Test organism	*MIC (mg/litre)*
Escherichia coli	75
Pseudomonas aeruginosa	3500
Staphylococcus aureus	35
Aspergillus niger	50
Aureobasidium pullulans	<20
Chaetomium globosum	75
Cladosporium herbarum	35
Cladosporium sphaerospermum	50
Coniophora cerebella	<20
Penicillium citrinum	20
Penicillium digitatum	<20
Penicillium funicolosum	35
Penicillium glaucum	10
Penicillium italicum	50
Polyporus versicolor	35
Trichophyton pedis	10
Trichoderma viride	50
Candida albicans	50
Torula rubra	50
Fresh water algae	20

5.7.7 Bis-(2-hydroxy-3,5-dichlorophenyl)-sulphide — 3,3',5,5'-Tetrachloro-2,2'-dihydroxy-diphenylsulphide — Bithionol

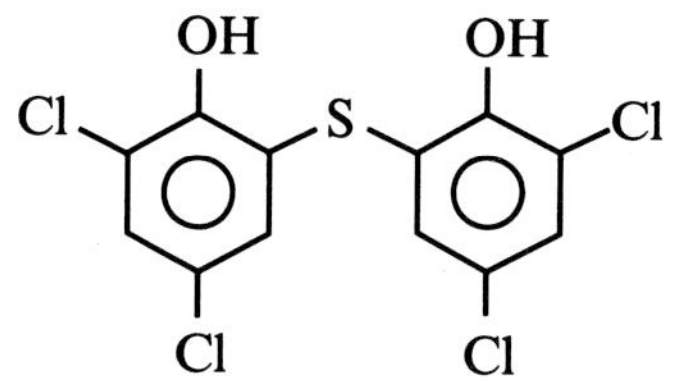

$C_{12}H_6Cl_4O_2S$ M_r 356·01
CAS-No. 97-18-7

Chemical and physical properties

Appearance	White, crystalline powder
Melting point	188°C
Solubility	Almost insoluble in water — soluble in alkaline solutions and in organic solvents

Toxicity data

LD_{50} oral	1·43 g/kg rat
	2·1 g/kg mouse

Antimicrobial effectiveness/applications

Due to its higher degree of chlorination Bithionol shows in comparison to Fentichlor even more activity against gram-positive bacteria, but has the same gap for Pseudomonades in its activity spectrum. There is also for bisphenols a typical discrepancy between microbistatic and microbicidal concentrations, differing from 1 to 1000 mg/litre, e.g. for *Staphylococcus aureus*.

With regard to the application of Bithionol as a microbicide for material protection what is said for Fentichlor (see Section 5.7.6) is valid for Bithionol.

5.8 Nitrophenols

The introduction of the nitro group into the phenol molecule leads to an increase in acidity and antimicrobial effectiveness, the latter being superior to most of the halophenols. However, the gain in activity is accompanied by an equal gain in toxicity. The yellow colour of nitrophenols is an additional handicap in the application of the compounds as microbicides for the protection of materials. In the meantime nitrophenols may be marked as old-line microbicides which are no longer of practical importance.

5.8.1 4-Nitro-phenol

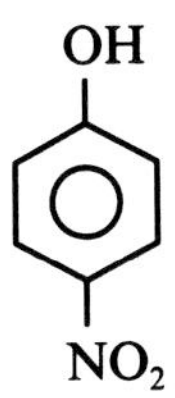

$C_6H_5NO_3$ M_r 139·11
CAS-No. 100-02-7

Chemical and physical properties

Appearance	Yellow crystals	
Melting point	113–115°C	
Boiling point	279°C	
Dissociation constant	$1{\cdot}38 \times 10^{-9}$	
Solubility	Sparingly soluble in water, soluble in organic solvents and alkaline solutions	
Sodium salt	M_r 161·09	
	melting point	>300°C
	flash point	90°C

Toxicity data

Toxic, irritant to skin and mucosa.

Antimicrobial effectiveness/applications

4-Nitrophenol is a broad spectrum microbicide, but out of use nowadays because of its toxicity and coloration effects. It was mainly used as a preservative in the leather industry.

5.9 Phenol-4-carbonic acid esters

These phenolic compounds are listed in Section 6, 'Acids' in Section 6.1.10 '*p*-Hydroxy-benzoates' because of their exceptional acidic character.

6 ACIDS

Among the large number of organic and inorganic acids there are only a few which are used as preservatives, especially in the food and cosmetic industry. In this section only those acids are listed which really exhibit antimicrobial activity; peroxycarboxylic acids are described in Section 19, 'Oxidizing agents'. That means, acids which act as acidulants, buffers, flavouring agents or synergists to antioxidants are not discussed here, e.g. phosphoric acid, citric acid, malic acid, adipic acid, succinic acid, glutaric acid, ascorbic acid, isoascorbic acid. Acids in general may be helpful in the combat against microbes, as low pH values facilitate the destruction of microbes, e.g. by heat, with the advantage that one needs shorter sterilization or pasteurization processing times thus saving energy and preserving product quality. In addition at low pH values bacterial growth generally is suppressed and spore germination is delayed or prevented. Finally acidulants lowering the pH of, for example, foods support the effect of antimicrobial active acids, as these are only effective in acidic preparations.

The microbicidal acids belong to the membrane-active substances. Typical is that acids display significant antimicrobial activity only when they are present in their undissociated state. A carboxylic acid for instance dissociates according to the following equation

$$\mathrm{R\text{-}\overset{\overset{\displaystyle O}{\|}}{C}\text{-}OH} \rightleftharpoons \mathrm{R\text{-}\overset{\overset{\displaystyle O}{\|}}{C}\text{-}O^- + H^+}$$

Undiss. state Diss. state

The acidity is characterized by the dissociation constant K; the greater is K, the higher is the concentration of H^+ ions or the stronger is the acid. With regard to the utility of acids as microbicides it is, however, as already said, important to have knowledge about the availability of the undissociated form. That is characterized by the pK_a value indicating the pH value at which 50% of the acid is present in the undissociated state. pK_a values of a selection of microbicides having an acidic character are listed in Table 65.

The antimicrobial active acids' mode of action is based on the ability of the undissociated forms to interact with or to pass through the membrane of the microbial cell which normally is negatively charged thus being a barrier for the negatively charged forms of the acids. In their undissociated state the acids may alter the membrane permeability of the microbial cell and interfere with many enzymatic processes in the cell. That means nutrient transport inhibition is mainly responsible for the antimicrobial effect of acids.

Table 65. pK_a Values of Acidic Compounds Used as Microbicides

Acidic microbicide	*p*K_a
Formic acid	3·8
Acetic acid	4·8
Propionic acid	4·9
Lactic acid	3·8
Dehydroacetic acid	5·4
Sorbic acid	4·8
Benzoic acid	4·2
Salicylic acid	3·0
Methyl *p*-hydroxybenzoate	8·5
Propyl *p*-hydroxybenzoate	8·1
2-Phenyl-phenol	11·6
4-Chloro-3-methyl-phenol	9·6
Sulphurous acid	1·8/6·9
Boric acid	9·1

6.1 Organic acids

6.1.1 Formic acid

```
   O
   ||
H-C-OH
```

CH_2O_2

M_r 46·03
CAS-No. 64-18-6

Chemical and physical properties

Appearance	Colourless fluid with a pungent odour
Melting point	8·6°C
Boiling point	100·8°C
Density (at 20°C)	1·220 g/cm^3
Refractive index at 20°C	n_D 1·3714
Flash point	68·9°C
Ignition temperature	601°C
Solubility	Miscible with water, alcohol, ether
pK_a value	3·75
Dissociation constant	1·77 × 10^4

Toxicity data

LD_{50} oral	1210 mg/kg rat
	1100 mg/kg mouse
intravenous	145 mg/kg mouse
	239 mg/kg rabbit

Concentrated formic acid irritates skin and mucous membranes.
Carcinogenic and teratogenic effects are not observed.
Acceptable daily intake (ADI): 3 mg/kg body weight.
MWC: 5 ml/m^3 (9 mg/m^3).

Antimicrobial effectiveness/applications

Formic acid exhibits its antimicrobial activity primarily in the undissociated state, i.e. at pH below 3·5 and therefore partly as an acidulant. The efficacy covers especially yeasts and some bacteria; lactic acid bacteria and moulds are relatively resistant.

In the form of its sodium or potassium salts (white, crystalline powders which are highly soluble in water) formic acid is used for the preservation of acidic foods (addition rates: 0·1–0·4%), frequently in combination with benzoic (Section 6.1.8) or sorbic acid (Section 6.1.5). In the EC list of preservatives permitted for cosmetics, formic acid is mentioned with a maximum authorized concentration of 0·5%.

6.1.2 Acetic acid

O
||
H_3C-C-OH $C_2H_4O_2$ M_r 60·05
CAS-No. 64-19-7

Chemical and physical properties

Appearance	Colourless fluid with a pungent odour and lachrymal effect
Melting point	16·5°C
Boiling point	118·1°C
Density at 20°C	1·0492 g/cm^3
Refractive index at 25°C	n_D1·3698
Vapour pressure at 20°C	15 hPa
Flash point	57·2°C
Solubility	Miscible with water, alcohol, ether, chloroform, glycerine and other organic solvents; insoluble in carbon disulphide
pK_a value	4·8

Toxicity data

LD_{50} oral	3·3 g/kg rat
intravenous	525 mg/kg mouse
inhalation	5620 ppm for mouse (exposure time: 1 h)

An ADI value is not fixed.
MWC: 10 ml/m^3 (25 mg/m^3)

Antimicrobial effectiveness/applications

At lower pH values (<4·5) acetic acid inhibits the growth of bacteria, yeasts and to a lesser extent the growth of moulds. The inhibitory effect is greater than due to pH alone. One assumes that undissociated acetic acid because of its good lipid solubility can penetrate the microbial cell and exert its antimicrobial effect. Bacteria of the genus Acetobacter can produce acetic acid by the oxidation of alcohol. In combination with heat acetic acid is especially effective.

Vinegar is one of the oldest preservatives and flavouring agents for food and also today acetic acid's application as a preservative is in food only.

6.1.3 Propionic acid

$H_3C\text{-}CH_2\text{-}COOH$ $C_3H_6O_2$ M_r 74·08
CAS-No. 79-09-4

Chemical and physical properties

Appearance	Colourless fluid with pungent odour
Boiling point	140·9°C
Solidification point	−22°C
Density at 20°C	0·992 g/cm^3
Refractive index at 19·9°C	n_D 1·38736
Vapour pressure at 20°C	3·2 hPa
Flash point	57·75°C
Solubility	Miscible with water, highly soluble in organic solvents
pK_a value	4·9

Toxicity data

LD_{50}	oral	4290 mg/kg rat
	intravenous	625 mg/kg rabbit

Irritant to skin and mucous membranes.
MWC: 10 ml/m^3 (30 mg/m^3).

According to experiments with the stomach mucous membranes of rats, which received a nutrition containing 4% propionic acid for 20 weeks, there is a suspicion that propionic acid is carcinogenic. On the other hand propionic acid is not foreign to the human body, as it is normally formed by decomposition of certain amino acids and the oxidation of fatty acids containing odd numbers of carbon atoms. The FAO and WHO therefore have not yet fixed an ADI value.

Antimicrobial effectiveness/applications

Propionic acid is an active microbicide in the undissociated form only; in consequence it exhibits optimum efficacy at pH values below 5. Propionic acid is primarily effective against moulds. The activity against yeasts is minimal and the activity against bacteria is poor. Microbes belonging to the genus Propionibacterium produce propionic acid during fermentation of Swiss Gruyère cheese. However, *Bacterium mesentericus* is an exception as it is very sensitive to the effect of propionic acid.

As a preservative propionic acid is used for the protection of acidic media; as the acid itself is corrosive one uses the calcium or the sodium salts, free flowable white powders, which are water soluble (solubility in water at 100°C in g/100 ml: sodium salt 150; calcium salt: 55·8). The main field of application for propionates is the baking industry where they can protect bread and cakes from the growth of moulds and the proliferation of *Bacterium mesentericus* which causes rope in bread. Addition rates: 0·15–0·3%. Propionates are also used for the protective treatment of cheese. In the EC list of preservatives those cosmetic products which may contain propionic acid and its salts are listed with a maximum authorized concentration of 2% (acid).

6.1.4 Lactic acid — α-Hydroxy-propionic acid

$H_3C\text{-}CH(OH)\text{-}COOH$	$C_3H_6O_3$	M_r 90·08 CAS-No. 50-21-5

Chemical and physical properties

Appearance	Colourless, odourless, highly viscous and hygroscopic fluid
Boiling point	119°C
Solidification point	18°C
Density at 20°C	1·21 g/cm^3
Solubility	Miscible with water, alcohol and glycerine; soluble in ether, insoluble in trichloromethane and benzene
pK_a value	3·79

Toxicity data

LD_{50} oral	3·73 g/kg mouse
intraperitoneal	0·288 g/kg mouse

Ames test and chromosome test negative.

The FAO and WHO have set no limit for the ADI value of lactic acid and its salts.

Antimicrobial effectiveness/applications

Lactic acid is a natural constituent of food; it is manufactured through microbial fermentations, e.g. by Lactobacilli. Synthetically lactic acid can be produced out of propylene and N_2O_4. Fermentation processes are the oldest methods for the protection of food. Lactic acid produced in such processes lowers the pH to levels unfavourable for the growth of spoilage organisms, that means it is mainly active as an acidulant. However, efficacy is observed against anaerobic bacteria, although at concentrations higher than 0·5%. Consequently for broader antimicrobial effectiveness lactic acid has to be combined with other microbicides active against yeasts and moulds, such as benzoic acid (Section 6.1.8) or sorbic acid (Section 6.1.5). Remarkable is the ability of lactic acid to inhibit specifically mycotoxin formation (Reiss, 1976; Lück, 1980).

6.1.5 Sorbic acid — 2,4-Hexadienoic acid

$H_3C\text{-}CH{=}CH\text{-}CH{=}CH\text{-}COOH$ $C_6H_8O_2$ M_r 112·13
CAS-No. 110-44-1

Chemical and physical properties

Appearance	White crystalline powder with a distinctive odour	
Melting point	134°C	
Boiling point	228°C (decomposition)	
Solubility (g/litre) at 20°C	In water	approx. 2
	ethanol	approx. 120
	isopropanol	approx. 85
	propylene glycol	approx. 5
	glycerine	approx. 3
	oils	approx. 5–10
	acetic acid	approx. 110
	contrary to the acid the potassium salt of sorbic acid is highly soluble in water (1390 g/litre)	
pK_a value	4·76	

Toxicity data

LD_{50} oral	7·36 g/kg rat

No mutagenic action.
No teratogenic action.
No carcinogenic action.

ADI value (FAO/WHO)	25 mg/kg body weight

Antimicrobial effectiveness/applications

In accordance with its pK_a value of 4·76 sorbic acid is most effective at pH 4 or below in its undissociated form; however, there is considerable antimicrobial activity of sorbic acid observed also at pH values up to 6–6·5. Sorbic acid inhibits especially the growth of moulds and yeasts; its efficacy against bacteria is not that equalized and consequently not reliable.

Sorbic acid is a naturally occurring compound; it may be characterized as one of the least toxic of all the preservative agents known. It is able to penetrate the semipermeable microbial cell membrane inhibiting nutrient transport and enzymes.

At addition rates between 0·05 and 0·2% sorbic acid is used as a preservative for foodstuffs (especially beverages, including wine), pharmaceuticals and cosmetics. The sodium, potassium and calcium salts are also available; the most frequently used salt form is, however, potassium sorbate. In the EC list of preservatives allowed for the protection of cosmetics, sorbic acid is listed with a maximum authorized concentration of 0·6%.

6.1.6 Undecylenic acid — Undecenoic acid

$CH_2{=}CH\text{-}(CH_2)_8\text{-}COOH$ $C_{11}H_{20}O_2$ M_r 184·28
CAS-No. 112-38-9

Chemical and physical properties

Appearance	Crystals or colourless fluid with a distinctive odour
Melting point	24·5°C
Boiling point	275°C (decomposition)
Density at 25°C	0·912 g/ml
Refractive index at 25°C	n_D 1·4478
Flash point	166°C
Solubility	Sparingly soluble in water, soluble in many organic solvents, e.g. alcohols, trichloromethane, ether

Toxicity data

LD_{50} oral	2·5 g/kg rat

Irritant to mucous membranes at concentrations >1%.

Antimicrobial effectiveness/applications

The antimicrobial activity of undecylenic acid is widely restricted to the inhibition of moulds including pathogenic fungi. The pH optimum is in the acidic range (4·5–6). Nevertheless undecylenic acid is listed among the preservatives allowed for the protection of cosmetics in the EC (maximum authorized concentration: 0·2%). Undecylenic acid may also be applied in dermatics as a fungicide (addition rates 2–10%). For preservation purposes it is recommended to use undecylenic acid only in combination with bactericides to complete the spectrum of effectiveness.

6.1.7 Dehydroacetic acid — 3-Acetyl-6-methyl-2[H]-pyran-2.4.[3H]-dione (DHA)

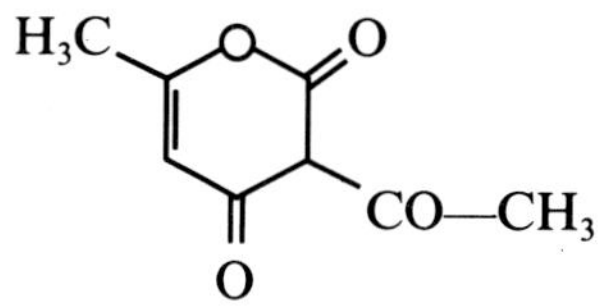

$C_8H_8O_4$ M_r 168·14
CAS-No. 520-45-6

Chemical and physical properties

Appearance	Colourless, crystalline, tasteless powder		
Melting point	109–111°C		
Boiling point	269·9°C		
Solubility (g/100 g solvent)		*DHA*	*DHA-Na*
	In water	<1	33
	ethanol (95%)	3	1
	methanol	5	14
	propylene glycol	1·7	48
	acetone	22	0·2
	ether	5	<0·1
	olive oil	1·6	<0·1
	tetrachloromethane	3	<0·1
	benzene	18	<0·1
Stability	Relatively stable; solutions of DHA tolerate heating up to 120–130°C for 1 h		

Toxicity data

LD_{50} oral Approx. 1 g DHA/kg rat
Approx. 0·5 g DHA-Na/kg rat

DHA does not cause primary skin irritation nor sensitization.
Ames test and chromosome test negative.
No-effect level (Cramer & Ford, 1978): > 50 mg/kg body weight.

Regarding all the toxicological data available it is, however, not justified to recommend DHA/DHA-Na as a food preservative.

Antimicrobial effectiveness/applications

DHA is more effective against mould producing fungi than against bacteria. The efficacy is restricted to media with acidic pH values; contrary to other acids used as preservatives DHA is, however, still effective at a pH as high as 6. Using

DHA/DHA-Na as a preservative for the in-can protection of functional fluids one has additionally to bear in mind that the active ingredient is not compatible with non-ionic detergents. In the EC list of preservatives allowed for the protection of cosmetics DHA is listed with a maximum authorized concentration of 0·6%. Addition rates for tooth-pastes: 0·05–0·15%.

6.1.8 Benzoic acid — Phenylformic acid

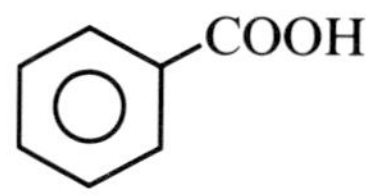

$C_7H_6O_2$ M_r 122·13
CAS-No. 65-85-0

Chemical and physical properties

Appearance	White crystalline powder		
Melting point	120–122°C		
Boiling point	250°C		
Flash point	121–131°C		
Solubility (g/litre) at 20°C	In water		2·9
		ethanol	435·0
Of sodium benzoate	in water		555·0
pK_a value	4·2		

Toxicity data

LD_{50} oral	2·37 g benzoic acid/kg mouse
	1·7 g benzoic acid/kg rat
	2·3 g sodium benzoate/kg rat
ADI value	0–5 mg/kg body weight

Ames test negative.

Antimicrobial effectiveness/applications

Table 66. Minimum Inhibition Concentrations (MIC) of Benzoic Acid in Nutrient Agar at pH 6 (Wallhäusser, 1984)

Test organism	*MIC (mg/litre)*
Staphylococcus aureus	20
Escherichia coli	160
Klebsiella pneumoniae	160
Pseudomonas aeruginosa	160
Pseudomonas fluorescens	160
Pseudomonas cepacia	160
Candida albicans	1200
Aspergillus niger	1000
Penicillium notatum	1000

Benzoic acid is a naturally occurring acid (detectable in several berries and fruits) which exhibits antimicrobial activity in its undissociated form only. It is therefore only significantly effective in acidic formulations at pH 4·5 or below (optimum pH 4–2·5). At pH 6 only 1·52% non-ionized benzoic acid is available. Apparently benzoic acid is more effective against bacteria and yeasts than against fungi; the antibacterial activity of benzoic acid is superior to that of sorbic acid (Section 6.1.5).

Sodium benzoate is an especially inexpensive preservative agent and probably the most extensively used food preservative in the world (addition rates 0·03–0·3%). A certain limitation in application as a food preservative is set only because of the astringent flavour of benzoate; it is therefore frequently used in combination with other active ingredients, e.g. sorbates or *p*-hydroxy-benzoates (Section 6.1.10). Benzoic acid, its salts and esters are permitted for the preservation of cosmetics (EC positive list); maximum authorized concentration: 0·5%. Sometimes benzoic acid or sodium benzoate are also applied as preservatives in acidic technical functional fluids, e.g. polymer emulsions, especially in those coming in contact with foodstuffs. In acidic two-phase systems (emulsions) it has to be taken into consideration that benzoic acid shows a marked tendency to migrate into the organic (oil) phase so that its activity in the water phase, where the microbes are vegetating, is decreased. Losses of activity are also observed in the presence of non-ionic detergents, QUACS, proteins and glycerine. Propylene glycol intensifies the efficacy of benzoic acid.

6.1.9 Salicylic acid — *o*-Hydroxy-benzoic acid

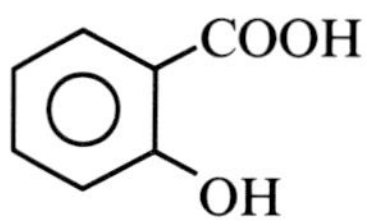

$C_7H_6O_3$ M_r 138·13
CAS-No. 69-72-7

Chemical and physical properties

Appearance	White, crystalline	
Melting point	157–159°C	
Boiling point	211°C	
Solubility (g/litre) at 20°C	In water	2·2
	ethanol	370
	glycerine	17
	oils	12
Stability	Sublimation at 76°C; light sensitive; thermal degradation into phenol and CO_2; coloration with iron salts	
pK_a value	2·97	
Dissociation constant	$1{\cdot}07 \times 10^{-3}$	

Toxicity data

LD_{50} oral 891 mg/kg rat

Percutaneous activity, cumulation possible.

Antimicrobial effectiveness/applications

The efficacy of salicylic acid is in first line directed against yeasts and moulds, but its activity against bacteria is also worth a mention as it is more pronounced than that of benzoic acid. However, because of its low pK_a value salicylic acid can only be used for the preservation of highly acidic products. As a food preservative it has been abandoned because of its toxicity. Still of some importance is the application of salicylic acid for the protection of cosmetics and pharmaceutical products. In the EC list of preservatives allowed for cosmetics, salicylic acid and its salts are mentioned with a maximum authorized concentration of 0·5%.

6.1.10 *p*-Hydroxy-benzoates

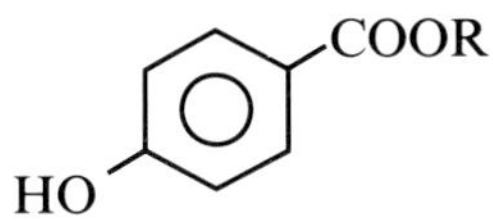

Hydroxy-benzoates, preferably *p*-hydroxy-benzoic acid alkyl esters, are listed under 'Acids' and not under 'Esters' because of their exceptional acidic character.

The starting product for the synthesis of *p*-hydroxy-benzoic acid alkyl esters, *p*-hydroxy-benzoic acid (PHBA) is a naturally occurring acid, e.g. in raspberries and blackberries. PHBA alkyl esters are relatively stable chemicals; hydrolysis takes places only under drastic conditions (e.g. pH 10 or pH 1). At pH 4 the PHBA methyl and ethyl ester are stable between 40 and 121°C. Lipid solubility and efficacy of PHBA alkyl esters increase with increasing length of the alkyl chain, reaching a maximum at the butyl ester. The PHBA methyl ester distinguishes itself with the optimal water solubility (0·25%), which decreases with increase in the alkyl chain. The lack of good water solubility is a certain limitation for the application of PHBA alkyl esters as preservatives. An advantage over benzoic acid and other preservative acids is the relatively high pK_a value (approx. 8·5) of the PHBA alkyl esters, which means that they exhibit antimicrobial activity even at neutral and slightly higher pH values (see Fig. 30).

PHBA alkyl esters are more effective against moulds and yeasts than bacteria, however, they are also useful for the control of bacterial growth. As membrane-active microbicides, their primary mode of action is based on nutrient transport

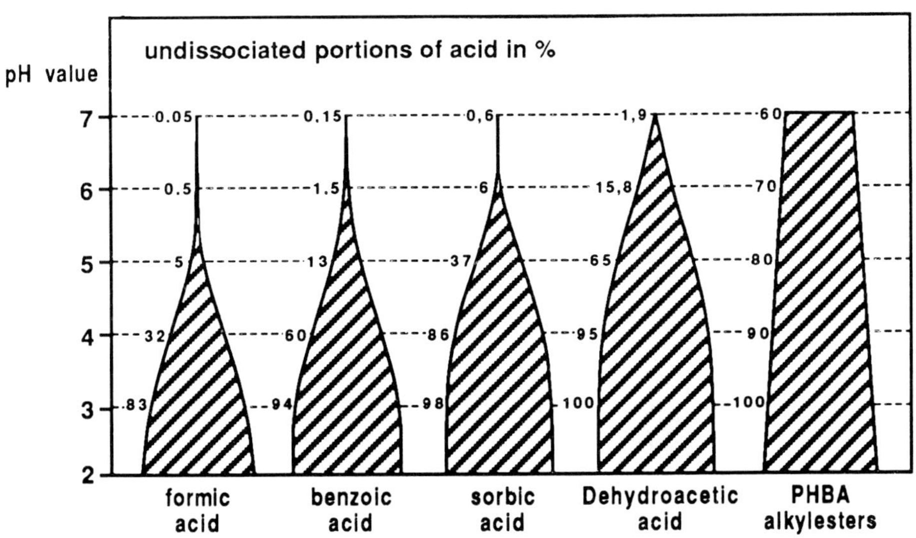

Fig. 30. Undissociated portions of different preservative acids at different pH values.

inhibition into the microbial cell. Because of the following favourable properties PHBA alkyl esters most closely approach the ideal preservative:

— broad spectrum of effectiveness covering moulds and bacteria,
— effective at low concentrations (0·05–0·2%),
— stable, odourless, colourless,
— not significantly influencing the taste of food,
— active over a wide range of pH and temperature,
— low toxicity, non-irritating parenterally.

Applications of PHBA alkyl esters

These are mainly in the preservation of foods, pharmaceuticals and cosmetics, although there is a certain sensitization potential. PHBA alkyl esters are also used for the in-can protection of technical functional fluids, when one looks for particular odourless and colourless preservatives. Using PHBA alkyl esters for the protection of non-ionic surfactant solutions one has to bear in mind that binding effects may occur which reduce the antimicrobial efficacy. This is valid for anionic alkaline detergent solutions, too. Macromolecules, e.g. saccharoseesters, polyoxy-40-stearate, polyvinyl-pyrrolidone, methylcellulose and carboxymethylcellulose, are also able to reduce the effectiveness of PHBA alkyl esters. In non-ionic formulations the activity of PHBA alkyl esters is increased by the addition of 2–5% propylene glycol. PHBA methyl and propyl esters are more widely used than the others and frequently in combination to broaden the spectrum of activity. The solubility of PHBA alkyl esters in preparations to be protected can be promoted by heating. In cases where preparations do not tolerate heating one can incorporate the sodium salts of PHBA alkyl esters which are highly soluble in water or aqueous formulations. An increase in pH caused through the alkaline reaction of the sodium salts can be corrected subsequently by the addition of an acidulant, e.g. citric acid or acetic acid.

PHBA esters are listed in the EC list of preservatives permitted for the in-can protection of cosmetics with a maximum authorized concentration of 0·4% for one ester and 0·8% for mixtures of esters.

Chemical and physical properties of PHBA alkyl esters

	6.1.10.1	*6.1.10.2*	*6.1.10.3*
PHBA ester	Methyl ester	Ethyl ester	Propyl ester
Structural formula	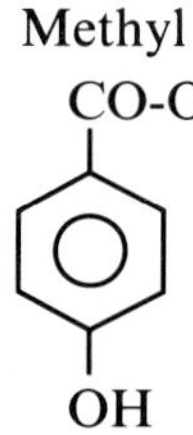	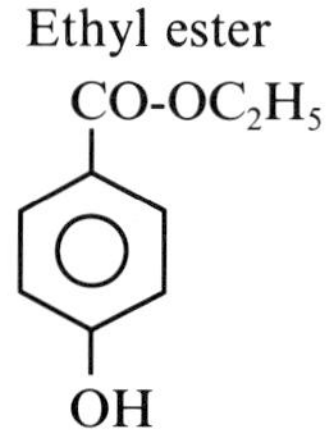	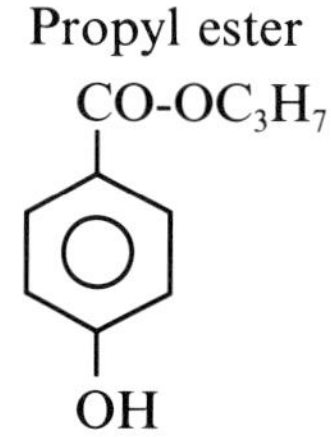
Empirical formula	$C_8H_8O_3$	$C_9H_{10}O_3$	$C_{10}H_{12}O_3$
CAS-No.	99-76-3	120-47-8	94-13-3

Molar mass		152·2	166·2	180·2
Appearance		White, crystalline powders, practically odourless		
Melting point		125–128°C	115–119°C	95–98°C
Boiling point		270–280°C	297–298°C	
Solubility (g/litre) in				
Water	(25°C)	2·5	1·2	0·4
	(80°C)	20·0	8·6	1·3
Methanol	(25°C)	550·0	700·0	1000·0
Ethanol	(25°C)	600·0	600·0	750·0
Diethylene	(20°C)	320·0	250·0	350·0
glycol	(60°C)	490·0	300·0	450·0
Propylene glycol	(25°C)	280·0	250·0	250·0
Glycerine	(25°C)	10·0	6·0	6·0
Olive oil	(25°C)	30·0	30·0	52·0
Mineral oil	(25°C)	10·0	10·0	14·0
Linseed oil	(25°C)	20·0	22·0	28·0
Paraffin oil	(25°C)	5·0	7·0	10·0
Stability in water at 20°C		Up to pH 8	Up to pH 8	Up to pH 8

Chemical and physical properties of sodium salts of PHBA alkyl esters

PHBA ester Na-salt	Na-salt of the methyl ester	Na-salt of the ethyl ester	Na-salt of the propyl ester
Empirical formula	$C_8H_7O_3Na$	$C_9H_9O_3Na$	$C_{10}H_{11}O_3Na$
Molar mass	174·1	188·2	202·2
CAS-No.	5026-62-0	35285-68-8	35285-69-9
Appearance	White, crystalline, hygroscopic powders		
pH of a 0·1% solution in H_2O	9·7–10·3	9·5–10·5	9·8–10·2
Solubility (g/litre) at 25°C in			
Water	Approx. 1000	Approx. 1000	Approx. 1000
Methanol	20	20	20
Ethanol	20	20	20
Propylene glycol	Approx. 250[a]	300[a]	400[a]
Glycerine	Approx. 500[a]	500[a]	500[a]

[a] after heating in a water bath.

Stability — Solutions in water which react alkaline (pH >8) are not storable as hydrolysis takes place with the formation of inactive *p*-hydroxybenzoic acid.

Toxicity data

LD_{50} oral	PHBA methyl ester	PHBA ethyl ester	PHBA propyl ester
For dogs	3 g/kg	5 g/kg	6 g/kg
For rats	>2 g/kg	>2 g/kg	>5 g/kg

Contact dermatitis caused through PHBA alkyl esters occurs occasionally, especially in cases where the esters are used at higher concentrations (5%). At concentrations of 0·5% allergic skin reactions in general fail to appear. As preservatives for pharmaceuticals, cosmetics and food, PHBA alkyl esters are normally used at an average concentration of 0·2%. Wallhäusser (1984) mentions an allergy quota of 0·3–0·8% for PHBA alkyl esters.

Cramer & Ford (1978) state that the no effect level for PHBA methyl, ethyl and propyl ester is 1200 mg/kg body weight. The ADI value is limited to 10 mg/kg body weight by the FAO/WHO Expert Committee on Food Additives.

Spectrum of antimicrobial effectiveness

Table 67. Minimum Inhibition Concentrations (MIC) of PHBA Alkyl Esters in Nutrient Agar

Test organism	*MIC (mg/litre)*		
	PHBA methyl ester	*PHBA ethyl ester*	*PHBA propyl ester*
Aerobacter aerogenes	2000	2000	1000
Escherichia coli	3000	1000	1000
Proteus vulgaris	2000	1000	300
Sarcina lutea	4000	3000	500
Staphylococcus aureus	4000	1500	700
Saccharomyces cerevisiae	1000	500	250
Saccharomyces pastorianus	500	500	200
Saccharomyces sp.	800	500	250
Candida albicans	2000	1000	500
Aspergillus niger	1000	500	400
Penicillium glaucum	1000	1000	500
Rhizopus nigricans	500	500	200

Apparently the a.m. PHBA alkyl esters are more effective against moulds and yeasts than against bacteria. The growth of lactic acid bacteria is only inhibited through PHBA butyl and benzyl esters (Section 6.1.10.4) and (Section 6.1.10.5). According to Schimmel & Husa (1956) the combination of PHBA alkylesters leads not only to additivity in effectiveness; very often synergism is observed, directed against moulds and bacteria. Proved combinations are:

PHBA methyl/propyl ester = 3/1 – 2/1
PHBA ethyl/propyl ester = 3/1 – 2/1
PHBA methyl/ethyl/propyl ester = 5/3/1
PHBA methyl/ethyl/propyl/butyl ester = 7/1/1/1

6.1.10.4 *p*-Hydroxy-benzoic acid butyl ester

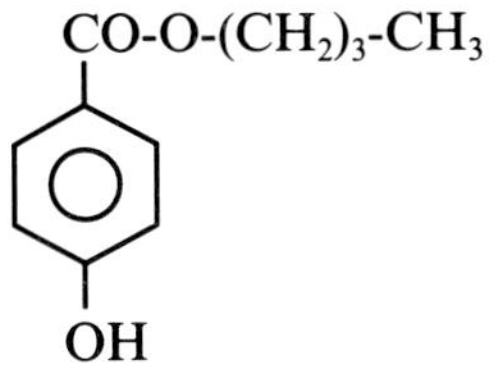

$C_{11}H_{14}O_3$ M_r 194·23
CAS-No. 94-26-8

Chemical and physical properties

Appearance	White crystalline powder	
Melting point	68–72°C	
Solubility (g/kg) at 25°C	In water	0·2
	ethanol	2100·0
	mineral oil	50·0
Stability	Stable and effective between pH 3 and 9·5; hydrolysis at lower or higher pH values, increasing with temperature	

Toxicity data

LD_{50} oral 6 g/kg dog

Ames test and chromosome test negative. PHBA butyl ester permeates *in vitro* through the human skin.

Antimicrobial effectiveness/applications

Table 68. Minimum Inhibition Concentrations (MIC) of PHBA Butyl Ester in Nutrient Agar according to Wallhäusser (1984)

Test organism	*MIC (mg/litre)*
Pseudomonas aeruginosa	120
Staphylococcus aureus	120
Candida albicans	125
Aspergillus oryzae	150
Penicillium chrysogenum	100

PHBA butyl ester is approximately 10 times more effective than the corresponding methyl and ethyl esters and additionally the spectrum of effectiveness is more equalized.

The application of PHBA butyl ester as a preservative in cosmetics and pharmaceuticals is, in spite of its high activity, very limited because of the low water solubility and is restricted to the application in combination with other PHBA alkyl esters (see Sections 6.1.10.1–6.1.10.3). PHBA butyl ester is permitted for use in cosmetics (EC list).

6.1.10.5 *p*-Hydroxy-benzoic acid benzyl ester

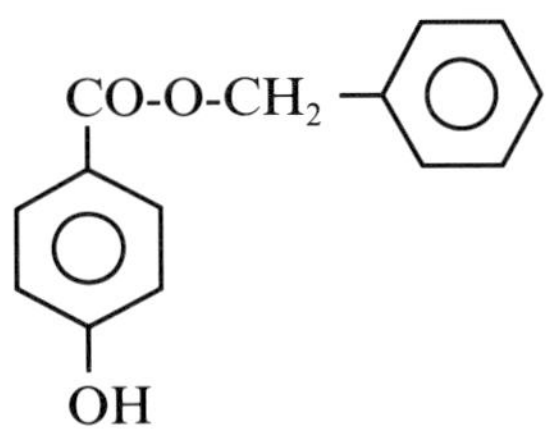

$C_{14}H_{12}O_3$ M_r 228·25
CAS-No. 94-18-8

Chemical and physical properties

Appearance	White crystalline powder	
Melting point	108–113°C	
Solubility (g/litre) at 25°C	In water	0·06
	ethanol	700
	mineral oil	5
Stability/compatibility	As indicated for PHBA alkyl esters in general	

Toxicity data

Although the CIR Expert Panel of CTFA in its final report states that PHBA benzyl ester in concentrations which are normally applied in cosmetics, can be regarded as safe (*J. Amer. Coll. Toxicol.*, **5** (1986) 301), the benzyl ester is not yet finally listed in the EC list of preservatives allowed for the in-can protection of cosmetics.

Antimicrobial effectiveness/applications

Table 69. Minimum Inhibition Concentrations (MIC) of PHBA Benzyl Ester in Nutrient Agar according to Wallhäusser (1984)

Test organism	*MIC (mg/litre)*
Escherichia coli	160
Klebsiella pneumoniae	160
Pseudomonas aeruginosa	160
Pseudomonas cepacia	160
Pseudomonas fluorescens	160
Staphylococcus aureus	120
Candida albicans	250
Aspergillus niger	1000
Penicillium notatum	500

Although highly effective the use of PHBA benzyl ester as a preservative is strongly limited because of its extremely poor water-solubility. Application is therefore only in combination with other PHBA esters. In the EC list of preservatives provisionally allowed for application in cosmetics, PHBA benzyl ester is mentioned with a maximum authorized concentration of 0·1% (acid).

6.1.11 Naphthenic acid copper salts — Copper naphthenate

Copper content 14–15% CAS-No. 1338-02-9

Chemical and physical properties

Appearance	Grey–green paste
Solubility	Virtually insoluble in water; soluble in organic solvents

Naphthenic acids are a by-product of oil refining. They are monocarboxylic acids of cyclic alkanes according to the following structure:

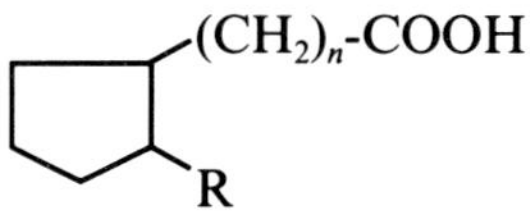

R = H, CH_3, cyclopentyl
$n = 0 - 60$

The salts of the alicyclic naphthenic acids are sparingly soluble in water and regarded as metallic soaps. Copper naphthenate is specifically active against a broad range of fungi and therefore mainly used in wood preservatives. It has been employed also for the fungicidal treatment of textiles (e.g. tents, sandbags, etc.), paper and cardboard. Copper naphthenate is applied in form of solutions in organic solvents and offers the advantage of not being leachable.

6.2 Inorganic acids

6.2.1 Boric acid

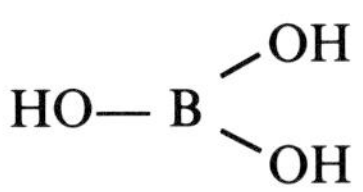

H_3BO_3 M_r 61·84
CAS-No. 10043-35-3

Chemical and physical properties

Appearance	White, crystalline, odourless powder
Melting point	171°C
Dissociation constant	7×10^{-10}
pK_a value	9·14
Solubility at 20°C	Approx. 55 g/litre water; Approx. 50 g/litre ethanol

Toxicity data

LD_{50} oral	5·14 g/kg rat
	3·45 g/kg mouse
	1–4 g/kg dog

Rapid absorption through the skin, slow excretion, danger of accumulation. Mutagenicity was not demonstrable.

Antimicrobial effectiveness/applications

Boric acid's spectrum of activity is rather selective and covers mainly yeasts with lower efficacy against moulds and bacteria. Although boric acid is only effective in its undissociated state, it is still active even at neutral pH because of its relatively high pK_a value (9·14). Boric acid acts as an enzyme inhibitor, especially blocking enzymes in the metabolism of phosphate.

In the EC list of preservatives, provisionally allowed for cosmetics, boric acid is listed with a maximum authorized concentration of 0·5% for products for oral hygiene and 3% for other products. As a food preservative boric acid is no longer of importance because of its toxicological properties.

For sapstain treatment (protection of freshly cut and sawn timber against staining fungi and superficial moulds) borax is used in mixture with other fungicides, in particular with sodium pentachlorophenate (Section 5.6.3), but with sodium phenylphenolate (Section 5.4.1) or with quaternary ammonium compounds (Section 16.1) too.

Mixtures of Borax with *o*-phenyl-phenol (OPP) possess synergistic effects against some species of wood staining fungi such as *Aureobasidium pullulans* and *Sclerophoma pityophila*. The marked synergism may be easily calculated from the minimum inhibition concentrations listed in Table 70 using the calculation method of Kull *et al.* (1961).

Borax is sodium borate: $Na_2B_4O_7 \cdot 10H_2O$; molar mass: 381·44; white crystalline, odourless powder; melting point: 75°C; soluble in H_2O (40 g/litre), insoluble in ethanol; pH of a 0·1 M solution in water: 9·2.

Table 70. Minimum Inhibition Concentrations (MIC) of Borax/OPP Mixtures in Nutrient Agar

Test organism	*MIC (mg/litre)*				
	Borax	*Borax/OPP*			*OPP*
		3/1 (x)[a]	*1/1* (x)	*1/3* (x)	
Aspergillus niger	≥5000	500	200	200	100
Aureobasidium pullulans	≥5000	625 (<0·5)	350 (<0·5)	200 (<0·4)	350
Chaetomium globosum	≥5000	200	150	100	75
Penicillium glaucum	≥5000	200	100	100	75
Sclerophoma pityophila	3500	875 (0·8)	500 (0·8)	350 (0·8)	350

[a] $x < 1$: synergism; $x = 1$: additivity; $x > 1$: antagonism.

6.2.2 Sulphurous acid/sulphur dioxide

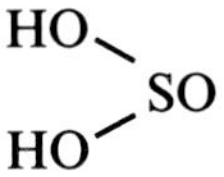

$H_2SO_3 \leftrightarrows SO_2 + H_2O$

M_r of SO_2: 64·06
CAS-No. 7446-09-5

Chemical and physical properties of SO_2

Appearance	Colourless, non-flammable gas with a pungent odour; forms a liquid under pressure
Boiling point	−10·08°C
Solidification point	−75·5°C
Solubility	Soluble in water forming sulphurous acid
pK_a values for H_2SO_3	1·8/6·9
Formation of salts	Water soluble salts sodium sulphite (Na_2SO_3; $Na_2SO_3 \cdot 7H_2O$) sodium bisulphite ($NaHSO_3$) sodium metabisulphite ($Na_2S_2O_5$) potassium metabisulphite ($K_2S_2O_5$) salt insoluble in water: calcium sulphite ($CaSO_3$)
Stability	Reactive and reductive

Toxicity data

LD_{50} oral	1–2 g SO_2/kg rat
	0·6–0·7 g SO_2/kg rabbit

No carcinogenic effects were detected.

ADI value 0·7 mg/kg body weight
MWC 2 ml/m^3 (5 mg/m^3)

Antimicrobial effectiveness/applications

SO_2/sulphurous acid is most effective in acid media (pH < 4); the activity decreases from $SO_2 \rightarrow H_2SO_3 \rightarrow HSO_3^- \rightarrow SO_3^{2-}$. At pH 3 only 6% of sulphurous acid is available in its effective undissociated state. The anhydride SO_2 is highly reactive and may interact with many constituents of the microbial cell, e.g. SH groups in structural proteins, enzymes, nucleic acids and lipids. Because of its reductive properties SO_2 also acts as an antioxidant.

Due to its properties sulphur dioxide has been used for many centuries mainly for the protection of acidic foodstuffs, especially in the wine industry. However its application is restricted owing to flavour problems occurring at concentrations above 5 mg/kg. It is applied as liquefied gas, a solution in water (H_2SO_3) or the a.m. salts. In the EC list of preservatives which cosmetics may contain inorganic sulphites and hydrogensulphites are mentioned with a maximum concentration of 0·2% expressed as free SO_2. But in finished products it is scarcely used, as these normally do not exhibit the low pH which is required for the antimicrobial effectiveness of sulphur dioxide or sulphurous acid.

7 CARBONIC ACID ESTERS

Although the microbicides described in this section all belong to the same substance class, namely to the acid esters, their antimicrobial activity is based on different mechanisms of action. The halogen containing esters (Sections 7.1–7.6) and dimethyl dicarbonate (Section 7.7) are substances with electrophilic character, a character which enables them to react with nucleophilic groups of the microbial cell and which equips the substances with a wide spectrum of antimicrobial effectiveness. The electrophilic character of the halogenated esters mentioned here is reduced due to the fact that the substances are bearing an activated halogen group in the α-position to an electronegative group (see Fig. 34, page 357). Bromo compounds are preferred, because due to the average electronegativity of the bromo atom in comparison to other halogen atoms (for scale of electronegativity values of halogen atoms see Table 71) these compounds are not too stable (not persistent) but also not too reactive (unstable), so that they are preferably used as slimicides not causing waste problems.

Table 71. Relative Electronegativity Values of Halogen Atoms according to Müller (1951).

Fluoro	(F)	4·0
Chloro	(Cl)	3·0
Bromo	(Br)	2·8
Iodo	(I)	2·4

Using these microbicides one has to bear in mind that in accordance with their reactivity these chemicals have skin and mucous membrane irritating properties.

The dicarbonate configuration in dimethyl dicarbonate (DMDC, Section 7.7) is responsible not only for the electrophilic character of the chemical but also for its distinguished reactivity (instability), which does not allow use of DMDC as a preservative, but as a powerful cold sterilizing agent.

The phenyl esters of long chain fatty acids (Sections 7.8 and 7.9) are membrane-active microbicides; they release phenolic compounds as the active ingredients. The advantage of these esters is that they are easier to handle and to apply than the phenolic compounds they are based on.

Finally it has to be pointed out that the antimicrobial esters of p-hydroxybenzoic acid (p-hydroxy-benzoates) are not described in this section; because of their high acidity they are listed under 'Acids' in Section 6.

7.1 2-Bromo-ethylacetate — Bromoacetic acid ethyl ester

$Br\text{-}CH_2\text{-}CO\text{-}O\text{-}C_2H_5$ $C_4H_7BrO_2$ M_r 167·01
CAS-No. 105-36-2

Chemical and physical properties

Appearance	Colourless, irritant fluid
Boiling point	168–169°C
Density at 20°C	1·499 g/ml
Refractive index	n_D^{20} 1·4510
Flash point	50°C
Solubility	Sparingly soluble in water, highly soluble in organic solvents, miscible with ethyl alcohol
Stability	Hydrolysis in water based solutions (increasingly with increasing pH values)

Toxicity data

LD_{50} oral	50 mg/kg rat
	100 mg/kg mouse

Highly toxic by inhalation and skin absorption.
Strongly irritant to mucous membranes and skin.
EPA genotox program 1988 — tumorigenic.

Antimicrobial effectiveness/applications

2-Bromo-ethylacetate is especially effective against yeasts, but at higher concentrations also against bacteria, fungi, slime forming micro-organisms and algae. The active ingredient has been used as a non-persistent preservative in drinks, e.g. wine and fruit juices. However, these applications are no longer permitted because of the toxicity and the irritant properties of the compound.

7.2 Bromo-benzylacetate

Br-CH_2-CO-O-CH_2-C_6H_5	$C_9H_9BrO_2$	M_r 299·08 CAS-No. 5437-45-6

Chemical and physical properties

Appearance	Colourless irritant fluid
Boiling point at 29 hPa	166–170°C
Density at 20°C	1·446 g/ml
Refractive index	n_D^{20} 1·5436
Flash point	>110°C
Solubility	Sparingly soluble in water, highly soluble in organic solvents
Stability	Hydrolysis in aqueous alkaline solutions

Toxicity data

Irritant to mucosa and skin.

Antimicrobial effectiveness/applications

The electrophilic active compound exhibits a broad spectrum of activity which covers bacteria, yeasts and fungi. It is used as a preservative for the in-can protection of water based functional fluids, e.g. paints. Due to its properties — irritant, moderate stability — the microbicide has been applied to a limited extent only.

7.3 2,3-Dibromo-(2-hydroxy-)ethylpropionate

CH_2Br-CHBr-CO-O-CH_2-CH_2-OH $C_5H_8Br_2O_3$ M_r 275·93
CAS-No. 68479-77-6

Chemical and physical properties (30% solution)

Appearance	Yellow to brown liquid
Density at 20°C	1·19 g/ml
Flash point	97°C
Stability	Hydrolytic cleavage in alkaline media
pH (100 mg/litre water)	5·6

No further data available.

Antimicrobial effectiveness/applications

Because of its distinguished efficacy against bacteria it is recommended to apply the compound as a preservative for water based paints, adhesives and polishes, but preferably together with a suitable fungicide. However, the suitability of the compound for that application is limited, due to the instability of the active ingredient in alkaline media. It is also recommended to use the compound as a slimicide in processing water circuits.

7.4 Bis-1,2-(bromoacetoxy)ethane

$Br\text{-}CH_2\text{-}CO\text{-}O\text{-}CH_2\text{-}CH_2\text{-}O\text{-}CO\text{-}CH_2\text{-}Br$ $C_6H_8Br_2O_4$ M_r 303·94
CAS-No. 3785-34-0

Chemical and physical properties

Appearance	Almost colourless irritant fluid
Boiling point at 18·5 hPa	176·5–177·5°C
Solubility	Sparingly soluble in water, highly soluble in alcohols, ether and benzene
Stability	Hydrolysis in water based solutions (increasingly with increase in pH and temperature)

Toxicity data

LD_{50} oral >400 mg/kg rat

Strongly irritant to skin and mucosa.

Antimicrobial effectiveness/applications

Due to its distinctive activity against slime forming micro-organisms bis-1,2-(bromoacetoxy)ethane has been an active ingredient in non-persistent slimicides for use preponderantly in the paper industry.

7.5 Bis-1,4-(bromoacetoxy)-2-butene

CH-CH_2-O-CO-CH_2-Br
||
CH-CH_2-O-CO-CH_2-Br $C_8H_{10}Br_2O_4$ *M*r 329·97
CAS-No. 20679-58-7

Chemical and physical properties

Appearance	Almost colourless fluid
Boiling point	130°C (decomposition)
Solubility	Sparingly soluble in water and polar solvents, highly soluble in other organic solvents
Stability	Hydrolysis in water based solutions, especially at pH values higher than 7

Toxicity data

LD_{50} oral	127 mg/kg rat
percutaneous	970 mg/kg rabbit

Strongly irritant and corrosive to the skin, eyes and mucous membranes.

Antimicrobial effectiveness/applications

The minimum inhibition concentrations of bis-1,4-(bromoacetoxy)-2-butene for fungi and some species of bacteria are in the range of 20 mg/litre only (see Table 72). The a.i. therefore was successfully introduced as a substitute for the persistent and highly toxic organomercurials (Sections 17.3–17.5) and pentachlorophenol (Section 5.6.3) in slimicides for the treatment of industrial water circuits, mainly in the paper industry.

Table 72. Minimum Inhibition Concentrations (MIC) of Bis-1,4-(Bromoacetoxy)-2-butene in Nutrient Agar

Test organism	*MIC (mg/litre)*
Aspergillus niger	<20
Botrytis cinerea	20
Chaetomium globosum	<20
Penicillium glaucum	35
Bacillus subtilis	20
Escherichia coli	200
Staphylococcus aureus	200

7.6 (2-Bromo-1,2-diiodoacryl-)ethylcarbonate

$$O{=}C\begin{matrix} \diagup O\text{-}C_2H_5 \\ \diagdown O\text{-}CH_2\text{-}CI{=}CBrI \end{matrix}$$

$C_6H_7BrI_2O_3$

M_r 460·84
CAS-No. 77352-88-6

Chemical and physical properties

Appearance	White crystals with a slight characteristic odour
Melting point	40°C
Vapour pressure at 25°C	$2{\cdot}4 \times 10^{-5}$ hPa
Solubility	0·119 g/litre water; soluble in organic solvents
Stability	Limited thermostability at temperatures >40°C; stable to UV light; hydrolyses in alkaline solutions

Toxicity data

LD_{50} oral	641–529 mg/kg rat
dermal	2858–2849 mg/kg rat
inhalation (4 h)	0·82–1·48 g/m^3 for rats
Irritation	Moderate to skin and mucosa

Several mutagenicity tests conducted in different biological systems demonstrated lack of genetic effects.

Antimicrobial effectiveness/applications

In view of the minimum inhibition concentrations listed in Table 73 it has to be stated that the compound's antimicrobial activity is above all directed against fungi. The efficacy of the active ingredient is not influenced by anionic, cationic or non-ionic components. Because of its effectiveness against wood-rotting fungi it has been recommended for use in wood preservatives, and for the protection of plywood by incorporating the fungicide into the glue in the production of plywood.

Table 73. Minimum Inhibition Concentrations (MIC) of (2-Bromo-1,2-diiodoacryl-)ethylcarbonate in Nutrient Agar

Test organism	*MIC (mg/litre)*
Alternaria alternata	50
Aspergillus niger	20
Aureobasidium pullulans	50
Chaetomium globosum	20
Cladosporium cladosporioides	10
Lentinus tigrinus	20
Penicillium glaucum	15
Sclerophoma pityophila	50
Trichoderma viride	100
Escherichia coli	350
Staphylococcus aureus	150

7.7 Dimethyl dicarbonate (DMDC)

H_3C-O-CO-O-CO-O-CH_3	$C_4H_6O_5$	M_r 134·09 CAS No. 4525-33-1

Chemical and physical properties

Appearance	Colourless liquid with a slightly pungent smell
Density at 20°C	1·25 g/ml
Viscosity at 20°C	2·1 mPa s
Vapour pressure at 20°C	0·7 hPa
Solidification point	17°C
Boiling point	172°C (decomposition)
Flash point	85°C
Ignition temperature	465°C
Solubility	In water approx. 35 g/litre, miscible with organic solvents, e.g. alcohol and toluene
Stability	Highly reactive carbmethoxylation agent, which means hydrolysis in water to methanol and carbon dioxide, reaction with N-H, S-H and O-H groups according to the following reaction scheme:

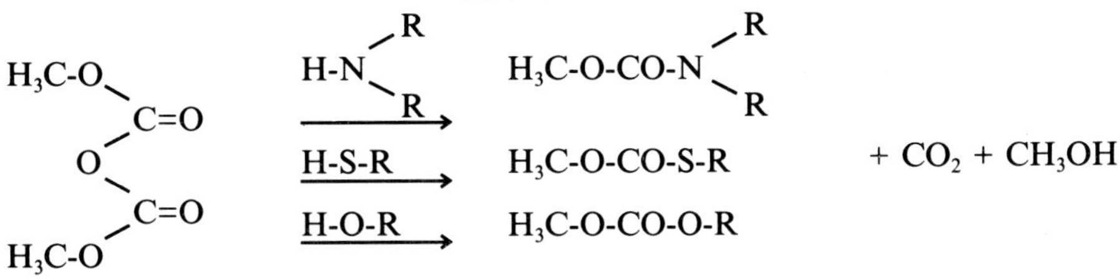

Half-life in water	At pH 2·8/10°C: 40 min pH 2·8/20°C: 15 min pH 2·8/30°C: 8 min

Toxicity data

LD_{50} oral	330–500 mg/kg rat
dermal	>1250 mg/kg rat; the dosage of 1250 mg/kg caused no symptoms of poisoning
LC_{50} inhalation	700 mg/m^3 air for rat (exposure: 4 h)

Severely irritant and corrosive to skin and mucous membranes.

LC_0 for fish (*Leuciscus idus*): 50 mg/litre (exposure: 48 h)

Remark: In contact with water DMDC decomposes completely to methanol and carbon dioxide within a few hours (see half-life).

Antimicrobial effectiveness/applications

The strong reactivity of DMDC is responsible for the antimicrobial action. When checking the microbicidal effectiveness of DMDC one has to bear in mind the short-half life of the chemical in water based media depending on temperature and pH. DMDC kills normal yeasts, mycoderma and fermentive bacteria at relatively low concentrations. At higher concentrations it also destroys other bacteria, wild yeasts and mould producing fungi. Minimal lethal concentrations of DMDC for a great number of individual microbe species are listed in Table 74.

Table 74. Minimum Lethal Concentrations (MIC) of DMDC (mg/litre)

Yeasts	
Saccharomyces carlsbergensis (non-flocculating yeast)	100
Saccharomyces carlsbergensis (flocculating yeast)	60
Saccharomyces diastaticus	200
Saccharomyces oviformis	100
Saccharomyces bailii	120
Saccharomyces cerevisiae	40
Saccharomyces uvarum	30
Saccharomyces pastorianus	100
Saccharomyces apiculatus	60
Saccharomyces globosum	40
Zygosaccharomyces priorianus	75
Rhodotorula mucilaginosa	50
Rhodotorula glutinosa	40
Rhodotorula rubra	200
Candida krusei	200
Pichia membranefaciens	40
Pichia farinosa	100
Torulopsis candida	100
Torulopsis versatilis	100
Torulopsis stellata	65
Torula utilis	240
Endomyces lactis	60
Kloeckera apiculata	40
Hansenula anomala	50
Bacteria	
Acetobacter pastorianum	80
Acetobacter xylinum	300
Escherichia coli	400
Staphylococcus aureus	100
Pseudomonas aeruginosa	100
Lactobacterium buchneri	40
Lactobacillus pastorianus	300
Lactobacillus brevis	200
Pediococcus cerevisiae	300
Moulds	
Penicillium glaucum	200
Byssochlamys fulva	100
Botrytis cinerea	100
Mucor racemosus	500
Fusarium oxysporum	100

The killing effect of DMDC bases on its irreversible reaction with microbe cells. In consequence DMDC destroys high cell numbers only at higher concentrations. The lethal concentrations in Table 74 were determined as follows: the microbe species concerned was introduced into an uncarbonated acidic drink (approx. pH 3) to give a viable cell count of 500 per ml; the effect of the treatment with DMDC was determined after the drink had been stored for 3 weeks at 28°C.

Although DMDC is highly effective, it cannot, because of its short half-life, be used as a preservative in water based media, when there is a risk of recontamination after the addition of DMDC. However, DMDC has found application for the cold sterilization of soft drinks and wine, and for the degermination of water, which is used for the production of drinks, cosmetics and pharmaceuticals. Once DMDC has decomposed there is no further sterilizing effect. It should therefore not be added until shortly before the drink is put into bottles or other containers and tightly closed. It has to be regarded as an important advantage that DMDC is not a persistent preservative and that its application does not influence either taste or quality of drinks. Addition rates range between 10 and 20 ml DMDC per 100 litre drink. But before DMDC is added the number of viable cells in the drink has to be reduced to approx. 500 per ml by filtration of flash pasteurization; the latter also inactivates enzymes which may decompose pectin. It is also recommendable to cool the drink before the addition of DMDC, preferably to 10–15°C; otherwise DMDC decomposes too fast not leaving time for sufficient antimicrobial action.

7.8 Pentachlorophenyl laurate (PCPL)

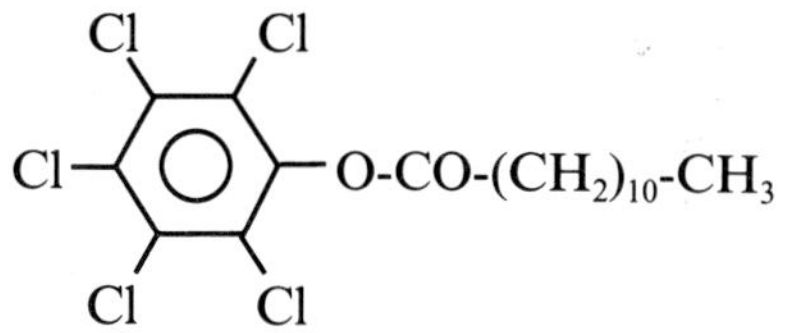

$C_{18}H_{23}Cl_5O_2$ M_r 448·65
CAS-No. 3772-94-9

Chemical and physical properties

Appearance	Oily brown, odourless liquid
Density at 20°C	1·25 g/ml
Solubility	Practically insoluble in water and alcohols, soluble in all portions of acetone, methyl ethyl ketone, trichloroethylene, carbon tetrachloride, toluene, white spirit, oils, fats and waxes.
Stability	Non-volatile, unaffected by dilute acids or alkalis, hydrolysis by high concentrations of alkali or by enzymatic action, sensitive to photochemical breakdown

Toxicity data

PCPL is of low toxicity and good skin compatibility, but its antimicrobial effectiveness is based on the liberation of pentachlorophenol (PCP) (see Section 5.6.3). PCPL releases approx. 57% PCP the toxicity data of which are listed in Section 5.6.3.

Antimicrobial effectiveness/applications

As can be seen from Table 75 the intact PCPL does not exhibit significant antimicrobial activity, especially not in comparison to PCP. Nevertheless PCPL has had considerable use as a microbicide for rot- and mould-proofing of various types

Table 75. Minimum Inhibition Concentrations (MIC) of PCPL and PCP in Nutrient Agar

Test organism	*MIC (mg/litre)*	
	PCPL	*PCP*
Aspergillus niger	>1000	50
Chaetomium globosum	>1000	20
Penicillium glaucum	>1000	50
Escherichia coli	>2500	500
Staphylococcus aureus	750	10

of materials, mainly textiles, ropes and cordage (addition rate approx. 2% calculated on the weight of the material to be protected) as in fact the active ingredient is PCP which is set free by enzymatic ester cleavage. In consequence PCPL is under pressure for substitution as is PCP and the application of PCPL is indeed in decline.

7.9 Fatty acid esters of 5,5'-dichloro-2,2'-dihydroxydiphenylmethane (Deiner, 1983)

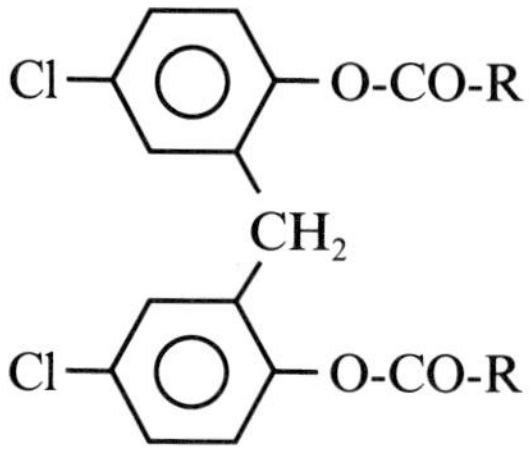

R = C_5–C_{19} alkyl

Chemical and physical properties

Appearance	Oily brown liquid of high viscosity
Solubility	Insoluble in water, highly soluble in organic solvents, preferably in non-polar solvents
Stability	Non-volatile, heat resistant (>200°C), hydrolyses in alkaline solutions

Toxicity data

As the active ingredient of the mixture of esters is Dichlorophen (see Section 5.7.3) which is liberated through enzymatic ester cleavage for antimicrobial action, one has to note the toxicity data of Dichlorophen.

Antimicrobial effectiveness/applications

The mixture of Dichlorophen fatty acid esters develops antimicrobial activity by the reconstitution of Dichlorophen through hydrolysis. The most important advantage of the ester mixture is the possibility of transferring it easily into stable emulsions which can be used for the impregnation of textile material together with water repellents without disturbing the effect of the water repellents. The application of Dichlorophen itself is fraught with difficulties. Stable emulsions of Dichlorophen for dilution with water are not available. Alkaline solutions of Dichlorophen are easy to apply on textile material by impregnation, but the alkali salts of Dichlorophen are not compatible with most of the water repellents. The application of Dichlorophen in solutions in organic solvents is not the solution to the problem, as the use of organic solvents is disliked in the textile industry.

7.10 *p*-Hydroxy-benzoates (see Section 6.1.10)

8 AMIDES

Carbonic acid amides do not generally belong to the substances with microbicidal effects. Toxophoric groups or toxophoric structural elements have to be introduced to obtain antimicrobial aliphatic carbonic acid amides. This possibility is exemplified by amides which, in the 2-position to the electronegative carbonamide grouping, possess a halogen atom, thus ranking among the electrophilic active microbicides. The addition of formaldehyde to such halogenated amides leads to antimicrobially effective *N*-hydroxymethyl amides, whose special feature is the presence of two toxophoric groups: an activated halogen atom and an activated hydroxymethyl group. Being formaldehyde-releasing compounds, they are treated in Section 3.4 ('Reaction products of amides with formaldehyde'). *N*-hydroxymethyl diamides of carbonic acid = *N*-hydroxymethyl ureas are formaldehyde releasing compounds, too, and therefore listed in Section 3.4.6.

Also salicylanilides (2-hydroxybenzanilides), long chain *N*-alkyl-salicylamides and carbanilides (urea derivatives) belong to the amide compounds with antimicrobial action. They are membrane-active substances, i.e. very small concentrations suffice to achieve microbistatic effects whereas microbicidal effects call for much higher addition rates.

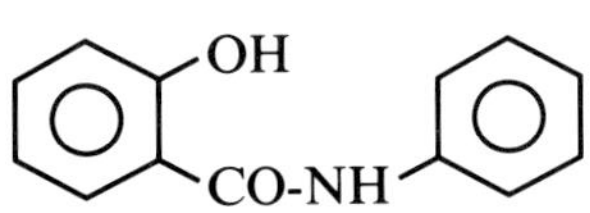

Salicylanilide

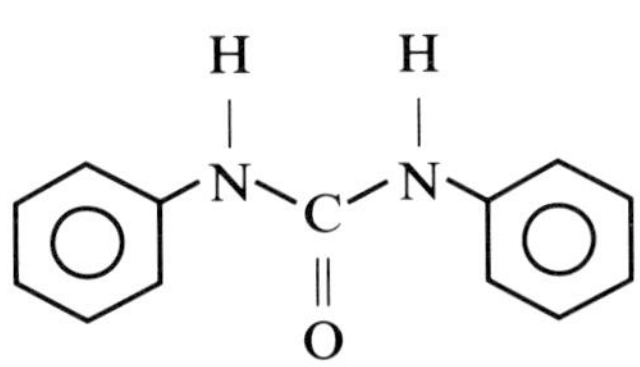

Carbanilide

As in the case of membrane-active phenol derivatives (see Section 5), the halogenation of salicylanilides or carbanilides increases the antimicrobial efficacy. The best results are obtained by means of di- to penta-chlorination or bromination, the halogen atoms being more or less evenly distributed on the two phenyl rings. On the other hand halogenated salicylanilides have photosensitizing properties, which has reduced their practical importance.

Also haloalkylthio amides, e.g. *N*,*N*-dimethyl-*N*'-phenyl-*N*'-dichlorofluoromethyl-thiosulphamide (Section 14.5) or *N*-triahalomethylthio-phthalimides (Sections 14.1 and 14.2) are well-known microbicides. They are electrophilic active agents with an activated N-S bond; their role as an important class of microbicides is described separately in Section 14 (*N*-haloalkylthio compounds).

For the sake of completeness, we conclude by mentioning a carbonic acid hydrazide, namely pyridine-4-carboxylic acid hydrazide (isonicotinic acid hydrazide, Isoniazide), an amide compound with a major potential for antimicrobial efficacy but also general toxicity, which is an obstacle to its practical application as a microbicide for the protection of materials.

8.1 2-Chloracetamide

$Cl\text{-}CH_2\text{-}CO\text{-}NH_2$ C_2H_4ClNO M_r 93·51
CAS-No. 79-07-2

Chemical and physical properties

Appearance	Colourless crystals	
Melting point	119–120°C	
Boiling point at 981 hPa	224°C (decomposition)	
Solubility (g/litre)	In water	at 20°C approx. 50;
		at 40°C approx. 150
	ethanol	at 20°C approx. 100
pH (1% solution in H_2O)	4·5	
Stability	Hydrolytic cleavage with acids and alkalis; relatively stable between pH 4 and 9 (half-life at pH 8, 2600 h = 108 days)	

Toxicity data

LD_{50} oral	350 mg/kg rat
	150 mg/kg mouse

Moderately irritant to skin and mucosa, contact dermatitis possible; potential skin sensitizer.
Not mutagenic (Ames test, micronucleus test, dominant lethal test negative).

Fish toxicity LC_0 for *Lebistes reticulatus*: 10 mg/litre (exposure: 10 days)

Tolerated concentration by activated sludge organisms: 150 mg/litre.

Antimicrobial effectiveness/applications

Among the three 2-halo-acetamides, chloro-, bromo- and iodo-acetamide, 2-chloracetamide is the one with the lowest antimicrobial efficacy, although because of the high electronegativity of the chloro atom, it is the more reactive compound. But its lipoid solubility is much lower than that of bromo- and iodo-acetamide. Comparison data of the three 2-halo-acetamides are listed in Table 76. Nevertheless 2-chloracetamide only has gained significant importance as a preservative for the protection of aqueous functional fluids because of its good water solubility and in consequence its favourable partition coefficient.

Other advantages of 2-chloracetamide are the following properties: colourless, odourless, effective over a wide range of pH values (4–9), compatible with anionic and non-ionic compounds. The antimicrobial activity increases even in the presence of anionic detergents. The activity of the 2-chloracetamide against mould producing fungi is more distinctive than the activity against bacteria, however in total, 2-chloracetamide cannot be listed among the strong microbicides (see Table 77). Accordingly the addition rates are relatively high; they vary between 0·2 and 0·6% for the in-can protection of water based paints, adhesives, glues and casein. 2-Chloracetamide may also be used as a preserving agent in the leather industry; for the protection of wet blues one needs approx. 0·2–2% active ingredient calculated on pelt weight.

Table 76. Minimum Inhibition Concentrations (MIC), M_r and half-life ($t_{½}$) of 2-Halo-acetamides

Halo-acetamide	*MIC (mg/litre) and (mM/litre) against*		M_r	$t_{½}$ *at pH 8 (h)*
	Escherichia coli	Staphylococcus aureus		
2-Chloro-acetamide	2000/21	1500/16	93·51	2600
2-Bromo-acetamide	200/1·5	100/0·75	137·97	5000
2-Iodo-acetamide	50/0·27	20/0·11	184·96	

Table 77. Minimum Inhibition Concentrations (MIC) of 2-Chloracetamide in Nutrient Agar

Test organism	*MIC (mg/litre)*
Aerobacter aerogenes	5000
Aeromonas punctata	2000
Bacillus subtilis	5000
Escherichia coli	2000
Proteus vulgaris	5000
Pseudomonas aeruginosa	1500
Pseudomonas fluorescens	5000
Staphylococcus aureus	1500
Alternaria alternata	100
Aspergillus niger	800
Aureobasidium pullulans	750

8.2 2-Bromacetamide

Br-CH_2-CO-NH_2 C_2H_4BrNO M_r 137·97
CAS-No. 683-57-8

Chemical and physical properties

Appearance	Colourless crystals
Melting point	88–90°C
Solubility	Sparingly soluble in water, soluble in organic solvents
Stability	Hydrolytic cleavage in acids or alkalis; relatively stable in water between pH 4 and 9 (half-life at pH 8, 5000 h = 208 days)

Toxicity data

Irritant and corrosive to skin and mucosa.

Antimicrobial effectiveness/applications

Table 78. Minimum Inhibition Concentrations (MIC) of 2-Bromacetamide in Nutrient Agar against Fungi

Test organism	*MIC (mg/litre)*
Alternaria alternata	20
Aspergillus niger	50
Aureobasidium pullulans	50
Chaetomium globosum	20
Coniophora puteana	5
Lentinus tigrinus	10
Penicillium glaucum	50
Polyporus versicolor	50
Sclerophoma pityophila	20
Trichoderma viride	100

2-Bromacetamide exhibits a broad and equalized spectrum of activity, somewhat more distinctive against fungi than against bacteria (see Tables 76 and 78). 5–10 mg bromacetamide/litre are sufficient to inhibit the growth of slime forming bacteria, e.g. those which can grow by utilizing ϵ-caprolactam or its cyclic dimer or trimer as the sole carbon source (Kato & Fukumura, 1962). 2-Bromacetamide exhibits antimicrobial activity also via the vapour phase. However, up to now bromacetamide has no significant importance as a preserving agent probably because for toxicity and cost reasons.

8.3 2-Iodacetamide

$I\text{-}CH_2\text{-}CO\text{-}NH_2$	C_2H_4INO	M_r 184·96 CAS-No. 144-48-9

Chemical and physical properties

Appearance	Colourless crystals
Melting point	91–93°C
Solubility	Sparingly soluble in water, highly soluble in organic solvents
Stability	Hydrolytic cleavage in acids and alkalis

Toxicity data

LD_{50} oral	74 mg/kg mouse
intraperitoneal	50 mg/kg mouse
intravenous	56 mg/kg mouse

Irritant to skin and mucosa.

Antimicrobial effectiveness/applications

The toxic 2-iodacetamide has no practical importance as a microbicide for the protection of materials and is mentioned here only for comparison purposes.

Table 79. Minimum Inhibition Concentrations (MIC) of 2-Iodacetamide in Nutrient Agar

Test organism	*MIC (mg/litre)*
Aspergillus niger	500
Chaetomium globosum	200
Penicillium glaucum	100
Rhizopus nigricans	200
Escherichia coli	50
Pseudomonas aeruginosa	20

8.4 *N*-(4-bromo-2-methylphenyl)-2-chloracetamide (BMPCA)

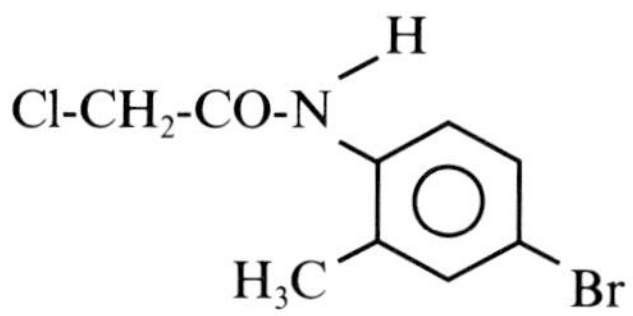

$C_9H_9BrClNO$ M_r 262·54
CAS-No. 96686-51-0

Chemical and physical properties

Appearance	White, odourless powder
Melting point	133–35°C
Vapour pressure at 24°C	<2 hPa
Solubility	Practically insoluble in water, soluble in organic solvents
Stability	Hydrolytic cleavage in acids or alkalis, stable in aqueous paints

Toxicity data

LD_{50}	oral	4044 mg/kg rat
	dermal	>21 g/kg rabbit
LC_{50}	inhalation	10 mg/litre for rats

No sensitization in the guinea pig test.
Slightly to moderately irritant to skin and mucosa.
Mutagenicity/carcinogenicity tests negative (Dalton, 1988).

Antimicrobial effectiveness/applications

Table 80. Minimum Inhibition Concentrations (MIC) of BMPCA in Nutrient Solution

Test organism	*MIC (mg/litre)*
Alternaria solani	31
Aspergillus niger	31
Aureobasidium pullulans	63
Chaetomium globosum	31
Cladosporium cladosporioides	50
Neurospora species	125
Penicillium glaucum	50
Sclerophoma pityophila	50

The antimicrobial activity of BMPCA is especially directed against a broad spectrum of fungi (see Table 80). As it is stable in aqueous formulations at pH values up to nine, not significantly leachable, non-volatile, not causing colorations (resistant to photo-oxidation), it has been recommended for use as a paint film fungicide (Dalton, 1988).

8.5 2,2-Dibromo-3-nitrilopropionamide (DBNPA) — 2,2-Dibromo-2-cyanoacetamide

NC-CBr_2-CO-NH_2 $C_3H_2Br_2N_2O$ M_r 241·86
CAS-No. 10222-01-2

Chemical and physical properties

Appearance	White crystalline powder with slightly antiseptic smell	
Melting point	123–125°C	
Solubility (g/litre)	In water	15
	acetone	350
	ethanol	250
	ethylene glycol	90
	propylene glycol	200
	isopropanol	350
	methyl ethyl ketone	500
	Poor solubility in non-polar organic solvents (<10 g/litre)	
Stability	Relatively stable in aqueous acidic media in the presence of stabilizers such as urea derivatives, sulphamide, oxamide, DMF, caprolactam, *N*-methyl-2-pyrrolidone, dimethylhydantoin, succinimide (Burk & Reineke, 1979); unstable to UV light, decomposition in media with pH values above 7 increasing with increase in pH and temperature	

Toxicity data

LD_{50} oral 126 mg/kg rat

Irritant and corrosive to skin and mucosa.

Fish toxicity LC_{50} for Rainbow trout, Blue gill: 0·9–1·8 mg/litre (exposure: 96 h)

Antimicrobial effectiveness/applications

DBNPA is especially effective against slime forming micro-organisms, e.g. those described by Kato & Fukumura (1962) are inhibited by 0·5–1·0 mg DBNPA/litre. Other minimum inhibition concentrations are listed in Table 81.

Because of its activity and its instability in aqueous systems DBNPA is applied as a potent but not persistent slimicide which does not cause waste water problems. Rates of hydrolysis of DBNPA at different pH values were determined by Exner *et al.* (1973) and are summarized in Table 82.

Table 81. Minimum Inhibition Concentrations (MIC) in Nutrient Agar[a] (Wolf & Sterner, 1972)

Test organism	*MIC (mg/litre)*
Enterobacter aerogenes ATCC 13048	100
Bacillus subtilis ATCC 8473	100
Desulfovibrio desulfuricans A.P.I., RP 38	10
Escherichia coli ATCC 11229	100
Pseudomonas aeruginosa ATCC 8709	100
Pseudomonas aeruginosa USDA, PRD 10	100
Salmonella typhosa ATCC 6539	100
Staphylococcus aureus ATCC 6538	250
Aspergillus terreus ATCC 10690	100
Candida albicans ATCC 10231	100
Candida pelliculosa ATCC 2149	100
Pullularia pullulans ATCC 9348	100
Rhizopus nigricans ATCC 6227 A	10

[a] The pH of the bacterial medium was 7·0–7·2, and that of the fungal medium was 5·0–5·5.

Table 82. Rates of Hydrolysis of DBNPA

pH	$r_{1/2}$ *(h)*	*Temp. (°C)*
3·9	2140	23
6·0	155	25
6·7	37·0	25
7·3	8·8	25
7·7	5·8	25
7·7	145	0
8·0	2·0	25
8·9	0·34	25
9·7	0·11	25
9·7	1·5	0

8.6 *N*-alkylsalicylamides — 2-Hydroxy-*N*-alkylbenzamides

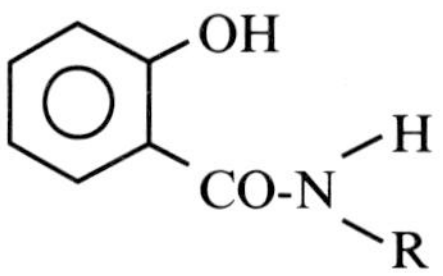

R = alkyl (C_4 – C_{14})

8.6.1 *N*-Butylsalicylamide

$C_{11}H_{15}NO_2$ — M_r 193·25, CAS-No. 57271-91-7 — viscous oil

8.6.2 *N*-Hexylsalicylamide

$C_{13}H_{19}NO_2$ — M_r 221·30, CAS-No. 67520-12-1 — melting point 42–43°C

8.6.3 *N*-Octylsalicylamide

$C_{15}H_{23}NO_2$ — M_r 249·36, CAS-No. 109972-90-9 — melting point 44–46°C

8.6.4 *N*-Decylsalicylamide

$C_{17}H_{27}NO_2$ — M_r 277·41, CAS-No. 116311-05-8 — melting point 59–60°C

8.6.5 *N*-Dodecylsalicylamide

$C_{19}H_{31}NO_2$ — M_r 305·46, CAS-No. 10586-70-6 — melting point 64–70°C

8.6.6 *N*-Tetradecylsalicylamide

$C_{21}H_{35}NO_2$ M_r 333·52 melting point 74–77°C
CAS-No. 109972-89-6

The long chain *N*-alkylsalicylamides are moderately soluble in water and fairly soluble in organic solvents. Their antimicrobial effectiveness is described by Leinen *et al.* (1988). Among the salicylamides mentioned here the *N*-octyl and *N*-decyl derivatives are the most effective ones. In concentrations of 2·5–10 mg/litre these derivatives inhibit the proliferation of Gram-positive bacteria such as *Staphylococcus aureus, Streptococcus mutans* or *Actinomyces viscosus*. *N*-alkylsalicylamides generally act at first as microbistats. Because of their selective activity they may be used as active ingredients in deodorant formulations, as it is known that the odour producing bacteria on the skin mainly belong to the group of Gram-positive organisms. *N*-alkylsalicylamides are also useful for incorporation into antimicrobial preparations as they strongly increase the activity of the preparations against cocci.

8.7 Salicylanilide — 2-Hydroxybenzanilide — *N*-Phenylsalicylamide

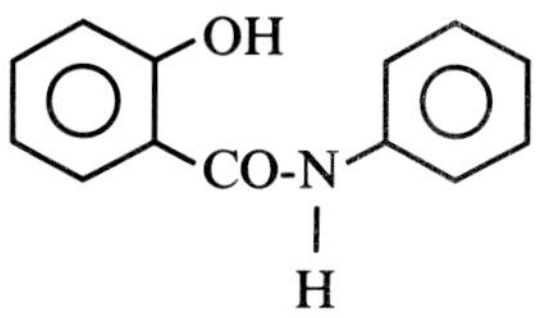

$C_{13}H_{11}NO_2$

M_r 213·24
CAS-No. 87-17-2

Chemical and physical properties

Appearance	White, odourless crystals
Melting point	136–138°C
Solubility	Very limited solubility in water; forms water soluble alkali salts; soluble in organic solvents
Stability	Relatively heat resistant, not volatile

Toxicity data

LD_{50} oral	2400 mg/kg mouse

Antimicrobial effectiveness/applications

Salicylanilide is especially active against fungi. As it is a practically odourless compound which does not cause coloration and is not hazardous in application, it has been used for the protection of textiles, leather, paper, plastic, paints, adhesives, etc. However, the compound is susceptible to leaching and therefore does not perform satisfactorily on material exposed outdoors. Effective addition rates are high, e.g. up to 10% in paints. Salicylanilide therefore has been widely substituted by more active and more economic microbicides. The sodium salt of salicylanilide may be used in fungicidal wall washes.

8.8 5,4'-Dibromosalicylanilide — Disanyl

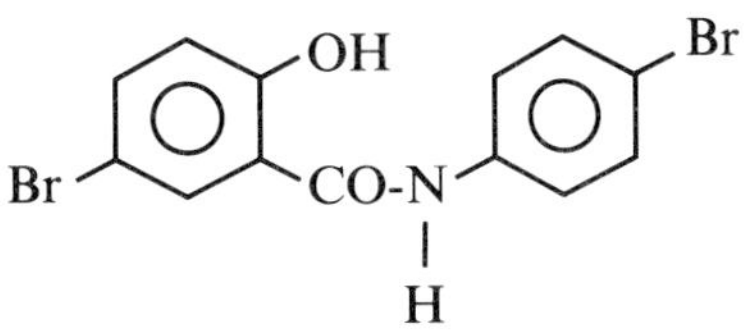

$C_{13}H_9Br_2NO_2$

M_r 371·03
CAS-No. 87-12-7

Chemical and physical properties

Appearance	White, odourless crystals
Melting point	238–245°C
Solubility	Practically insoluble in water, poorly soluble in ethanol, soluble in acetone and other organic solvents
Stability	Light and heat resistant, not compatible with proteins

Toxicity data

LD_{50} oral 1700 mg/kg mouse

Only moderately irritant to skin and mucosa — photosensitization is possible.

Antimicrobial effectiveness/applications

Although dibromosalicylanilide offers a broad activity spectrum its application as a microbicide for the protection of materials is on the decline, as there are available more effective and more economic substitutes.

8.9 3,5,4'-Tribromosalicylanilide — Trisanyl

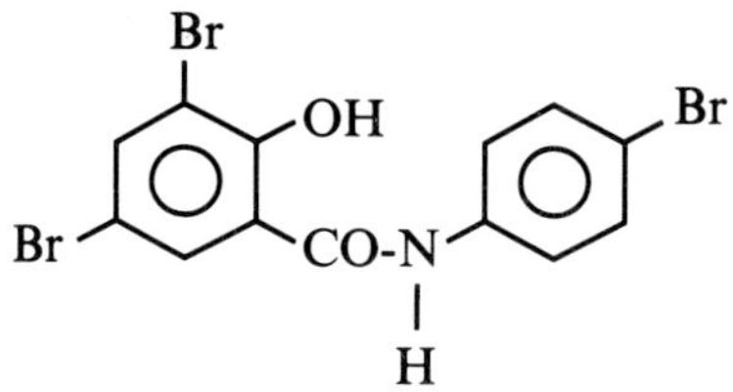

$C_{13}H_8Br_3NO_2$

M_r 449·96
CAS-No. 87-10-5

Chemical and physical properties

Appearance	White, odourless crystals
Melting point	227–228°C
Solubility	Practically insoluble in water, soluble in aqueous alkalis, hot acetone, dimethylformamide
Stability	Light and heat resistant, not compatible with proteins, compatible with anionic, cationic and non-ionic detergents

Toxicity data

LD_{50} oral 570 mg/kg rat

Only moderately irritant to skin and mucosa — photosensitization possible.

Antimicrobial effectiveness/applications

Broad spectrum microbicide the application of which for the protection of materials is on the decline.

8.10 Dithio-2,2'-bis(benzmethylamide)

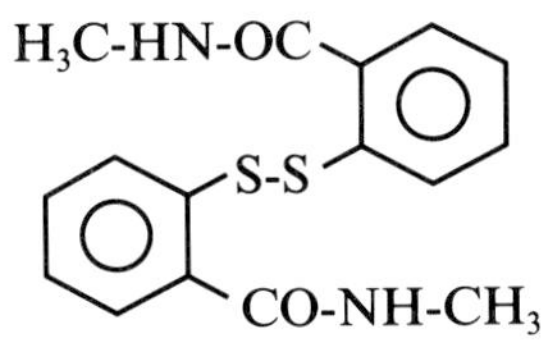

$C_{16}H_{16}N_2O_2S_2$ M_r 332·45
CAS-No. 2527-58-4

Chemical and physical properties

Appearance	Brownish powder	
Solubility (g/litre)	In water	0·20
	acetic acid ester	0·25
	methyl ethyl ketone	1·00
	toluene	0·02
	white spirit	0·01
Stability	Hydrolyses in alkaline media to *N*-methyl-benzisothiazolinone and *N*-methylthiosalicylamide; excellent heat stability	

Toxicity data

LD_{50} oral 15·000 mg/kg rat

Irritant to skin and mucosa, skin sensitizer.

Ecotoxicity

LC_{50} against japanese rice fish (*Oryzias latipes*): 5 mg/litre (exposure: 48 h)

Activated sludge organisms tolerate 40 mg/litre.

Antimicrobial effectiveness/applications

As is shown by the minimum inhibition concentrations in Table 83, dithio-2,2'-bis(benzmethylamide) is a broad spectrum fungicide which is also effective against bacteria. Apparently *Trichoderma viride* is more resistant to the compound than other fungal species.

The compound is recommended for application in water based coatings as a paint film fungicide, which at higher concentrations also exhibits algicidal effectiveness. It is active between pH 3·5 and 11. One of the degradation products in alkaline media is *N*-methyl-benzisothiazolinone which is an antimicrobial compound, too, acting as a preservative. Water based paints containing Dithio-2,2'-bis(benzmethylamide) as a paint film fungicide do not need an additional active ingredient for in-can protection. The microbicide may also be used for the protection of wet blues against microbial deterioration.

Table 83. Minimum Inhibition Concentrations (MIC) of Dithio-2,2'-bis(benzmethylamide) in Nutrient Agar

Test organism	*MIC (mg/litre)*
Alternaria alternata	20
Aspergillus niger	75
Aureobasidium pullulans	5
Chaetomium globosum	50
Coniophora puteana	1
Lentinus tigrinus	5
Penicillium glaucum	35
Polyporus versicolor	5
Sclerophoma pityophila	1
Trichoderma viride	100
Escherichia coli	25
Staphylococcus aureus	10

8.11 2,5-Dimethyl-*N*-cyclohexyl-*N*-methoxy-furan-3-carbonamide — Furmecyclox

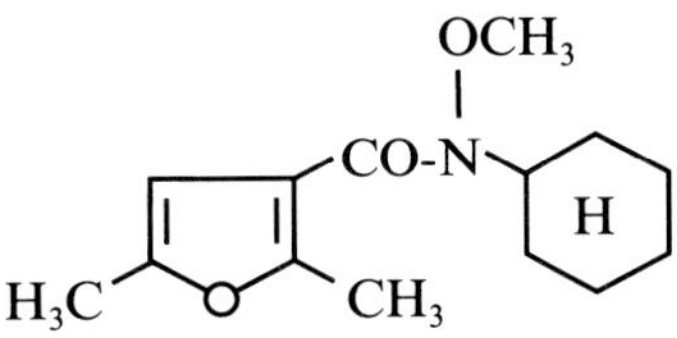

$C_{14}H_{21}NO_3$ M_r 251·33
CAS-No. 60568-05-0

Chemical and physical properties

Appearance	Yellow, slightly viscous fluid
Density at 20°C	1·081 g/cm^3
Boiling point at 0·7 hPa	135–140°C
Solidification point	Approx. 30°C
Vapour pressure at 20°C	9·6 × 10^{-5} hPa
Solubility	In water <1 g/litre; highly soluble in organic solvents
Stability	Hydrolyses in strong acids and alkalis

Toxicity data

LD_{50}	oral	3780 mg/kg rat; approx. 3500 mg/kg rabbit
	dermal	>5000 mg/kg rat
LC_{50}	inhalative	5600 mg/m^3 air against rats (exposure: 4 h)

Antimicrobial effectiveness/applications

Furmecyclox is a fungicide which is especially active against wood-rotting fungi, e.g. *Coniophora puteana, Coriolus versicolor, Lentinus lepideus* and *Lenzites abietina*. It has been used therefore as an active ingredient in preservatives for the protection of wood, generally in combination with other fungicides which are active against wood staining fungi, e.g. *Aureobasidium pullulans*. However, in the meantime the production of Furmecyclox has been discontinued and azole fungicides (see Section 12) are becoming important as substitutes.

8.12 *N*-(2-methylnaphthyl)maleinimide

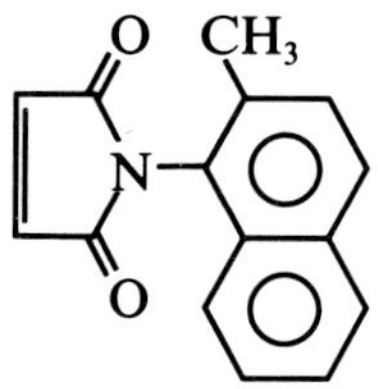

$C_{15}H_{11}NO_2$

M_r 237·26
CAS-No. 70017-56-0

Chemical and physical properties

Appearance	White powder		
Density	1·25–1·28 g/ml		
Melting point	163–165°C		
Solubility at 25°C (g/100 ml)	In dimethylsulphoxide	17–19	
	acetone	9–11	
	acetonitrile	9·5	
	dimethyl phthalate	6– 7	
	tetrahydrofuran	5·0	
	ethyl acetate	3– 4	
	toluene	2·0	
	xylenes	1– 2	
	carbon tetrachloride	0·7	
	ethanol	0·4	
	propylene glycol	0·3	
	octanol	0·2	
	isopropanol	0·2	
	n-heptane	0·03	
	water	0·0025	
Stability	Heat resistant (≥190°C)		

Toxicity data

LD_{50}	oral	>7000 mg/kg rat
	dermal	>7000 mg/kg rat

Acute inhalation: lethal concentration >10·5 mg/litre.

Slightly irritant to the skin, moderately irritant to mucous membranes; did not produce any sensitization in the guinea pig test.

No indications of mutagenic potential in the Ames test.
Not teratogenic in tests with rats and rabbits.

Ecotoxicity

LC_{50} for trouts	34 ppb
for bluegill	169 ppb
for daphnia	2·6 ppm

Antimicrobial effectiveness/applications

N-(2-methylnaphthyl)maleinimide has been found to be a microbicide capable of protecting fabrics, plastics, paints, etc., from fungal and bacterial attack (Becker & Gurnee, 1979). As can be seen from the minimum inhibition concentrations in Table 84, it is, however, particularly effective against fungi and yeasts. This activity of the product in combination with its heat stability and low water solubility makes the activity especially recommendable for the antimicrobial treatment of plastics.

Table 84. Minimum Inhibition Concentrations (MIC) of *N*-(2-methylnaphthyl)maleinimide in Nutrient Agar

Test organism	*MIC (mg/litre)*
Aspergillus niger	2
Stv. reticulum (Pink stain)	2–5
Brewer's yeast	10–25
Candida guilliermondii	1–10
Candida lipolytica	1–10
Escherichia coli	100
Klebsiella pneumoniae	20
Pseudomonas aeruginosa	1000
Staphylococcus aureus	20
Streptococcus faecalis	500–1000

8.13 3,4,4'-Trichlorocarbanilide (TCC) — *N*-(4-dichlorophenyl)-*N*'-(3,4-dichlorophenyl)-urea — Triclocarban

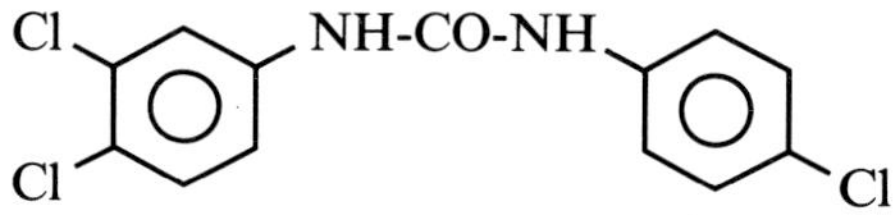

$C_{13}H_9Cl_3N_2O$ M_r 315·59
CAS-No. 101-20-2

Chemical and physical properties

Appearance	White, odourless, crystalline powder
Density at 20°C	Approx. 1·56 g/cm^3
Bulk density	Approx. 650 g/litre
Vapour pressure at 50°C	<1 hPa
Melting point	250–255°C
Flash point	>300°C
Solubility at 20°C	In water (<1 g/litre); soluble in organic solvents (10–100 g/litre in glycols)
Stability	Light stable and also chemically very stable; does not cause colorations in the presence of traces of heavy metals (e.g. Fe^{2+} ions); compatible with detergents

Toxicity data

LD_{50} oral	3600 mg/kg rat
	600 mg/kg mouse
LD_{50} intraperitoneal	2100 mg/kg mouse

Not irritant to skin and eyes.
Non-mutagenic (Salmonella microsome test).

Ecotoxicity

The compound degrades readily.

Antimicrobial effectiveness/applications

In standard concentrations TCC primarily inhibits the growth of Gram-positive bacteria; *Staphylococcus aureus* is especially sensitive to TCC. Microbicidal activity results only if the application concentration is many times the minimum inhibition concentration (MIC).

MIC	for *Staphylococcus aureus*	<1 mg/litre
	for *Pseudomonas aeruginosa*	1000 mg/litre

TCC is used as an active ingredient in deodorant soaps, other deodorant products such as sprays, roll-ons and sticks, shampoos, skin creams and shaving foams. The compound has an affinity for the skin which means that a longer-lasting antimicrobial effect is achieved by using a product containing TCC. In EC countries Triclocarban is one of the preservatives approved for use in cosmetics (permissible concentration is 0·2%). The Cosmetic, Toiletry and Fragrance Association (CTFA) in the USA includes Triclocarban in its Cosmetic Ingredient Dictionary.

8.14 *N'*-(3,4-dichlorophenyl)-*N,N*-dimethyl urea — Diuron

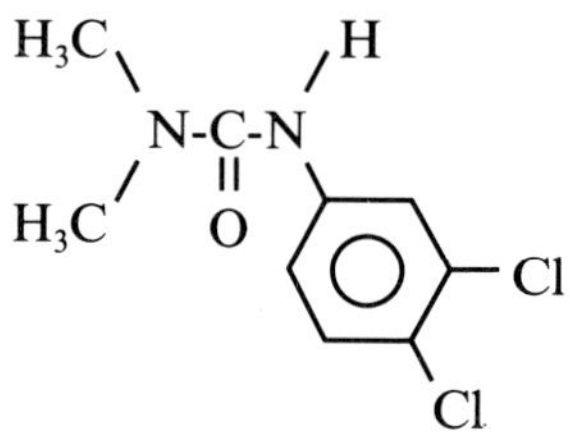

$C_9H_{10}Cl_2N_2O$ M_r 249·1
CAS-No. 330-54-1

Chemical and physical properties

Appearance	Pale yellow, crystalline powder	
Bulk density	Approx. 450 g/litre	
Melting point	154–159°C	
Vapour pressure at 20°C	Approx. $2{\cdot}3 \times 10^{-9}$ hPa	
Solubility at 20°C	In water	0·035 g/litre
27°C	acetone	53 g/kg
27°C	benzene	1·2 g/kg
20°C	butylglycol	60·0 g/litre
20°C	cyclohexanone	90·0 g/litre
Stability	Decomposition begins at 200°C; hydrolyses in strong acids and alkalis	

Toxicity data

LD_{50}	oral	4150 mg/kg rat
	dermal	>5000 mg/kg rat

Diuron is not irritant to skin and mucosa.
No sensitization was observed in the guinea pig test.

Ecotoxicity

LC_{50} against	*Carassius auratus*	5·8 mg/litre (exposure: 48 h)
	Cyprinus carpio	3·2 mg/litre (exposure: 48 h)
	Lepomis machrochirus	7·4 mg/litre (exposure: 48 h)
	Poecilia reticulata	25·0 mg/litre (exposure: 96 h)

Abiotic degradation

Half-life (22°C)	approx. 5 years at pH 4
	approx. 6 years at pH 7
	approx. >1 year at pH 9

Antimicrobial effectiveness/applications

Diuron is a broad spectrum algicide which is highly effective against both sea-water and freshwater algae by interrupting the photosynthetic electron transport. There is a big difference between the minimum algistatic concentrations of Diuron ($\ll 1$ mg/litre) and its minimum algicidal concentrations (100 mg/litre), indicating that the compound acts as an inhibitor of photosynthesis. Its favourable properties — non-volatile, low solubility in water, good toxicological properties, low ecotoxicity — make Diuron ideal for use as an algicide in masonry paints and lately in antifouling coatings. Unlike organotin compounds, it is light stable and, more importantly, it has low tolerable ecotoxicity. Although the antimicrobial effect of Diuron covers some species of mould fungus, the overall fungicidal activity of the compound is not very strong. As coatings for exterior applications are subjected to both algal attack and fungal infestation, an effective and suitable fungicide must also be added to the coating together with Diuron, unless one does not prefer to use a microbicide which has both a fungicidal and algicidal effect.

8.15 4-Trifluoromethylphenylsulphonic acid amide

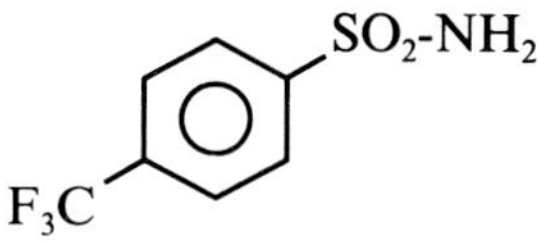

$C_7H_6F_3NO_2S$ M_r 225·19
CAS-No. 830-43-3

Chemical and physical properties

Appearance	Colourless crystals
Melting point	181–183°C
Solubility	Poor solubility in water, soluble in aqueous alcohols and organic solvents
Stability	Stable in slightly acid or alkaline media

Toxicity data

LD_{50} oral	270 mg/kg rat

Antimicrobial effectiveness/applications

4-Fluoromethylphenylsulphonic acid amide is described as a broad spectrum microbicide, recommended mainly as a paint film fungicide especially in water based paints, where it achieves simultaneously in-can protection. This property of the microbicide points to the water solubility of the substance and accordingly to leachability out of paint films.

9 CARBAMATES

Carbamates comprise the esters and the salts of carbamic acids. This is a systematic classification from chemical aspects. Microbicides of this class of substances widely differ in terms of both efficacy and mechanisms of activity. The carbamic and dithiocarbamic acids, the basis of the carbamates, lack chemical stability; they occur only intermediarily and disintegrate instantly to form carbon dioxide (CO_2), carbon disulphide (CS_2) and amine (see Fig. 31).

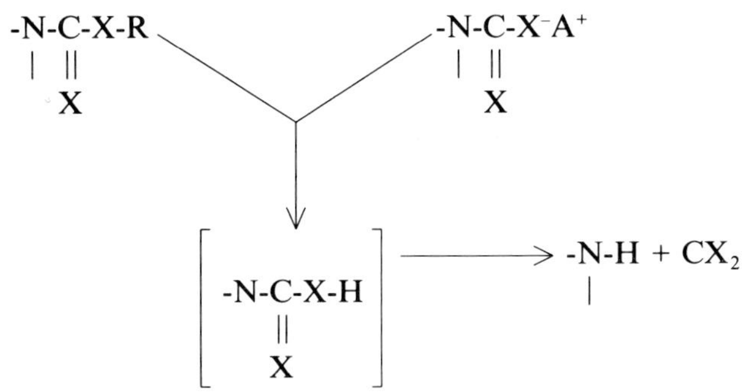

X = O or S; A^+ = metal kation

Fig. 31. Hydrolysis/degradation of carbamates.

The present section also deals with such microbicides as contain the carbamate grouping (-N-C-X-; with =X on C; X = O or S) within the molecule as a structural element, e.g. dithiocarbamic acid disulphides such as thiram. Other compounds of that nature, with a hydroxymethyl group on the N atom and thus able to release formaldehyde, are listed in Section 3. ‘Formaldehyde releasing compounds’:

3.4.10.1 = Sodium *N*-hydroxymethyl-*N*-methyldithiocarbamate
3.4.10.2 = *N*-hydroxymethyl-benzothiazoline-2-thione
3.4.10.3 = *N*-hydroxymethyl-5,6-dichlorobenzoxazolinone
3.4.10.4 = *N*-hydroxymethyl-5-chlorobenzoxazoline-2-thione.

Dazomet contains the dithiocarbamate grouping as well; as a reaction product of amine and formaldehyde it is however described in Section 3.1.16 = 3,5-Dimethyl-tetrahydro-1,3,5-2H-thiadiazine-2-thione.

Before dithiocarbamates gained importance as microbicides for the protection of materials they were known as potent, non-systemic fungicides for plant and crop protection. They are characterized as fungitoxic metal binding agents. A survey on the chemistry and mode of action of dithiocarbamate fungicides is given by Ludwig & Thorn (1960).

9.1 3-Iodopropargyl-*N*-butylcarbamate (IPBC)

$H_3C\text{-}CH_2\text{-}CH_2\text{-}CH_2\text{-}\underset{H}{\underset{\vert}{N}}\text{-}\underset{O}{\underset{\Vert}{C}}\text{-}O\text{-}CH_2\text{-}C{\equiv}C\text{-}I$	$C_8H_{12}INO_2$ M_r 281·07 CAS-No. 55406-53-6

Chemical and physical properties

Appearance	White odourless crystals
Melting point	65–66°C
Solubility at 20°C	156 mg/litre water; highly soluble in aromatic and polar organic solvents, low solubility in aliphatic hydrocarbon solvents and alkyd/hydrocarbon resins
Stability	Relatively stable in water and certain solvent based formulations; hydrolyses in strong alkaline media. Thermal degradation at 192°C

Toxicity data

LD_{50} oral	1470 mg/kg rat
dermal	>2000 mg/kg rat
inhalation	>6·89 mg/litre for rats (exposure: 4 h)

IPBC is not a primary skin irritant, but is irritant and corrosive to the eyes. No skin sensitization effects were observed in the guinea pig maximization test. Ames test and micronucleus assay did not show signs of mutagenicity.

Fish toxicity (exposure: 96 h)

LC_{50} for rainbow trout	0·124 mg/litre
for blue gill sunfish	0·5 mg/litre

Antimicrobial effectiveness/applications

IPBC is highly effective against a wide variety of fungal species, e.g. blue stain fungi (*Aureobasidium pullulans, Sclerophoma pityophila*), sapstainers (*Diplodia natalensis, Ceratocystis virescens, Ceratocystis pluriamulata*), wood rotting fungi, (*Coniophora puteana, Polyporus versicolor, Poria monticola, Gloeophyllum trabeum, Lenzites trabea*).

IPBC is used as a paint film fungicide and algicide in both solvent or water based surface coatings. It serves as an active ingredient in non-film forming decorative wood stains and can be formulated in wood preservative formulations suitable for the protection of freshly cut and sawn timber against staining fungi and moulds and for combating wood rotting fungi.

Table 85. Minimum Inhibition Concentrations (MIC) of IPBC in Nutrient Agar

Test organism	*MIC (mg/litre)*
Alternaria alternata	2
Aspergillus niger	2
Aureobasidium pullulans	1
Cephaloascus fragrans hanava	2
Ceratocystis pilifera	1
Chaetomium globosum	5
Cladosporium cladosporioides	2
Gliocladium virens	5
Lentinus tigrinus	2
Penicillium glaucum	1
Phialophora fastigiata	2
Sclerophoma pityophila	1
Trichoderma pseudokoningii	500
Trichoderma viride	100
Rhodotorula rubra	20
Sporobolomyces roseus	7·5
Escherichia coli	1000
Staphylococcus aureus	200

9.2 3-Iodopropargyl-*N*-phenylcarbamate (IPPC)

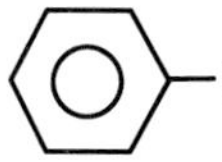–NH-CO-O-CH_2-C≡C-I $C_{10}H_8INO_2$ M_r 301·09
CAS-No. 22618-38-8

Chemical and physical properties

Appearance	White crystalline powder
Melting point	144–146°C
Solubility	Practically insoluble in water, highly soluble in aromatic and polar organic solvents (in acetone 100 g/litre, in dimethylformamide 200 g/litre), low solubility in aliphatic hydrocarbon solvents
Stability	Stable in both solvents and water based formulations, hydrolyses in strong alkaline media

Toxicity data

Not available.

Antimicrobial effectiveness/applications

Table 86. Minimum Inhibition Concentrations (MIC) of IPPC in Nutrient Agar

Test organism	*MIC (mg/litre)*
Alternaria alternata	3·5
Aureobasidium pullulans	5
Aspergillus niger	1
Chaetomium globosum	1
Coniophora puteana	0·5
Lentinus tigrinus	0·75
Penicillium glaucum	1·5
Polyporus versicolor	1
Sclerophoma pityophila	5
Trichoderma viride	35
Candida albicans	10
Rhodotorula rubra	10
Sporobolomyces roseus	5
Torula utilis	3·5
Escherichia coli	200
Staphylococcus aureus	1000

IPPC is a broad spectrum fungicide as is IPBC (Section 9.1). As can be seen from Table 86 the antimicrobial activity is also against yeasts and algae. IPPC may be used in fields of application which are described for IPBC. In exposure tests with paint films IPPC containing coatings appeared to be of somewhat longer lasting effectiveness in comparison to IPBC, which apparently is a little more leachable. However, up to now IPPC is not a commercially available microbicide.

9.3 3-Iodopropargylcarbamate (IPC)

$H_2N\text{-}CO\text{-}O\text{-}CH_2\text{-}C{\equiv}C\text{-}I$	$C_4H_4INO_2$	M_r 224·99
		CAS-No. 129348-50-1

Chemical and physical properties

Appearance	White solid
Melting point	85–87°C
Solubility	Sparingly soluble in water, soluble in organic solvents
Stability	Hydrolytically stable between pH 3 and 11.

Antimicrobial effectiveness/applications

IPC is reported (Rayudu, 1988) to have a number of advantages over IPBC (Section 9.1) and IPPC (Section 9.2). It is an excellent microbicide and its antimicrobial activity covers both fungi and bacteria including slime forming microorganisms. These properties open a variety of application fields to the microbicide IPC which are closed for instance to IPBC and IPPC. The compound may be used as a preservative for aqueous functional fluids, e.g. polymer emulsions, latex paints and metal working fluids. A further recommended use for IPC is the addition to pulp slurry or cooling water to inhibit the formation of slime. Additionally IPC protects coatings from mould growth after application. However, in accordance with its chemical structure it should be more leachable than IPBC and IPPC and therefore be able to act as an in-can preservative of aqueous functional fluids, which has been confirmed.

9.4 Methyl-*N*-benzimidazol-2-ylcarbamate — *N*-benzimidazol-2-ylcarbamic acid methylester (BCM) — Carbendazim

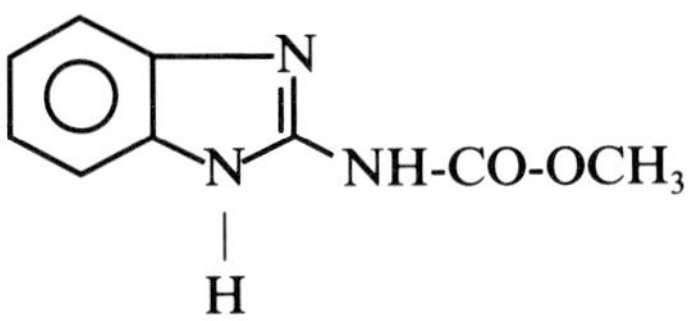

$C_9H_9N_3O_2$ M_r 191·19
CAS-No. 10605-21-7

Chemical and physical properties

Appearance	Pale grey crystalline powder	
Density at 20°C	Approx. 1·44 g/ml	
Bulk density	Approx. 500 g/litre	
Vapour pressure at 20°C	Approx. $0{\cdot}1 \times 10^{-6}$ hPa	
Melting point	Approx. 310°C (decomposition)	
Solubility (g/litre)	In water	0·008
	ethanol	approx. 0·3
	methyl glycol acetate	approx. 0·3
	acetone	approx. 0·3
	trichloromethane	approx. 0·1
	Solvesso 100	approx. 0·09;
	Soluble in strong acids — salts are formed	
Stability	Stable between pH 2 and 12; at pH > 13 slow hydrolysis to 2-aminobenzimidazole; endothermal decomposition starts at 185°C	

Toxicity data

LD_{50} oral	>15 000 mg/kg rat;
	>15 000 mg/kg mouse
dermal	>2000 mg/kg rat
intraperitoneal	>4000 mg/kg rat

Not irritant to the skin, but causes slight irritation of the eyes.
No sensitization.

Ecotoxicity

Biotic and abiotic degradation in water; microbial degradation in soil (half-life approx. 5–6 weeks).

LC_{50} for *Salmo gairdneri*	2·3 mg/litre
Lepomis gibbosus	3·2 mg/litre
LC_0 for *Salmo gairdneri*	0·56 mg/litre (exposure: 96 h)
LC_{50} for *Daphnia magna*	0·46 mg/litre (exposure: 48 h)

Antimicrobial effectiveness/applications

The antimicrobial activity of BCM and Benomyl (see Section 9.5) is based on a severe inhibition of DNA synthesis, which takes place before other systems are affected (Clemons & Sisler, 1971).

BCM exhibits a striking activity against a variety of fungal species and yeasts, but is more or less ineffective against bacteria and algae. However, there are deep gaps in BCM's spectrum of efficacy (see Table 87). Another handicap of BCM is the poor solubility of the fungicide in solvents, making it difficult to formulate the active ingredient and additionally inhibiting its penetration, for example, into wood. The activity of BCM against wood staining fungi is excellent; it is therefore especially suitable for use in the formulation of anti-blue stain agents. In contrast to many other fungicides for the protection of materials BCM is also highly effective against *Trichoderma* species. The main faults are against species of *Alternaria, Mucor, Geotrichum, Streptoverticillium, Cephaloascus* and *Candida*. In cases where complete protection is required, e.g. in fungicidal coatings, impregnating agents, sealants and plastics, it makes sense to combine BCM with another fungicide which has the relevant spectrum of activity. Examples of fungicides which are able to fill in the gaps in the activity spectrum of BCM are *N*-trihalomethylthio compounds (Section 14), dithiocarbamates (Sections 9.10.2 and 9.12), methylsulphonyl-tetrachloropyrimidine (Section 11.3), azole fungicides (Section 12) and others. In some cases of such combinations not only additivity of effectiveness is observed but synergism too (Brake, 1974; Paulus & Genth, 1986).

Table 87. Minimum Inhibition Concentrations (MIC) of BCM in Nutrient Agar

Test organism	*MIC (mg/litre)*
Alternaria alternata	>1000
Aureobasidium pullulans	0·1–0·5
Aspergillus niger	5–10
Cephaloascus fragrans hanava	>1000
Ceratocystis pilifera	0·5
Chaetomium globosum	0·5
Cladosporium cladosporioides	0·5
Gliocladium virens	1·0
Lentinus tigrinus	>1000
Penicillium glaucum	0·5
Phialophora fastigiata	0·5
Sclerophoma pityophila	0·5
Streptoverticillium reticulum	>1000
Trichoderma viride	1–2
Candida albicans	>1000
Candida krusei	>1000
Rhodotorula mucilaginosa	7·5
Rhodotorula rubra	5
Sporobolomyces roseus	1–2

The applications of BCM as a microbicide for the protection of materials are determined by the fact that its useful properties — it is non-volatile, heat resistant, non-leaching and colour-fast and has low toxicity — are counter-effected by its bad properties — it has an unequalized spectrum of activity and poor solubility.

9.5 Methyl-*N*-(1-butylcarbamoyl-)benzimidazol-2-ylcarbamate — Benomyl

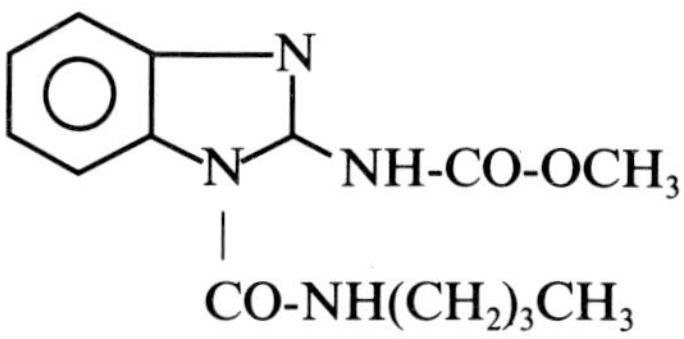

$C_{14}H_{18}N_4O_3$ — M_r 290·33

CAS-No. 17804-35-2

Chemical and physical properties

Appearance	Pale grey powder	
Melting point	Decomposes	
Solubility (g/litre) at 25°C	In water	0·004
	acetone	18
	heptane	400
Stability	Unstable in aqueous media, generation of BCM (see Section 9.4); at pH ≥ 13 quantitative conversion to 3-butyl-2,4-dioxo-s-triazino[1,2-α] benzimidazole (STB) (Chiba & Singh, 1986); decomposition when heated (release of methanol and formation of STB — see Section 9.5.1)	

Toxicity data

LD_{50} oral	5600 mg/kg mouse
	10 000 mg/kg rat

Mildly irritant to skin and mucosa.

Antimicrobial effectiveness/applications

As Benomyl in aqueous media always splits into BCM and *N*-butylcarbamic acid which is very reactive and unstable (decomposition to butylamin and CO_2), the antimicrobial efficacy of the fungicide corresponds to BCM (see Section 9.4). Even in solutions of Benomyl in organic solvents there exists an equilibrium between Benomyl on one side and BCM and butyl isocyanate on the other side; precipitation of BCM due to poor solubility in organic solvents shifts the equilibrium further to the side of BCM. Also STB (Section 9.5.1) which is formed from Benomyl at pH 13 or on heating exhibits an antimicrobial activity spectrum similar to that of BCM. As Benomyl apparently is a BCM (or STB) releasing compound having no remarkable advantages over BCM, it has not found much interest as a microbicide for the protection of materials. Benomyl has gained importance for plant protection as a fungicide which is systemic in plant tissue.

9.5.1 3-Butyl-2,4-dioxo-*s*-triazino[1,2-α]benzimidazole (STB)

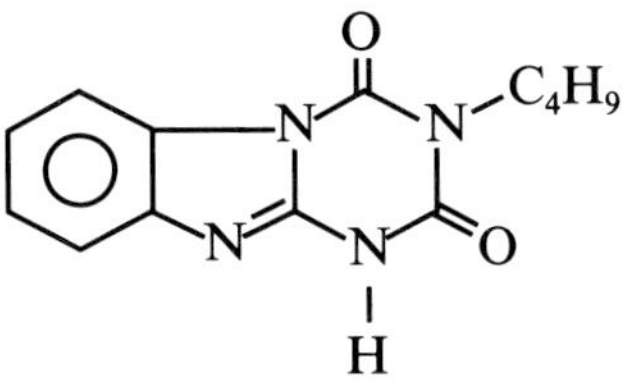

$C_{13}H_{14}N_4O_2$ M_r 258·28
CAS-No. 41136-38-3

Chemical and physical properties

Appearance	Crystalline powder
Melting point	280°C
Solubility	Practically insoluble in water, moderately soluble in acetonitrile, DMF, DMSO; in alkalis conversion to soluble alkali salts
Stability	At pH ≥ 13 slow degradation to 1-(2-benzimidazolyl)-3-*n*-butylurea (Chiba & Singh, 1986)

Antimicrobial effectiveness/applications

Table 88. Minimum Inhibition Concentrations (MIC) of STB in Nutrient Agar

Test organism	*MIC (mg/litre)*
Aspergillus niger	10–20
Chaetomium globosum	0·5
Penicillium glaucum	5
Rhizopus nigricans	>2000

9.6 5,6-Dichlorobenzoxazolinone

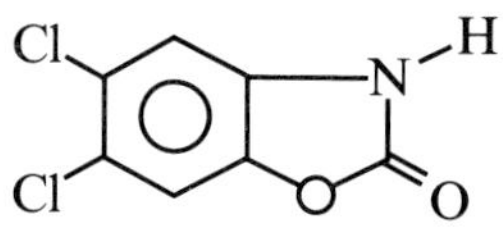

$C_7H_3Cl_2NO_2$ M_r 204·01
CAS-No. 5285-41-6

Chemical and physical properties

Appearance	Greyish powder with a characteristic odour	
Melting point	205–206°C	
Bulk density	240 g/litre	
Solubility (g/litre) at 25°C	In water	0·25
	ethanol	61
	isopropanol	56
	propylene glycol	60
	polyethylene	285
	acetone	134
	methyl ethyl ketone	120
	toluene	4·8
Stability	Conversion to soluble salts in alkaline media; on boiling in alkaline solutions hydrolysis to 3,4-dichloro-6-aminophenol	

Toxicity data

LD_{50} oral 1100 mg/kg mouse

No skin irritation.

Antimicrobial effectiveness/applications

5,6-Dichlorobenzoxazolinone is a very potent fungicide which had been used in the past mainly for the protection of textile material against biodeterioration. The minimum inhibition concentrations of the compound for a wide variety of fungal species are in the range of 10–100 mg a.i./ml nutrient agar. However, 5,6-dichlorobenzoxazolinone is no longer of practical importance.

9.7 3-(3-Iodopropargyl)-benzoxazol-2-one

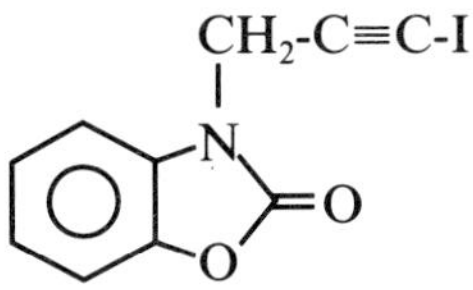

$C_{10}H_6INO_2$ M_r 299·07

CAS-No. 135306-48-8

Melting point 162–164°C

Antimicrobial effectiveness, see Table 89.

9.8 3 (3–Iodopropargyl)-6-chloro-benzoxazol-2-one

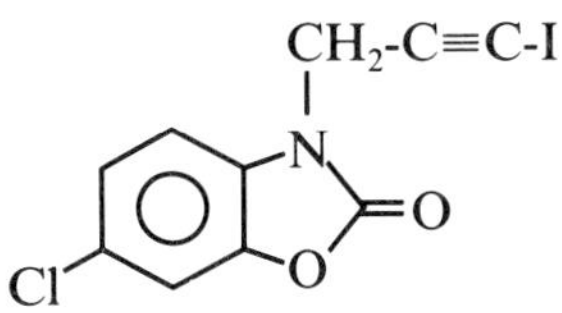

$C_{10}H_5ClINO_2$ M_r 333·52

CAS-No. 135306-49-9

The microbicides given in Sections 9.7 and 9.8 are novel benzoxalone compounds. Their antimicrobial activity, mainly directed against fungi and their utility as microbicides is described by Chi-Tung Hsu (1991).

Table 89. Minimum Inhibition Concentrations (MIC) of 3-(3-Iodopropargyl)-benzoxazol-2-one (9.7) and 3-(3-Iodopropargyl)-6-chloro-benzoxazol-2-one (9.8) in Nutrient Agar

Test organism	*MIC (mg/litre)*	
	9.7	*9.8*
Aspergillus niger	<0·13	<0·13
Aureobasidium pullulans	0·25	≤1
Cladosporium resinae	≤5	≤1
Gloeophyllum trabeum	≤0·2	≤1
Penicillium funicolosum	≤1	≤1
Rhodotorula rubra	≤1	≤1
Saccharomyces cerevisae	≤1	≤5
Escherichia coli	>250	>250
Pseudomonas aeruginosa	>250	>250
Pseudomonas fluorescens	>250	16
Staphylococcus aureus	>250	16

9.9 Sodium *N*-methyldithiocarbamate

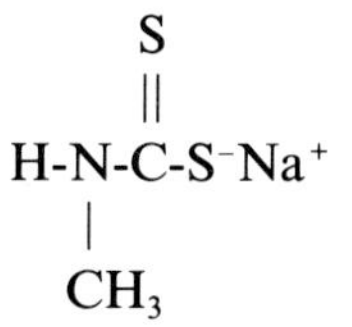

$C_2H_4NS_2Na$

M_r 129·18
CAS-No. 137-42-8

Chemical and physical properties

Appearance	Colourless crystals
Solubility	Highly soluble in water and mixtures of water and alcohols; solutions react alkaline
Stability	Decomposes when heated and in acid media; can cause coloration on contact with heavy metals

Toxicity data

LD_{50} oral	300–800 mg/kg rat

Strongly irritant to skin and eyes.

Antimicrobial effectiveness/applications

For practical application — controlling bacterial and fungal slime — in general 40% solutions of the sodium or potassium salt in water are available.

9.10 Salts of *N*-dimethyldithiocarbamic acid

9.10.1 Sodium *N*-dimethyldithiocarbamate

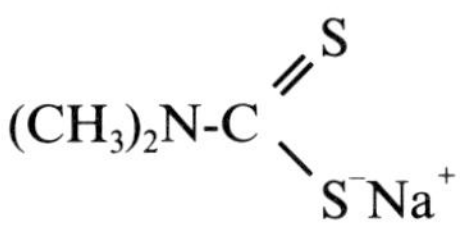

$C_3H_6NS_2Na$ M_r 143·21
CAS-No. 128-04-1

Chemical and physical properties (42% solution in water)

Appearance	Yellow liquid
Density at 20°C	Approx. 1·18 g/ml
Viscosity at 20°C	Approx. 4·9 mPa s
Vapour pressure	Approx. 20 hPa at 20°C Approx. 100 hPa at 50°C
Solidification point	Approx. −5°C (crystallization starts at 5°C)
Ignition temperature	>500°C
Solubility	Miscible with water and mixtures of alcohols and water
pH (1% solution in H_2O)	Approx. 9
Stability	Stable between pH 7 and 13; degradation in acid media with liberation of CS_2 and dimethylamine; may form deeply coloured, insoluble salts with Fe, Cu, Sn, Pb, and Co

Toxicity data

LD_{50} oral	3600–5100 mg/kg rat
dermal	>5000 mg/kg rat (exposure: 24 h)

Irritant to skin and eyes.

Ecotoxicity

LC_0 for *Leuciscus idus*	Approx. 1 mg/litre (exposure: 48 h)
LC_{100} for *Leuciscus idus*	Approx. 5 mg/litre (exposure: 48 h)

Activated sludge organisms in biological sewage treatment plants tolerate 50 mg/litre.

Antimicrobial effectiveness/applications

Table 90. Minimum Inhibition Concentrations (MIC) of a 42% Solution of Sodium *N*-dimethyldithiocarbamate in Nutrient Agar

Test organism	*MIC (mg/litre)*
Alternaria tenuis	200
Aspergillus niger	>1000
Aureobasidium pullulans	500
Chaetomium globosum	500
Cladosporium cladosporioides	500
Lentinus tigrinus	200
Penicillium glaucum	500
Sclerophoma pityophila	150
Trichoderma viride	>1000
Candida albicans	500
Candida krusei	500
Rhodotorula mucilaginosa	500
Saccharomyces bailii	200
Saccharomyces cerevisiae	500
Torula rubra	500
Torula utilis	500
Aerobacter aerogenes	150
Aeromonas punctata	200
Bacillus mycoides	100
Bacillus subtilis	200
Escherichia coli	500
Leuconostoc mesenterioides	500
Proteus mirabilis	>1000
Pseudomonas aeruginosa	>1000
Pseudomonas fluorescens	>1000
Staphylococcus aureus	500

The solution of sodium dimethyldithiocarbamate is a broad spectrum microbicide for controlling the growth of bacteria and fungi in pulp and paper manufacture and in related systems. It is employed not only as a slimicide but also for the protection of formulations used for papermaking, e.g. glues, starch and clay slurries and coating preparations. But when adding sodium dimethyldithiocarbamate to such formulations one has to bear in mind, that traces of heavy metals can cause colorations. For slime control one usually adds 50–200 mg/litre water, in pulp and paper mills 20–80 kg/100 t of dry pulp or paper per day.

N-Dimethyldithiocarbamate is highly effective against lactic acid bacteria. The sodium salt solution therefore is used in sugar production to inhibit the bacterial degradation of sugar to lactic acid. It is added to the sugar juice (10–20 mg/litre). It decomposes relatively quickly under these conditions especially with increasing temperature and accordingly is not retained by the sugar.

Potassium salt solutions of *N*-dimethyldithiocarbamic acid are also available.

9.10.2 Zinc dimethyldithiocarbamate — Ziram

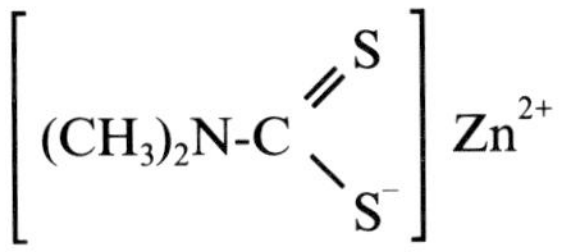

$C_6H_{12}N_2S_4Zn$

M_r 305·83
CAS-No. 15465-14-2

Chemical and physical properties

Appearance	White to grey odourless powder
Density at 25°C	1·66 g/ml
Melting point	248–250°C
Ignition temperature	> 300°C
Solubility at 25°C	0·065 g/litre water; sparingly soluble in organic solvents
Stability	Stable between pH 5 and 10, decomposition in acid media with release of CS_2 and dimethylamine; causes only coloration when coming into contact with traces of heavy metals, e.g. Cu, Fe; disturbs/inhibits the oxidative drying of resins (alkyds)

Toxicity data

LD_{50} oral	1400 mg/kg rat
	1900 mg/kg mouse
LD_{50} cutaneous	>5000 mg/kg rat (exposure: 24 h);
inhaled for rats	>0·5 mg/litre (exposure: 4 h)

Does not cause skin irritation, but irritates mucous membranes.
No sensitization was observed in the guinea pig test.

Antimicrobial effectiveness/applications

Ziram displays an unusually broad spectrum of activity covering fungi, yeasts, bacteria and algae. As it also has a low toxicity and is non-volatile, is non-leaching and relatively heat resistant, it is a suitable microbicide for use in coatings, plaster, adhesives, rubber and sealing compounds, cement jointing filler, particle and fibre board. The addition rates range between 0·2 and 1%.

Table 91. Minimum Inhibition Concentrations (MIC) of Ziram in Nutrient Agar

Test organism	*MIC (mg/litre)*
Alternaria alternata	200
Aspergillus flavus	200
Aspergillus niger	200
Aspergillus terreus	750
Aureobasidium pullulans	100
Chaetomium globosum	75
Cladosporium herbarum	100
Coniophora puteana	<20
Lentinus tigrinus	200
Paecilomyces variotii	200
Penicillium citrinum	500
Penicillium glaucum	150
Polyporus versicolor	200
Rhizopus nigricans	75
Trichoderma viride	200
Trichophyton pedis	100
Candida albicans	350
Candida krusei	200
Rhodotorula mucilaginosa	150
Rhodotorula rubra	200
Sporobolomyces roseus	50
Torula rubra	200
Torula utilis	200
Bacillus subtilis	150
Escherichia coli	50
Pseudomonas aeruginosa	75
Staphylococcus aureus	200
Algae	1

9.11 Zinc ethylenebisdithiocarbamate — Zineb

$$\left[\begin{array}{c} H_2C\text{-}NH\text{-}CS\text{-}S^- \\ | \\ H_2C\text{-}NH\text{-}CS\text{-}S^- \end{array}\right] Zn^{2+}$$

$C_4H_6N_2S_4Zn$ *M*r 275·76
CAS-No. 142-14-3

Chemical and physical properties

Appearance	Faintly coloured powder with an unpleasant odour
Solubility	Practically insoluble in water, only moderately soluble in organic solvents
Melting point	240°C (decomposition)
Stability	Decomposes when heated; unstable in acid media

Toxicity data

LD_{50} oral	5200 mg/kg rat
	7600 mg/kg mouse
	4450 mg/kg rabbit
intraperitoneal	1940 mg/kg mouse

Irritant to skin and mucous membranes.

Antimicrobial effectiveness/applications

Table 92. Minimum Inhibition Concentrations (MIC) of Zineb in Nutrient Agar

Test organism	*MIC (mg/litre)*
Alternaria tenuis	10
Aspergillus niger	50
Aureobasidium pullulans	5
Chaetomium globosum	200
Coniophora puteana	2
Lentinus tigrinus	5
Penicillium glaucum	20
Polyporus versicolor	50
Sclerophoma pityophila	5
Trichoderma viride	500
Escherichia coli	100
Staphylococcus aureus	100
Algae	20–100

Although Zineb displays a broad spectrum of effectiveness and considerable antimicrobial activity, it has not often been employed as a microbicide for the protection of materials. The stable Ziram (see Section 9.10.2) is the preferred zinc dithiocarbamate for that application. The disodium salt of ethylenebisdithiocarbamic acid (Nabam) is also available.

9.12 Bis-(dimethylthiocarbamoyl)-disulphide — Tetramethylthiuramdisulphide — Thiram

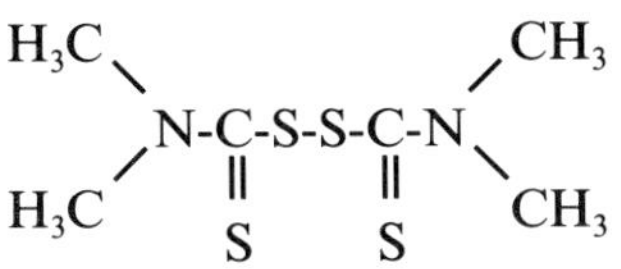

$C_8H_{12}N_2S_4$

M_r 240·43
CAS-No. 137-26-8

Chemical and physical properties

Appearance	Colourless, odourless crystals	
Melting point	146–148°C	
Density	1·29 g/ml	
Solubility (g/litre)	In water	0·03
	acetone	25
	ethanol	38
	dimethylformamide	75
	Moderately soluble in benzene, trichloro-methane, highly soluble in carbon disulphide	
Stability	Stable between pH 4·5 and 9; decomposition in acid media with release of carbon disulphide (CS_2) and dimethylamine ($(CH_3)_2NH$); Thiram causes staining on contact with metals, particularly iron, copper, bronze or brass	

Toxicity data

LD_{50} oral	Approx.	2500 mg/kg rat
dermal		>1000 mg/kg rat

Only moderately irritant to the skin, but strongly irritant to the eyes.

MWC 5 mg/m^3

Ecotoxicity

LC_0 for *Brachydanio rerio* 0·01–0·03 mg/litre (exposure: 96 h)

Thiram causes no BOD under the strict conditions of the closed bottle test. This result does not exclude the possibility that the product may be degraded in the environment.

Antimicrobial effectiveness/applications

Table 93. Minimum Inhibition Concentrations (MIC) of Thiram in Nutrient Agar

Test organism	*MIC (mg/litre)*
Alternaria alternata	35
Aspergillus niger	150
Aspergillus ustus	50
Aureobasidium pullulans	25
Cephaloascus fragrans	50
Chaetomium globosum	50
Cladosporium cladosporioides	25
Cladosporium herbarum	5
Coniophora puteana	1
Fusarium culmorum	75
Fusarium moniliforme	50
Fusarium solani	50
Lentinus tigrinus	25
Paecilomyces variotii	2·5
Penicillium citrinum	1000
Penicillium glaucum	50
Phialophora fastigiata	100
Polyporus versicolor	50
Rhizopus nigricans	25
Rhizopus stolonifer	100
Sclerophoma pityophila	10
Stachybotris chartarum	500
Trichoderma viride	175
Trichophyton pedis	250
Candida albicans	150
Candida krusei	100
Rhodotorula mucilaginosa	100
Saccharomyces bailii	50
Saccharomyces cerevisiae	50
Sporobolomyces roseus	50
Torula rubra	100
Torula utilis	100
Aerobacter aerogenes	50
Aeromonas punctata	75
Bacillus mycoides	25
Bacillus subtilis	100
Desulfovibrio desulfuricans	5
Escherichia coli	50
Leuconostoc mesenterioides	50
Proteus mirabilis	>2500
Proteus vulgaris	>2500
Pseudomonas aeruginosa	>2500
Pseudomonas fluorescens	>2500
Staphylococcus aureus	35

With regard to fungi and yeasts Thiram's spectrum of activity is considerably equalized. But deep gaps occur in the activity spectrum for bacteria; pseudomonades especially resist Thiram (see Table 93). The compound therefore is not appropriate for the in-can protection of functional fluids. However, Thiram is an important paint film fungicide and algicide, as it is virtually insoluble in water, light-stable and non-volatile. It can be used in emulsion paints and in solvent based coatings as long as they contain no binders which dry by oxidation (Thiram disturbs/inhibits the drying of such binders). Thiram containing coatings should not be used for outdoor applications as staining may be caused by splashes of water transporting traces of heavy metals, e.g. rainwater from gutters. Mould resistant coatings based on Thiram are mainly applied in locations with high humidity, e.g. in the food industry, in bathrooms etc. Paper, cardboard, shoe insoles and glues are other examples of materials which may be protected by Thiram against biodeterioration.

10 DIBENZAMIDINES

Dibenzamidines of the following general structure

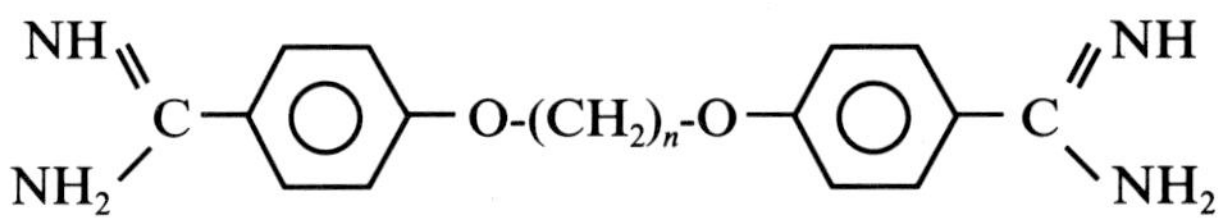

possess striking antimicrobial properties (Wien *et al.*, 1948). They have a cationic character and in consequence belong to the membrane-active microbicides. Their main application as microbicides is in antiseptics; additionally they are used to some extent as preservatives for cosmetics. They have not gained importance as microbicides for the protection of materials, although their activity spectrum covers not only bacteria but fungi, too. Dibenzamidines in general are of poor water-solubility, but because of their aminic character they are able to form water soluble salts with acids. The dibenzamidine salts of 2-hydroxyethanesulphonic acid, the isethionates, are of practical importance.

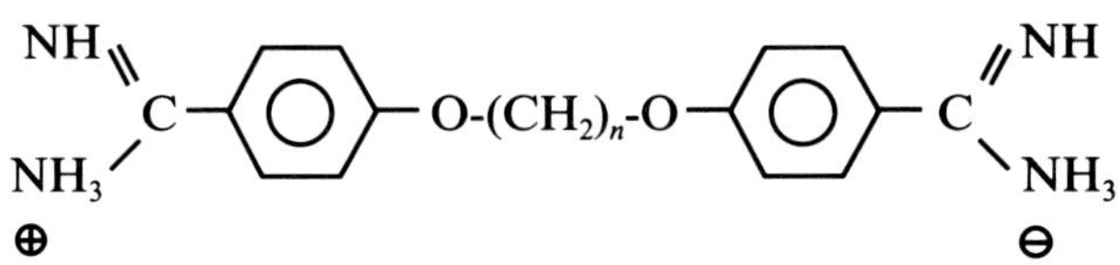

$^{\ominus}O_3S\text{-}CH_2\text{-}CH_2\text{-}OH$ $\qquad$ $HO\text{-}H_2C\text{-}H_2C\text{-}SO_3^{\ominus}$

10.1 4,4'-(Trimethylenedioxy)-dibenzamidine — 4,4'-Diamidinophenoxypropane — Propamidine

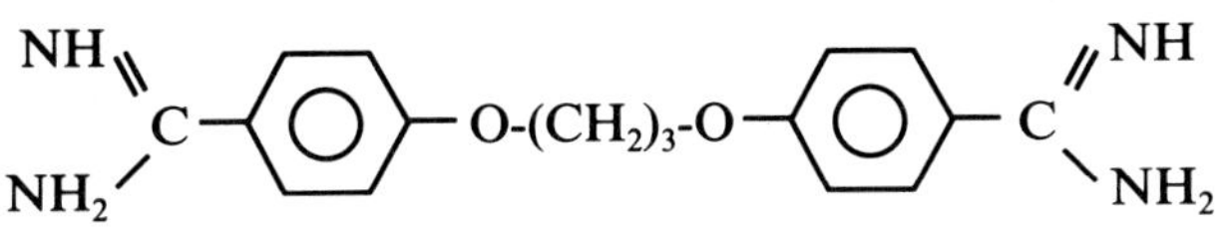

$C_{17}H_{20}N_4O_2$

M_r 312·38
CAS-No. 104-32-5

10.1.1 4,4'-(Trimethylenedioxy)-dibenzamidine-bis(β-hydroxy-ethanesulphonate) — Propamidineisethionate

$C_{21}H_{32}N_4O_{10}S_2$ M_r 564·65

Chemical and physical properties (isethionate)

Appearance	White, crystalline, hygroscopic, odourless powder
Melting point	235°C
Solubility	In water 200 g/litre; highly soluble in alcohol
pH (5% solution in H_2O)	4·5–6·5
Stability	Precipitation and inactivation in the presence of phosphates

Toxicity data

LD_{50} subcutaneous	42 mg/kg mouse
intravenous	55 mg/kg mouse

Antimicrobial effectiveness/applications

The minimum inhibition concentrations vary from 5 mg/litre against *Staphylococcus aureus* to 75 mg/litre against *Escherichia coli*, to 250 mg/litre against *Pseudomonas aeruginosa* and to 500 mg/litre against *Aspergillus niger*. The activity is reduced by blood or serum and at low pH values.

Propamidine is a component of different antiseptics and sometimes a preservative in cosmetics (concentration in 0·1%).

10.2 4,4'-Diamidino-2,2'-dibromodiphenoxypropane — Dibromopropamidine

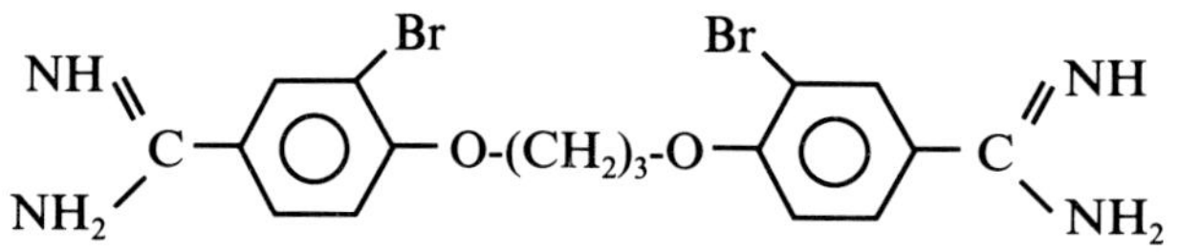

$C_{17}H_{18}Br_2N_4O_2$

M_r 470·17
CAS-No. 496-00-4

10.2.1 4,4'-Diamidino-2,2'-dibromodiphenoxypropane-bis(β-hydroxy-ethanesulphonate) — Dibromopropamidineisethionate

$C_{21}H_{30}Br_2N_4O_{10}S_2$ M_r 722·47

Chemical and physical properties

Appearance	White, crystalline powder
Melting point	226°C
Solubility	Highly soluble in water and alcohol, soluble in glycerine, insoluble in oils
Stability	Tolerates heating to 100°C for 30 min; slight decomposition during storage, not compatible with chloride and sulphate ions and anionic detergents

Toxicity data

LD_{50} subcutaneous	300 mg/kg mouse
intravenous	10 mg/kg mouse

Antimicrobial effectiveness/applications

The bromination of Propamidine to Dibromopropamidine is accompanied by a considerable increase in antimicrobial efficacy. However the difference in activity against Gram-positive and Gram-negative bacteria remains unchanged. In the EC list of preservatives permitted for the in-can protection of cosmetics, Dibromopropamidine is mentioned with a maximum authorized concentration of 0·1%.

10.3 4,4'-(Hexamethylenedioxy)-dibenzamidine — 4,4'-Diamidinophenoxyhexane — Hexamidine

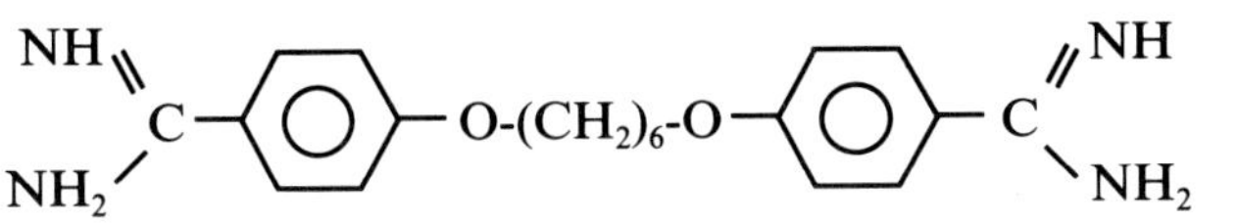

$C_{20}H_{26}N_4O_2$

M_r 354·46
CAS-No. 3811-75-4

10.3.1 4,4'-(Hexamethylenedioxy)-dibenzamidine-bis(β-hydroxy-ethansulphonate) — Hexamidineisethionate

$C_{24}H_{38}N_4O_{10}S_2$ M_r 606·72

Chemical and physical properties

Solubility	Fairly soluble in water and alcohol, insoluble in oils
Stability	Not compatible with chloride and sulphate ions and anionic detergents

Toxicity data

LD_{50} oral	500–750 mg/kg rat, mouse, rabbit

Moderately irritant to skin and mucosa; sensitization and photosensitization effects were not observed.

Antimicrobial effectiveness/applications

Hexamidine exhibits bactericidal, bacteriostatic and fungistatic activity similar to Propamidine (Section 10.1). The activity is reduced in the presence of protein. In the EC list of preservatives permitted for cosmetics Hexamidine is mentioned with a maximum authorized concentration of 0·1%.

10.4 4,4'-Diamidino-2,2'-dibromodiphenoxyhexane — Dibromohexamidine

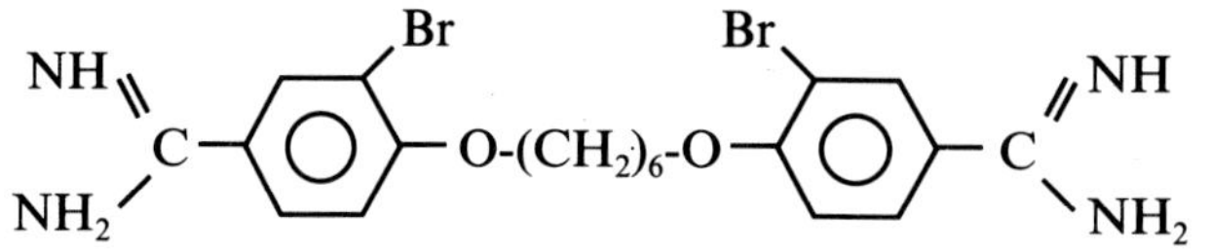

$C_{20}H_{24}Br_2N_4O_2$

M_r 512·25
CAS-No. 93856-82-7

10.4.1 Dibromohexamidine isethionate

$C_{24}H_{36}Br_2N_4O_{10}S_2$ M_r 764·55

Chemical and physical properties

Solubility	Soluble in water, alcohol, glycerine, insoluble in benzene

Toxicity data

LD_{50} oral and cutaneous	>4000 mg/kg rat
intravenous	71 mg/kg rat

Irritant to skin and mucosa. No sensitization.

Antimicrobial effectiveness/applications

The brominated Hexamidine is more active than Hexamidine. The EC list of preservatives for cosmetics indicates an application concentration of 0·1%.

11 PYRIDINE DERIVATIVES AND RELATED COMPOUNDS (BENZOPYRIDINES = QUINOLINES)

Among other pyridine derivatives 2-hydroxy-pyridine-*N*-oxides, 2-mercapto-pyridine-*N*-oxides and 8-hydroxyquinolines are described in this section which may be looked at as membrane-active microbicides with chelating properties. The following pyridine compounds with antimicrobial activity but without significant importance for the protection of materials shall only be mentioned:

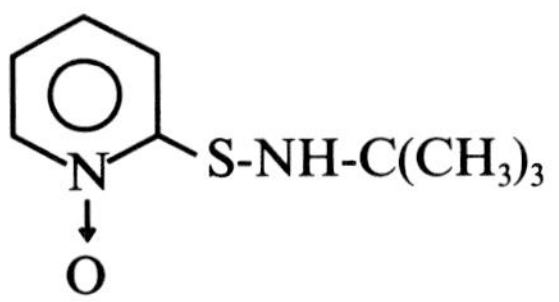

2-(tert.-butylaminothio)pyridine-*N*-oxide

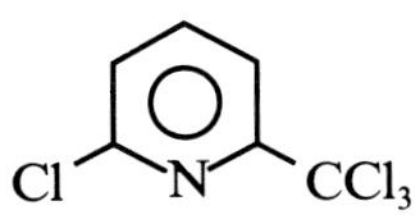

2-chloro-6-(trichloromethyl)pyridine — Nitrapyrin

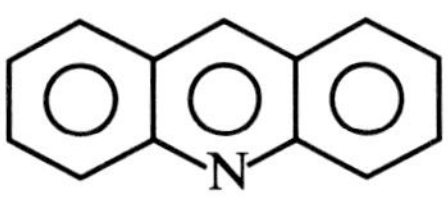

dibenzpyridine — Acridine

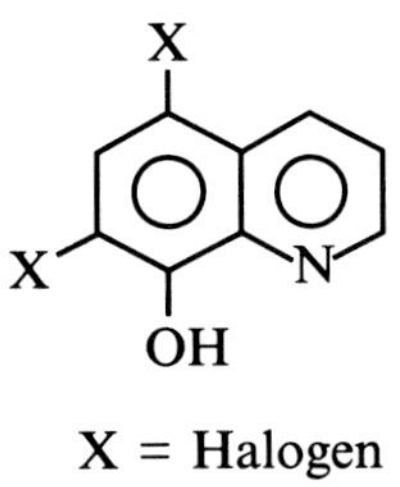

Halo-8-hydroxyquinolines, especially 5,7-dihalo-8-hydroxyquinolines transgress the parent compound (Oxine, Section 11.4) in antimicrobial activity

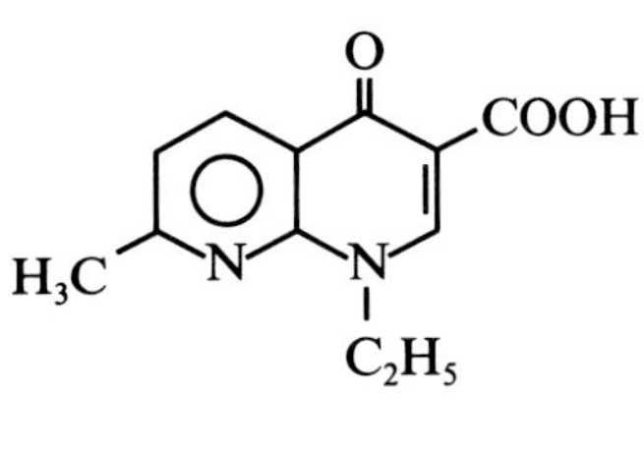

1-ethyl-1,4-dihydro-7-methyl-4-oxo-1,8-naphthyridine-3-carboxylic acid — Nalidixic acid

Quaternary pyridinium and isoquinolinium salts are mentioned in Section 16, 'Surface active agents'.

11.1 Pyridine-*N*-oxides

The most important compound is 2-mercapto-pyridine-*N*-oxide = Pyrithione. According to Albert (1968) its mechanism of antimicrobial activity is based on chelation complex formation. But there are findings (Cooney & Felix, 1972; Chandler & Segel, 1978) which demonstrate that other modes of action are involved, too, for example, influence on ATP levels, nutrient transport, interference with protein synthesis.

Pyrithiones are used to prevent biodeterioration in aqueous functional fluids, e.g. adhesives, latex paints, polymer emulsions, including cosmetics. Pyrithiones are used very successfully in metal working fluids for the control of fungi; here they assist formaldehyde releasing compounds (see Section 3), e.g. hexahydro-*s*-triazines (Section 3.3.12), which are especially toxic for bacteria (Rossmoore *et al.*, 1978). Another application of Pyrithiones is in fuel storage to inhibit microbial growth in entrained water and the resulting microbial induced corrosion.

11.1.1 2-Hydroxypyridine-*N*-oxide (A) — 1-Hydroxypyridine-2-one (B)

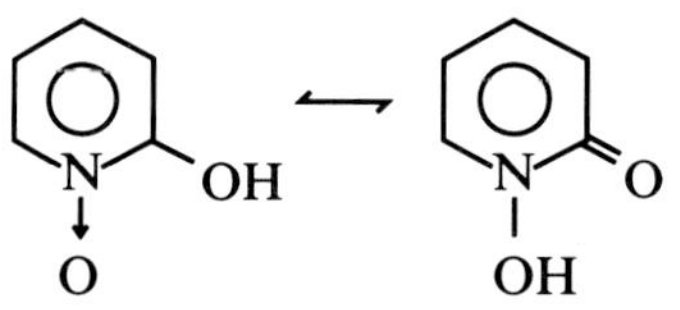

A B

$C_5H_5NO_2$

M_r 111·10
CAS-No. 822-89-9

Chemical and physical properties

Appearance	White crystalline powder
Melting point	151°C
Solubility	Soluble in water, alkalis and alcohols
Stability	Strong reducing agents attack the *N*-oxide and reduce the compound to 2-hydroxypyridine

Toxicity data

LD_{50} oral	920 mg/kg rat

Characteristics for a 20% solution of 2-hydroxypyridine-*N*-oxide in water:

Appearance	Yellow-brown liquid
Density at 20°C	1·135 g/ml
Refractive index	n_D^{20} 1·3885
pH	7–8
Solubility	Soluble in water and polar organic solvents

Antimicrobial effectiveness/applications

Table 94. Minimum Inhibition Concentrations (MIC) of 2-hydroxypyridine-*N*-oxide in Nutrient Agar

Test organism	*MIC*[a] *(mg/litre)*
Bacillus subtilis	0·1
Enterobacter aerogenes	3·3
Escherichia coli	2·5
Lactobacillus acidophilus	0·2
Proteus vulgaris	1·8
Pseudomonas aeruginosa	3·5
Serratia marcescens	1·2
Staphylococcus aureus	0·4
Streptococcus haemolyticus	0·2

[a] According to Pyrion-Chemie for Oxy-Pyrion.

Oxy-Pyrion is mainly used as a preservative in cosmetics.

11.1.2 1-Hydroxy-4-methyl-6-(2,4,4-trimethyl-pentyl)-2(1H)pyridone ethanolamine salt — Piroctone olamine

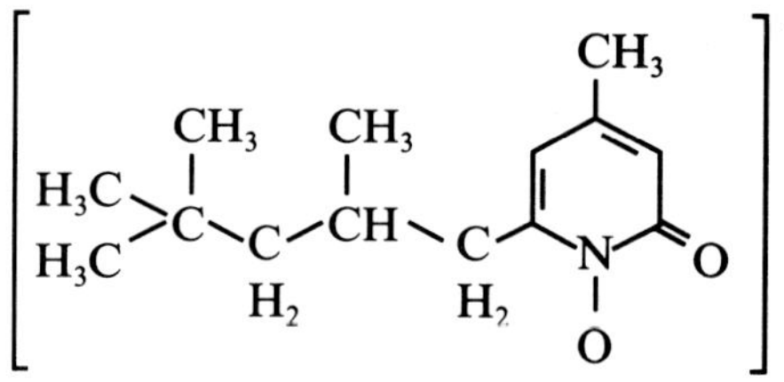

$H_3\overset{+}{N}$-CH_2-CH_2-OH $\qquad C_{16}H_{30}N_2O_3$

M_r 298·46
CAS-No. 50650-76-5

Chemical and physical properties

Appearance	White to slightly yellowish fine powder — odourless	
Melting point	136°C (decomposition)	
Solubility	In water	approx. 2 g/litre
	alcohols	approx. 100 g/litre
	oils	approx. 0·5–1 g/litre
Stability	Stable between pH 5 and 9; heat resistant (80°C)	
pH (1% aqueous solution)	9–9·8	

Toxicity data

LD_{50} oral	8·1 g/kg rat
	5·0 g/kg mouse

No-effect level for rats and dogs: 100 mg/kg/day.

Good skin and mucous membrane tolerance (rabbits).

No sensitization.

No phototoxic properties.

Non-teratogenic, non-mutagenic.

Antimicrobial effectiveness/applications

Fungi and Pseudomonades are 10–20 times more resistant than other bacteria and yeasts. The main advantage of Piroctone olamine is its low toxicity — it is practically non-toxic. As the microbicide is additionally compatible with anionic, cationic and amphoteric compounds it serves mainly as an active ingredient in antidandruff hair tonics, and as a preservative in shampoos. The EC positive list of preservatives for cosmetics authorizes a maximum concentration of 1% for products rinsed off and 0·5 for other products. As a preservative for other functional fluids Piroctone olamine has up to now not gained importance.

Table 95. Minimum Inhibition Concentrations (MIC) of Piroctone olamine in Nutrient Agar according to Wallhäusser (1984)

Test organism	*MIC (mg/litre)*
Escherichia coli	64
Klebsiella pneumoniae	32
Pseudomonas aeruginosa	625–1250
Staphylococcus aureus	32
Candida albicans	64
Penicillium notatum	625

11.1.3 Sodium salt of 2-mercaptopyridine-*N*-oxide — Sodium 2-pyridinethiol-1-oxide — Sodium Pyrithione

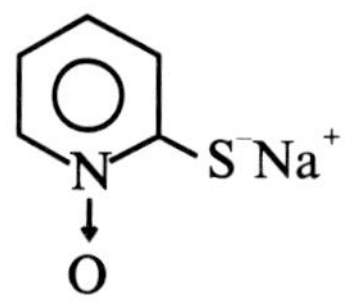

C_5H_4NOSNa M_r 149·15
CAS-No. 116503-44-7

Chemical and physical properties

Appearance	White to yellow crystalline powder with a mild odour
Melting point	250°C (decomposition)
Solubility	Highly soluble in water (>500 g/litre), soluble (>100 g/litre) in ethanol, propylene glycol, polyethylene glycol, DMSO, moderately soluble in isopropanol (approx. 8 g/litre)
pH (2% aqueous solution)	8
Stability	In solution stable between pH 4·5 and 9·5; below pH 4·5 conversion to free Pyrithione which is very unstable in the presence of oxygen or light; above pH 9·5 slow conversion to the sodium salt of Pyrithione sulphonic acid. Oxidizing agents such as peroxides and hypohalites will convert Pyrithione first to 2,2'-dithio-bis-pyridine-1,1'-dioxide (Section 11.1.5) and finally to inactive pyrithione sulphinic or sulphonic acid. Strong reducing agents attack the N-oxide group and reduce Pyrithione to 2-mercaptopyridine which is less active than the parent compound.

Toxicity data

LD_{50} oral	875 mg/kg rat
	1172 mg/kg mouse

It is neither an allergic skin sensitizer nor a photosensitizer. Good skin compatibility is attested for use dilutions.

Characteristics (40% solution in water)

Appearance	Amber, mild odour
Density at 25°C	1·27 g/ml
pH at 25°C	9·2
Solubility	Soluble in water and alcohols
Stability	Heat resistant up to 100°C; precipitation of free pyrithione below pH 6·5

Toxicity data

LD_{50} oral	1500 mg/kg rat

Antimicrobial effectiveness/applications

Sodium Pyrithione is a widely used preservative for water based functional fluids, cosmetics included. In consequence of its activity spectrum it is preferably applied when problems due to the growth of fungi have to be overcome, e.g. in metal working fluids. As is demonstrated by the MIC in Table 96 Sodium Pyrithione is a highly effective microbicide; the addition rates therefore are relatively low: 0·02–0·06%. However, users of Sodium Pyrithione have to pay attention to the fact that it is a chelating agent which in the presence of, for example, Fe^{2+} ions or Cu^{+} ions is converted to the corresponding chelates. These are sparingly soluble and highly coloured compounds. That means that they can cause colorations and precipitation, thus withdrawing active ingredients from the functional fluid to be protected. The ferric complex is blue, for example, and only a few ppm in a formulation can cause a noticeable discoloration. Between pH 7 and 9, chelating agents such as salts of diethylenetriaminepentaacetic acid or hydroxyethylene-diaminetriacetic acid are of some use at higher than theoretical amounts to inhibit coloration, if the chelating agent is added before the Sodium Pyrithione. The optimum scope of pH for the application of Sodium Pyrithione is between 7 and 10. Non-ionic detergents may partly inhibit the antimicrobial effectiveness of the compound.

Table 96. Minimum Inhibition Concentrations (MIC) of Sodium Pyrithione in Nutrient Agar (Wallhäusser, 1984)

Test organism	*MIC (mg/litre)*
Aspergillus niger	2
Penicillium notatum	2
Trichophyton mentagrophytes	0·5
Candida albicans	4
Escherichia coli	8
Pseudomonas aeruginosa	512
Salmonella typhimurium	64
Staphylococcus aureus	1
Streptococcus faecalis	2

11.1.4 Zinc salt of 2-mercaptopyridine-*N*-oxide — Zinc-bis-(2-pyridine-thiol-1-oxide) — Zinc Pyrithione

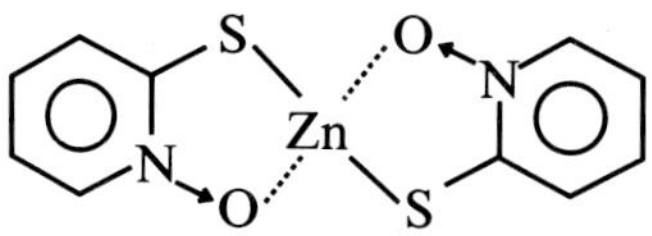

$C_{10}H_8N_2O_2S_2Zn$ M_r 317·71
CAS-No. 3138-01-0

Chemical and physical properties

Appearance	White to yellow crystalline powder of a mild odour	
Melting point	240°C (decomposition)	
Solubility (g/litre) at 20°C	In water	0·02
	acetone	0·7
	ethanol	0·1
	polyethylene glycol	2·0
	DMF	80·0
pH (10% solution in water)	6·5	
Stability	Stable between pH 4 and 8; below pH 3·5 conversion to free Pyrithione which in its undissociated form is sparingly soluble in water and precipitates; above pH 8 Zinc Pyrithione is converted to highly soluble alkali salts (e.g. Section 11.1.3); transchelation and coloration in the presence of heavy metal ions. Sensitive to strong oxidizing and reducing agents and light (see Section 11.1.3).	

Toxicity data

LD_{50} oral	180–200 mg/kg rat
	300 mg/kg mouse

Irritant to skin and mucous membranes.
No sensitization.

Reproduction and teratology studies were performed by Nolen & Dierkman (1979).

Antimicrobial effectiveness/applications

Zinc Pyrithione is mainly used as a preservative in cosmetics; the addition rates vary between 0·025 and 0·1%. In the EC list of preservatives allowed the microbicide is mentioned with a maximum authorized concentration of 0·5%, however, with the limitation authorized only in products rinsed off, forbidden in products for oral hygiene. Another application for Zinc Pyrithione is the in-can/in-tank protection of metal working fluids, polymer emulsions, latex paints, etc., especially against mould growth. Non-ionic detergents can reduce the antimicrobial activity of Zinc Pyrithione.

Table 97. Minimum Inhibition Concentrations (MIC) of Zinc Pyrithione in Nutrient Agar

Test organism	*MIC (mg/litre)*
Alternaria alternata	7·5
Aspergillus niger	100
Aureobasidium pullulans	15
Chaetomium globosum	20
Cladosporium cladosporioides	5
Penicillium brevicaule	50
Sclerophoma pityophila	5
Trichoderma viride	50
Pseudomonas aeruginosa	400
Staphylococcus aureus	<10

11.1.5 2,2'-Dithio-bis(pyridine-*N*-oxide) — Bis-(2-pyridyl-*N*-oxide)-disulphide — Pyrithionedisulphide

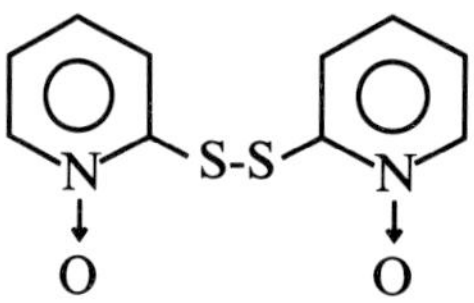

$C_{10}H_8N_2O_2S_2$ M_r 252·32
CAS-No. 3696-28-4

Chemical and physical properties

Appearance	Off-white to tan crystalline solid with a faint odour, slightly hygroscopic	
Melting point	205°C (decomposition)	
Solubility at 21°C	In water (pH 4·5)	11·3 g/litre
	alcohol	8·1 g/litre
	acetone	1·1 g/litre
	chloroform	28·1 g/litre
	hexane	<10 mg/litre
pH (solution in water)	Approx. 7	
Stability	Stable in mineral or acetic acid solutions (formation of soluble acid salts without cleavage of the sulphur–sulphur bond); in alkaline solutions slow cleavage of the disulphide linkage under release of the alkali salts of 2-mercaptopyridine-*N*-oxide (see Section 11.1.3) and to a certain extent Pyrithione sulphinate	

Toxicity data

LD_{50} oral	1640 mg/kg male rat
	1240 mg/kg female rat
	543 mg/kg mouse

Mildly irritating to the skin (rabbit); a 0·8% solution in water did not cause irritation on rabbit eyes.

Antimicrobial effectiveness/applications

The antimicrobial activity of the Pyrithione disulphide corresponds to the activity of Pyrithione (Section 11.1.3) which is released when the disulphide bond splits up; that means also the disulphide is especially toxic for mould producing fungi and Gram-positive bacteria and not very effective for Gram-negative bacteria.

It may be used as a preservative for the in-can/in-tank protection of aqueous functional fluids including jet fuel and cosmetics. Also with the application of Pyrithione disulphide coloration risks as described under Section 11.1.3 for Sodium Pyrithione have to be taken in consideration. The EC positive list of preservatives for cosmetics mentions Pyrithione disulphide for use in products rinsed off with a maximum authorized concentration of 0·2%. The addition rates for the preservative vary between 0·06 and 0·2%.

11.2 Pyridine-4-carboxylic acid hydrazide — Isonicotinic acid hydrazide — Isoniazide

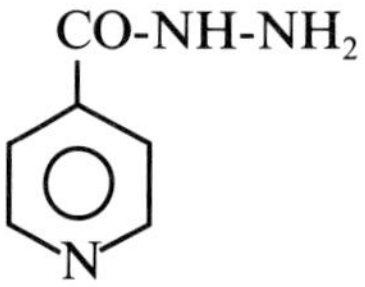

$C_6H_7N_3O$ M_r 137·14
CAS-No. 54-85-3

Chemical and physical properties

Appearance	Colourless crystals
Melting point	170–173°C
Solubility	Moderately soluble in water, soluble in polar solvents, practically insoluble in ether, benzene

More information is given in Section 8 ‘Amides’.

11.3 2,3,5,6-Tetrachloro-4-(methylsulphonyl)pyridine

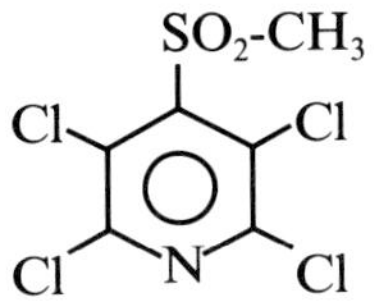

$C_6H_3Cl_4NO_2S$ M_r 294·28
CAS-No. 13108-52-6

Chemical and physical properties

Appearance	Off-white to fawn crystalline powder	
Melting point	152°C	
Solubility at 25°C	In water	0·025 mg/litre
	DMF	343 g/litre
	methanol	20 g/litre
	acetone	220 g/litre
	methylethyl ketone	180 g/litre
	dichloromethane	70 g/litre
	toluene	50 g/litre
	white spirit	< 5 g/litre
Stability	Stable and effective over a pH range of 4–8·5; hydrolysis in alkaline media; negligible degradation on heating to 290°C	

Toxicity data

LD_{50} oral 1·4 g/kg rat

Skin irritant, can cause skin sensitization.
Exotoxic — may inhibit sewage treatment processes; is dangerous to fish.

Antimicrobial effectiveness/applications

Tetrachloro-4-(methylsulphonyl)pyridine exhibits a broad spectrum of high antifungal activity and additionally antialgal efficacy. In accordance with its physical and chemical properties it is used as a paint film fungicide and algicide in aqueous and solvent based surface coatings requiring protection from biological growth, in PVC wall coverings, shower curtains, silicone sealants, polypropylene films and wood stains. The addition rates vary between 0·1 and 0·5%. Hydrolysis and inactivation of the active ingredient occurs at a pH in excess of 8·5 and in the presence of organic amines, e.g. morpholine, triethanolamine, diethanolamine; such conditions should be avoided.

Table 98. Minimum Inhibition Concentrations (MIC) of Tetrachloro-4-(methyl-sulphonyl)pyridine in Nutrient Agar

Test organism	*MIC (mg/litre)*
Alternaria alternata	7·5
Aspergillus flavus	50
Aspergillus niger	10
Aspergillus terreus	150
Aspergillus ustus	10
Aureobasidium pullulans	7·5
Chaetomium globosum	10
Cladosporium cladosporioides	5
Cladosporium herbarum	10
Coniophora puteana	<1
Paecilomyces variotii	10
Penicillium citrinum	10
Penicillium glaucum	10
Sclerophoma pityophila	5
Stachybotris chartarum	10
Trichoderma viride	50
Candida albicans	10
Candida krusei	5
Rhodotorula mucilaginosa	10
Saccharomyces bailii	5
Saccharomyces cerevisiae	50
Torula rubra	10
Torula utilis	50
Aerobacter aerogenes	500
Aeromonas punctata	150
Bacillus mycoides	100
Bacillus subtilis	200
Escherichia coli	5000
Pseudomonas aeruginosa	5000
Staphylococcus aureus	50

11.4 8-Hydroxyquinoline — Oxine

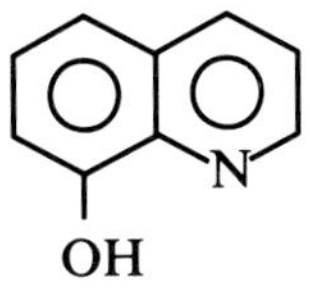

C_9H_7ON M_r 145·16
CAS-No. 148-24-3

Chemical and physical properties

Appearance	Almost colourless needles	
Melting point	73–75°C	
Boiling point	267°C	
Solubility (g/litre)	In water	0·52 at 18°C
		3·6 at 90°C
	Soluble in alcohol, acetone, trichloromethane, benzene, forms water soluble salts with mineral acids, e.g. HCl, H_2SO_4	
Stability	Light sensitive; chelating agent	

Toxicity data

LD_{50} oral	730 mg/kg rat
	1200 mg/kg guinea pig

Evaluations of Gocke *et al.* (1981) strengthen the suspicion that Oxine is mutagenic. It is irritant to skin and mucous membranes.

Ecotoxicity

Oxine is biodegradable (test performed according to OECD Guideline for Testing of Chemicals, 301 D), but toxic for fish.

Antimicrobial effectiveness/applications

The antimicrobial activity described here is only valid for 8-hydroxyquinoline and not for the seven other possible hydroxyquinolines which are widely inactive and do not, as does 8-hydroxyquinoline, chelate metal cations, e.g. copper, iron and zinc. Contrary to Oxine the corresponding chelation complexes, e.g. copper 8-hydroxyquinoline (Section 11.5) possess considerable lipid solubility enabling the complex to pass through the cell membrane and then to dissociate into the toxic 8-hydroxyquinoline. However, according to the findings of Block (1983) the chelation itself may already potentiate the antifungal activity of Oxine in addition to the resultant increase in lipid solubility.

As is demonstrated in Table 99, Oxine is a broad spectrum microbicide, although it is much more toxic for Gram-positive bacteria than for Gram-negative bacteria. But unfavourable solubility properties and coloration risks have prevented Oxine from becoming an important microbicide for material protection. More popular are the chelate complexes, above all copper 8-hydroxyquinoline (Section 11.5).

Table 99. Minimum Inhibition Concentrations (MIC) of Oxine in Nutrient Agar

Test organism	*MIC (mg/litre)*
Alternaria alternata	5
Aspergillus niger	50
Aureobasidium pullulans	10
Chaetomium globosum	10
Cladosporium cladosporioides	5
Coniophora puteana	7·5
Lentinus tigrinus	10
Penicillium glaucum	10
Polyporus versicolor	20
Sclerophoma pityophila	5
Trichoderma viride	50
Aerobacter aerogenes	15
Aeromonas punctata	15
Bacillus mycoides	35
Bacillus subtilis	7·5
Escherichia coli	100
Leuconostoc mesenterioides	75
Proteus mirabilis	100
Pseudomonas aeruginosa	500
Pseudomonas fluorescens	500
Staphylococcus aureus	20

11.5 Copper 8-hydroxyquinoline — Oxine copper

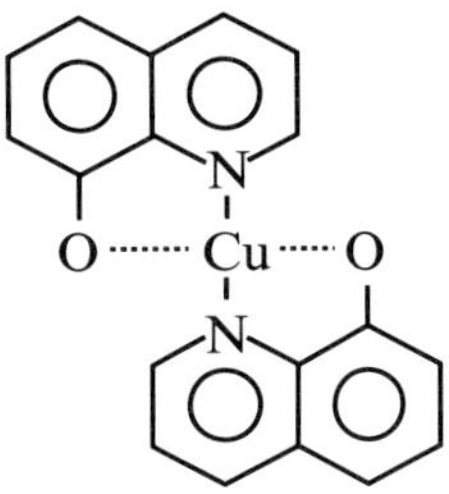

$C_{18}H_{12}CuN_2O_2$ M_r 351·85
CAS-No. 132-71-8

Chemical and physical properties

Appearance	Crystalline greenish powder; non-hygroscopic
Copper content	18%
Solubility	Insoluble in water, moderately soluble in xylene, trichloromethane, 0·4 g/ litre in 1,2-dichlorobenzene
Stability	Stable between pH 3 and 12; light stable; heat resistant up to 270°C

Toxicity data

LD_{50} oral 4500 mg/kg rat

Antimicrobial effectiveness/applications

Table 100. Minimum Inhibition Concentrations (MIC) of Oxine Copper in Nutrient Agar

Test organism	*MIC (mg/litre)*
Aspergillus flavus	2
Aspergillus niger	2
Aureobasidium pullulans	2
Chaetomium globosum	5
Penicillium citrinum	5
Penicillium funicolosum	2
Trichoderma viride	2

Oxine copper is distinctively more effective than Oxine (Section 11.4) and a little bit more active than the corresponding zinc chelate which offers, however, the advantage of being colourless. The activity spectrum of Oxine copper covers bacteria, yeasts, fungi and algae. Because of its high efficacy against wood staining and mould-producing fungi it is recommended for the protection of freshly sawn timber and wooden products, further for the protection of paper, cardboard, textile and plastic material, but only as far as the coloration caused through the incorporation of Oxine copper does not disturb. Another disadvantage of Oxine copper is the fact that the compound is not easily transferred into formulations appropriate for the homogeneous distribution of the active ingredient in the material to be protected from biodeterioration.

12 AZOLES

Azole fungicides represent a new class of microbicides with favourable toxicological property patterns and low ecotoxicity; some of them are suitable for the fungicidal treatment of various materials, especially wood and polymers.

The antimicrobial action of the azole derivatives described in this section is based primarily on the inhibition of ergosterol biosynthesis, e.g. in fungi, through blocking the biosynthetic reactions in the conversion of lanosterol to ergosterol (Kato *et al.*, 1986). More precisely, the azole fungicides act as inhibitors for C-14 demethylase which is responsible for the 14-α-demethylation of lanosterol leading to ergosterol (Fig. 32).

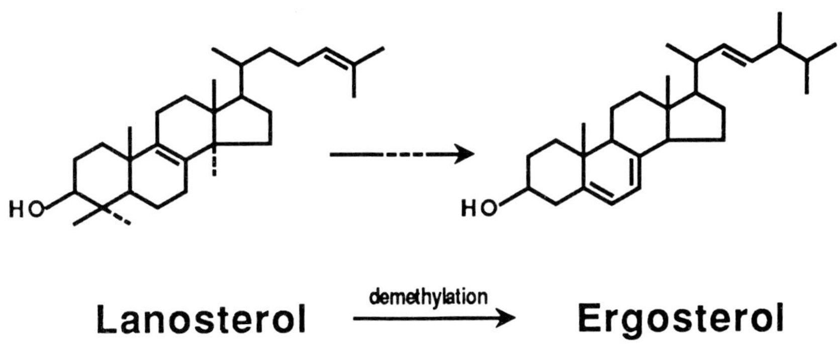

Fig. 32. Demethylation of lanosterol to ergosterol.

Azole derivatives with antifungal activity, e.g. those listed here, can be characterized as substituted aromatic heterocycles (imidazole, triazole) with an unsubstituted nitrogen atom at the *m*-position and one nitrogen atom bearing lipophilic groups including a benzene ring (see Fig. 33). They are able to attack cytochrome P-450 involved in the 14-α-demethylation of lanosterol. The presence of the *m*-nitrogen atom is absolutely necessary for the antifungal activity, respectively, the inhibition of C-14 demethylase. It is likely that the *m*-N atom of the aromatic heterocycles binds coordinatively to the protohaem iron of cytochrome

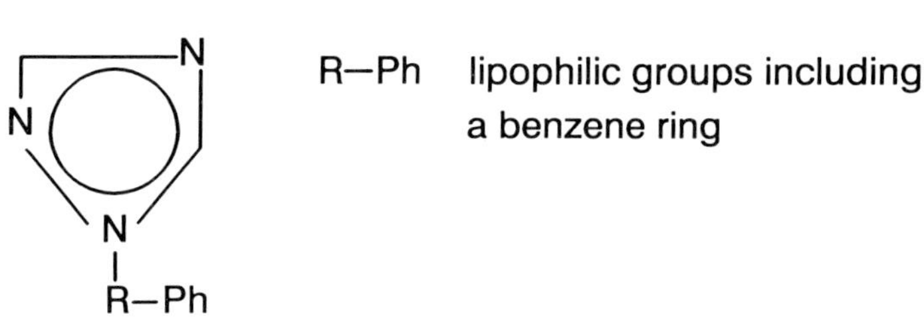

Fig. 33. Azole derivatives with antifungal activity.

Table 101. Minimum Inhibition Concentrations (MIC) (mg/litre) of three Azole Fungicides in Nutrient Agar

Test organism	*Tebuconazole*	*Azaconazole*	*Propiconazole*
Alternaria alternata	200	1000	350
Aspergillus niger	10	250	100
Aureobasidium pullulans	35	35	5
Cephaloascus fragrans	5	25	5
Ceratocystis pilifera	20	50	10
Chaetomium globosum	15	75	10
Cladosporium cladosporioides	50	35	
Gliocladium virens	75	350	35
Lentinus tigrinus	10	50	10
Penicillium glaucum	15	200	100
Phanerochaete sanguinea	10		50
Phialophora fastigiata	50	1000	100
Sclerophoma pityophila	5	5	1
Stereum sanguinolentum	5		15
Trichoderma viride	>1000	>1000	>1000

P-450. The lipophilic part of the azole inhibitor, which is structurally variable to a great extent, may bind to cytochrome P-450 at various positions through the formation of hydrophobic bonds and in some cases hydrogen bonds. As ergosterol is a principle sterol in fungi and an indispensable compound in the membrane structures, the azole inhibitors affect membrane permeability; at lower inhibitor concentrations this ultimately results in retardation of fungal growth. At higher concentrations the azoles act fungicidally; they disrupt fungal membranes immediately, because they probably act as cationic surface active ingredients (membrane-active microbicides) (Kato *et al.*, 1986).

They were developed as fungicides for the protection of plants and seeds. Not only their comparatively high efficacy (which permits lower concentrations in application), but also their selectivity and generally very good environmental acceptance are undeniable assets. At first, however, the azole active ingredients were not regarded as interesting fungicides for the protection of materials because of their unbalanced effect spectrum (see Table 101). But in view of a demand for microbicides of very low toxicity and an activity spectrum graduated according to the intended use it became apparent that there were fields of application for which the azole fungicides' limited effect spectrum seemed to be tailored. That applies to the field of wood preservation: here the azole fungicides can perform partial functions, namely the control of wood-decaying fungi. Other fungicides which have also a limited effect spectrum but on the other hand an extremely high degree of efficacy, e.g. dichlofluanide, are able to control the wood-staining fungi. And since a new generation of insecticides has also been developed, it is today possible to formulate wood preservatives on the basis of active substances of low toxicity and ecotoxicity. These modern wood preservatives are able to replace those of doubtful toxic and ecotoxic characteristics, especially formulations with pentachlorophenol or organotin compounds.

12.1 α-[2-(4-Chlorophenyl)ethyl]-α-(1,1-dimethylethyl)-1H-1,2,4-triazolyl-(1)-ethanol — α-tert.Butyl-α(*p*-chlorophenylethyl)-1H-1,2,4-triazole-1-ethanol — Tebuconazole

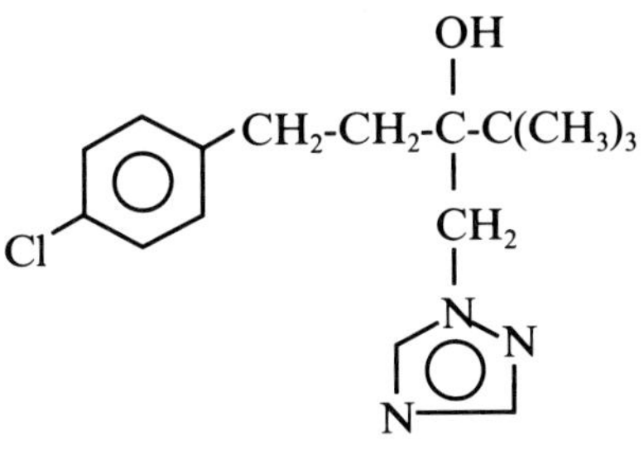

$C_{16}H_{22}ClN_3O$ M_r 307·83
CAS-No. 107.534-96-3

Chemical and physical properties

Appearance	White to yellow crystalline powder	
Melting point	103–105°C	
Vapour pressure (hPa)	$7{\cdot}2 \times 10^{-9}$ at 20°C	
	$4{\cdot}6 \times 10^{-6}$ at 60°C	
Solubility (g/litre) at 20°C	In water	0·032
	ethylene glycol	270
	propylene glycol	50
	white spirit	3
	xylene	40
Stability	Stable in acid and alkaline solutions; light stable; heat resistant up to 320°C according to differential thermal analysis	

Toxicity data

LD_{50} oral	3900 mg/kg rat
dermal	>5000 mg/kg rat
LC_{50} (inhalation, 4 h)	>5000 mg/m^3 air for rats

Not irritant to skin and mucosa, not sensitizing.

Several tests conducted in different biological systems demonstrated lack of mutagenicity, embryotoxicity or teratogenicity.

Ecotoxicity

LC_{50} for *Leuciscus idus melanotus*	8·7 mg/litre/96 h
Salmo gairdneri	6·4 mg/litre/96 h
Daphnia magna	25·0 mg/litre/24 h

Antimicrobial effectiveness/applications

Tebuconazole is a fungicide with some gaps in its activity spectrum (see Table 101), but together with other microbicides filling the gaps in the spectrum of effectiveness it may be used for the fungicidal treatment of materials as described by Büchel *et al.* (1988). Small amounts of BCM (Section 9.4) are able to fill the Trichoderma gap. Evaluating a test report of Gründlinger & Exner (1990) it will be noticed that Tebuconazole is particularly effective against wood decaying fungi, but less effective against blue-staining fungi. It has been demonstrated that Tebuconazole is resistant to both leaching and evaporation and therefore is suitable for long lasting protection of wood against decay fungi.

12.2 1-[(2-(2',4'-Dichlorophenyl)-4-propyl-1,3-dioxolan-2-yl-methyl]-1H-1,2,4-triazole — Propiconazole

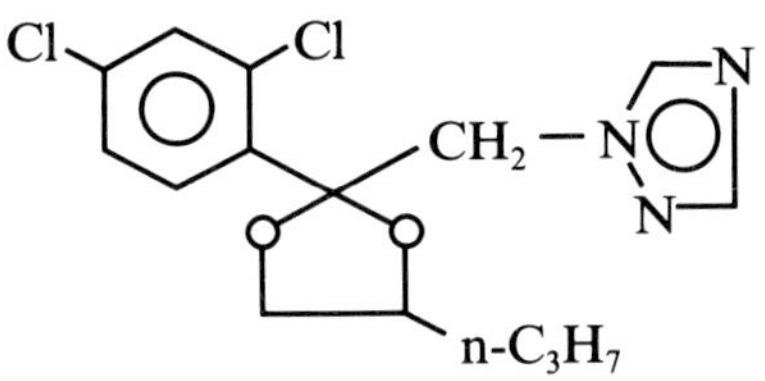

$C_{15}H_{17}Cl_2N_3O_2$

M_r 342·24
CAS-No. 60207-90-1

Chemical and physical properties

Appearance	Yellow, viscous liquid
Boiling point at 0·13 hPa	180°C
Density at 20°C	1·27 g/ml
Vapour pressure (hPa)	$1 \cdot 3 \times 10^{-6}$ at 20°C
	$1 \cdot 5 \times 10^{-5}$ at 40°C
Viscosity (mPa s)	74·300 at 20°C
	9·100 at 35°C
	1·400 at 50°C
Volatility	0·019 mg/m³ at 20°C
pK_a	Approx. 1
Flash point	>150°C
Ignition temperature	430°C
Solubility (g/litre)	In water 0·11
	acetone >500
	ethylene glycol 260
	propylene glycol >500
	toluene >500
	mineral spirit >100
Stability	Stable in acid and alkaline solutions; heat and light stable

Toxicity data

LD_{50} oral	1517 mg/kg rat
dermal	>4000 mg/kg rat
LC_{50} (inhalation, 4 h)	>5800 mg/m³ air for rats

Slightly irritant to skin and eyes (rabbits), not sensitizing. In sub-chronic, chronic, oncogenicity and reproduction studies, performed in rats, mice, dogs and rabbits, propiconazole proved to be a safe chemical under normal use conditions. Several mutagenicity tests, conducted in different biological systems, demonstrated lack of genetic effects (Valcke, 1989).

Ecotoxicty

LC_{50} for Brown trout	11·5 mg/litre/48 h
Daphnia magna	20·0 mg/litre/96 h

Antimicrobial effectiveness/applications

The activity spectrum of Propiconazole which is demonstrated by the MICs in Table 101, is similar to that of Tebuconazole (Section 12.1). Although Propiconazole shows a somewhat better water solubility and higher vapour pressure than Tebuconazole it may according to extensive tests performed by Valcke (1989) be regarded as a most interesting wood preserving fungicide, being resistant to leaching and evaporation stresses and having as good stability in treated wood and in formulations. A more thorough analysis of the test results (Gründlinger & Exner, 1990; Valcke, 1989) discloses that Tebuconazole and Propiconazole differ in their effect on differing wood decaying fungi but complement each other: Tebuconazole is more effective against *Gloephyllum trabeum* and Propiconazole more effective against species of Poria.

12.3 1-[2-(2,4-Dichlorophenyl)-1,3-dioxolan-2-yl-methyl]-1H-1,2,4-triazole — Azaconazole

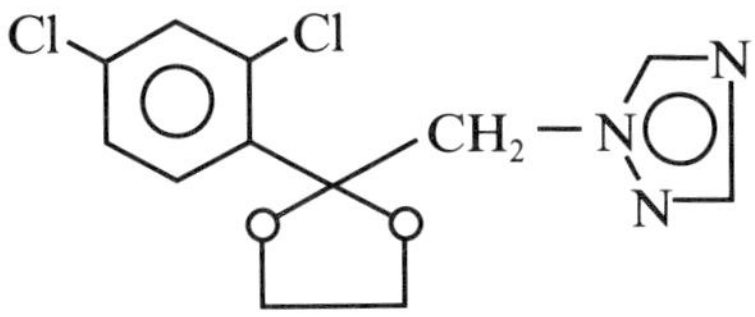

$C_{12}H_{11}Cl_2N_3O_2$

M_r 300·14
CAS-No. 60207-31-0

Chemical and physical properties

Appearance	White to brownish powder	
Melting point	100–110°C	
Boiling point	450°C	
Vapour pressure (hPa)	$5{\cdot}3 \times 10^{-9}$ at 20°C	
	$2{\cdot}7 \times 10^{-6}$ at 70°C	
Solubility (g/litre)	In water	0·3
	methanol	158
	n-propanol	36
	ethylene glycol	35
	propylene glycol	66
	acetone	156
	toluene	111
	xylene	45
	cyclohexanone	345
Stability	Stable up to 220°C; stable in acid and alkaline solutions (pH 1–9) when stored at temperatures up to 40°C	

Toxicity data

LD_{50} oral		308 mg/kg rat;
		1123 mg/kg mouse
	dermal	>2500 mg/kg rat

Moderately irritant to skin and mucosa.

In sub-chronic, chronic, oncogenicity and reproduction studies performed in rats, mice, dogs and rabbits, Azaconazole proved to be a very safe chemical. Several mutagenicity tests conducted in different biological systems demonstrated the lack of genetic effects.

Ecotoxicity

LC_{50} for Bluegill sunfish	18–23 mg/litre/96 h
Brown trout	22 mg/litre/96 h
Daphnia magna	86 mg/litre/48 h

Antimicrobial effectiveness/applications

According to Valcke & Goodwine (1985) Azaconazole is effective against *Basidiomycetes* and staining fungi (blue stain, sapstain and mould species). Because of its stability Azaconazole may be formulated in aqueous and solvent based wood preservatives or into emulsion concentrates to be used for solid wood and modified wood out of ground contact. However, in the meantime the other azole derivatives, Tebuconazole (Section 12.1) and Propiconazole (Section 12.2) are regarded as the more interesting fungicides for use in wood preservatives.

12.4 1-[2-(2,4-Dichlorophenyl)-2-(2-propenyloxyethyl)]-H-imidazole — Imazalil

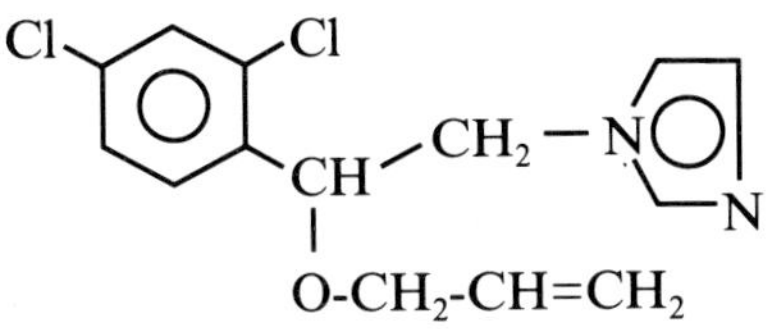

$C_{14}H_{14}Cl_2N_2O$ M_r 297·18
CAS-No. 35554-44-0

Chemical and physical properties

Appearance	Slightly yellow to brown solidified oil
Density at 23°C	1·243 g/ml
Vapour pressure at 20°C	$9{\cdot}3 \times 10^{-4}$
Refractive index	n_D^{20} 1·5643
Melting point	50°C
Solubility	Poor solubility in water (approx. 1 g/litre), very soluble in organic solvents
Stability	Stable to temperatures up to 285°C; stable to hydrolysis in dilute acids and alkalis at room temperature, in the absence of light

Toxicity data

LD_{50} oral	320 mg/kg rat;
	>640 mg/kg dog
dermal	4200–4880 mg/kg rat
on inhalation	>16 g/m^3 for rats

Causes irreversible eye damage.
No indication of teratogenicity in rats.
Toxic to fish.

Antimicrobial effectiveness/applications

The azole fungicide Imazalil is listed here for completeness sake. It is used for the protection of citrus fruits against mould infestation.

Table 102. Minimum Inhibition Concentrations (MIC) of Imazalil in Nutrient Agar

Test organism	*MIC (mg/litre)*
Alternaria alternata	50
Aspergillus niger	50
Aureobasidium pullulans	5
Chaetomium globosum	10
Cladosporium cladosporioides	50
Coniophora puteana	50
Lentinis tigrinus	50
Penicillium digitatum	1
Penicillium glaucum	100
Penicillium italicum	1
Polyporus versicolor	100
Rhizopus nigricans	200
Sclerophoma pityophila	5
Trichoderma viride	500

13 HETEROCYCLIC N,S COMPOUNDS

These are cyclic organic microbicides containing N and S atoms in the ring system, e.g. thiazole or isothiazole derivatives.

thiazole

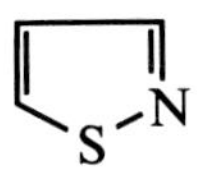

isothiazole

The heterocyclic N,S compounds Dazomet (Section 3.3.16) and Taurolin (Section 3.5.2.) can be regarded as formaldehyde releasing compounds; therefore they are listed in Section 3. The *N*-hydroxymethyl derivative of 2-mercaptobenzothiazole (Section 3.4.10.2) is a formaldehyde releasing compound too and is described under Section 3.4, 'Reaction products of amides and formaldehydes'.

13.1 2-Methyl-3(2H)-isothiazolone — 2-Methyl-4-isothiazolin-3-one (MIT)

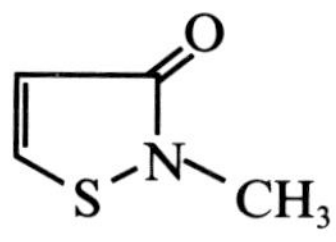

C_4H_5NOS M_r 115·16
CAS-No. 2682-20-4

Chemical and physical properties

Appearance	Colourless, extremely hygroscopic crystals
Melting point	50–51°C
Boiling point at 0·04 hPa	93°C
Solubility	30 g/litre H_2O; highly soluble in organic solvents
Stability	At exposure to air conversion into an oily compound; decomposition on heating starts at 55°C; reacts under inactivation with sodium bisulphite, SH compounds, amines

For toxicity data, antimicrobial effectiveness and application see Section 13.3.

Biodegradation in river water (indications of Rohm & Haas Company, Philadelphia, USA).

Concentration tested	0·6 mg/litre.	
Active ingredient remaining after	2 days	98%
	4 days	52%
	7 days	28%
	14 days	0%

13.2 5-Chloro-2-methyl-3(2H)-isothiazolone — 5-Chloro-2-methyl-4-isothiazolin-3-one (CMIT)

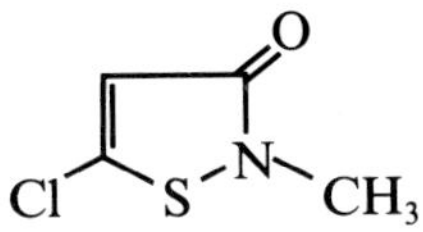

C_4H_4ClNOS M_r 149·60
CAS-No. 26172-55-4

Chemical and physical properties

Appearance	Slightly yellow crystals
Melting point	54–55°C
Solubility	Approx. 5 g/litre H_2O; moderately soluble in organic solvents
Stability	Decomposition on heating starts at 55°C; sensitive to sodium bisulphite, SH compounds, amines, alkalines (pH > 8)

Toxicity data, antimicrobial effectiveness and applications see Section 13.3. Strongly irritant to skin and mucous membranes; potent skin sensitizer.

Biodegradation in river water (indications of Rohm & Haas Company, Philadelphia, USA).

Concentration tested	0·6 mg/litre.	
Active ingredient remaining after	2 days	92%
	4 days	83%
	7 days	72%
	14 days	61%
	35 days	25%

13.3 Mixture of 5-Chloro-2-methyl-4-iosothiazolin-3-one and 2-methyl-4-isothiazolin-3-one (Sections 13.1 and 13.2)

Such mixtures of 2-substituted (3(2H))-isothiazolones are formed in high yield by cyclization of *N*-methyl-3-3'-dithiopropioamide induced through chlorine or sulphurylchloride (Lewis *et al.*, 1971).

Chemical and physical properties

Appearance	Amber liquid of mild odour
Total active ingredient	13·9% (typical value) (3·8% MIT + 10·1% CMIT)
Inert ingredients	86·1% (min. 55% water, stabilizers, e.g. magnesium salts)
Solubility	Miscible with water and alcohols, limited solubility in aliphatic and aromatic solvents
Density at 20°C	1·25–1·30 g/ml
pH	2–5
Stability	Decomposition on heating starts at 55°C; sensitive to amines, SH compounds, alkalis (pH > 8); strong reducing or oxidizing agents, sodium bisulphites deactivate rapidly; heat resistance is considerably increased in the presence of phenoxyalkanol (Willingham & Mattox, 1990). Thermal or storage degradation is also reduced by the addition of formaldehyde, formaldehyde-releasers or metal salts

Toxicity data

LD_{50} oral	457 mg/kg albino rat
dermal	660 mg/kg albino rabbit

Acute inhalation: LC_{50} for rat 2·36 mg/litre aerosol (exposure: 4 h at 23°C).
Severe skin irritant, corrosive to the eye.
Skin sensitization possible, especially at concentrations greater than 15 mg active ingredients per litre.

MWC for a 3/1 mixture of CMIT/MIT: 0·5 mg/m^3.

Ecotoxicity

LC_{50} for Bluegill sunfish	0·54 mg active ingredients/litre (exposure: 6 days)

For biodegradation see Sections 13.1 and 13.2.

Antimicrobial effectiveness/applications

3-(2H)-isothiazolones are electrophilic active ingredients containing an activated N-S bond which enables them to react with nucleophilic cell entities, thus exerting antimicrobial activity. The 5-chloro-2-methyl-3(2H)-isothiazolone (Section 13.2) contains additionally a vinyl activated chloro atom and can therefore be characterized as a molecule with two toxophoric structural elements (Paulus, 1988). As a result CMIT should have greater antimicrobial efficacy than the halogen-free MIT. This is indeed the case, as shown by Schmitt (1987) who found in a comparison test considerably lower minimum inhibition concentrations of CMIT for fungi and bacteria (see Table 103).

Table 103. Minimum Inhibition Concentrations (MIC) of CMIT and MIT in Nutrient Agar

Test organism	*MIC (mg/litre)*	
	CMIT	*MIT*
Aspergillus niger	<20	500
Chaetomium globosum	<20	500
Penicillium glaucum	<20	500
Escherichia coli	<20	35
Staphylococcus aureus	<20	50

The isothiazolinone mixture (Section 13.3) contains the chloro compound and non-chlorinated compound in a ratio of approx. 2·7:1 which is an optimum with regard to the antimicrobial effectiveness; higher concentrations of the non-chlorinated compound in such mixtures decrease the antimicrobial activity considerably. The mixture of active ingredients in Section 13.3 can be really considered as a broad spectrum microbistat. There is no significant difference in activity against Gram-positive and Gram-negative bacteria. The MIC for bacteria, fungi and yeasts range between 2·5 and 10 mg active ingredients per litre. The algistatic concentrations for green and blue–green algae and diatoms vary between 0·1 and 1·0 mg a.i./litre. At concentrations 5–10 times higher than the microbistatic concentrations the mixture of active ingredients exhibits also microbicidal effectiveness if only slowly (incubation period 24 h).

With regard to the antimicrobial activity it is not surprising that the mixture of isothiazolinones has become one of the most important preservatives for the in-can/in-tank protection of aqueous functional fluids, including detergents and cosmetics. A big advantage of the active ingredient mixture is the fact that it does not show a Pseudomonas gap. The EC positive list of preservatives for cosmetics mentions the mixture with a maximum authorized concentration of 0·003% (30 mg/litre) active ingredients, although with regard to the sensitization potential of mainly 5-chloro-2-methyl-4-isothiazolin-3-one it its nowadays recommended not to use concentrations higher than 10–15 mg/litre. The compatibility with non-ionic, anionic and cationic compounds is faced with the lack of heat resistance and limited stability, above all in the presence of amines, SH compounds and alkalis. Especially the main active ingredient, CMIT, is due to its chemical nature, sensitive to hydrolysis in media with pH values > 8. Riha *et al.* (1990) have found that the use of Cu^{2+} with CMIT may effectively reduce the required dose of CMIT and minimize its toxicological potential; apparently Cu^{2+} protects CMIT from nucleophilic attack in the extracellular environment and additionally synergistically enhances its activity at the cellular level. The properties of the isothiazolinone mixture have also made possible its successful use as a water treatment microbicide.

13.4 2-*n*-Octyl-4-isothiazolin-3-one (OIT)

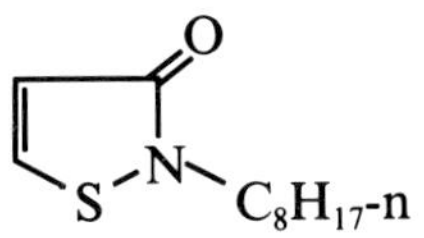

$C_{11}H_{19}NOS$

M_r 213·34
CAS-No. 26530-20-1

Chemical and physical properties

Appearance	The a.i. is normally available as a concentrate in organic solvents, e.g. 45–48% in propylene glycol (amber liquid)
Solubility of the a.i.	0·48 g/litre water at 25°C; infinitely soluble in methanol, ethanol, propylene glycol, acetone, ethyl ether, ethyl acetate, chloroform, corn and mineral oil; approx. 250 g/litre toluene; insoluble in heptane
Vapour pressure (hPa)	$2{\cdot}7 \times 10^{-7}$ at 20°C $5{\cdot}3 \times 10^{-6}$ at 40°C
Stability	Inactivation by ammonia, primary and secondary amines increasing with increasing pH; strong reducing or oxidizing agents such as hypochlorites, bisulphites, metabisulphites, H_2S, may also effect the efficacy of the active ingredient. Where possible zinc oxide may be used to increase the stability of OIT in the presence of amines

Toxicity data

LD_{50} oral	270 mg/kg rat
dermal	311 mg/kg rabbit
Primary skin irritation	
45% a.i. solution	severely irritant
Primary eye irritation	
45% a.i. solution	Corrosive
5% a.i. solution	Slightly to moderately irritant
0·5% a.i. solution	Not irritant

Skin sensitizer.
Non-teratogenic, non-mutagenic, non-carcinogenic.

Ecotoxicity

LC_{50} for fish (exposure 96 h):	(mg a.i./litre)
Bluegill sunfish (*Lepomis macrochirus*)	0·196
Rainbow trout (*Salmo gairdneri*)	0·065
Chamel catfish (*Ictaluras punctatus*)	0·177

Biodegradation — In both river and pond water at an active level of 0·1 mg/litre, 90% degradation is achieved within 7 days. At higher concentrations of 0·5 and 1·0 mg/litre, 14 days were required to reach the 90% degradation level

Antimicrobial effectiveness/applications

Table 104. Minimum Inhibition Concentrations (MIC) of 2-*n*-Octyl-4-isothiazolin-3-one in Nutrient Agar

Test organism	*MIC of a.i. (mg/litre)*
Aspergillus niger	5–10
Aureobasidium pullulans	0·5
Chaetomium globosum	10
Penicillium glaucum	2·5
Rhizopus nigricans	75
Candida albicans	2·5
Rhodotorula rubra	5
Aerobacter aerogenes	50
Aeromonas punctata	50
Bacillus mycoides	15
Bacillus subtilis	10
Escherichia coli	125
Leuconostoc mesenterioides	20
Proteus mirabilis	100
Pseudomonas aeruginosa	500
Staphylococcus aureus	10

OIT is also effective against algae; it prevents algal growth at concentrations of 0·5–5·0 mg a.i./litre.

In accordance with its extraordinary high toxicity for fungi the active ingredient is mainly used as a fungicide, e.g. for paint film protection, in non-film forming decorative wood stains, in the leather industry for the protection of wet blues, in adhesives and sealants, pulp, paper and cardboard, etc. Surface coatings treated with the fungicide may lose mould resistance when exposed to leaching with water because of the still relatively high water solubility of OIT. Nevertheless the length of a straight chain alkyl group at the N-atom of the isothiazolinone ring has a significant effect on water solubility. The substitution of the methyl group in MIT (Section 13.1) by the *n*-octyl group causes approx. 100-fold decrease in water solubility. In the search for a non-leachable *N*-alkyl–isothiazolinone, OIT represents an optimum with regard to antimicrobial activity and poor water-solubility.

13.5 4,5-Dichloro-2-(*n*-octyl)-4-isothiazolin-3-one (DCOIT)

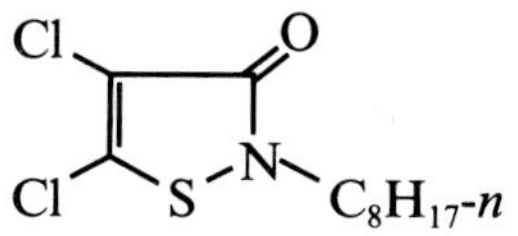

$C_{11}H_{17}Cl_2NOS$

M_r 282·24
CAS-No. 64359-81-5

The development of DCOIT is based on the findings of Miller & Lovegrove (1980) who systematically examined the effect of substitution on the isothiazolinone ring with regard to antimicrobial activity and water solubility. This investigation has shown DCOIT to be the preferred candidate, having high biological activity and very low water-solubility (2 ppm). DCOIT is more a biocide than a microbicide; due to the fact that the molecule contains several toxophoric structural elements — a reactive N-S bond, an activated chloro atom in the α-position to a (electronegative) carbonyl group, a vinyl activated chloro atom — it is toxic for fungi, yeasts, bacteria, algae, sea animals and insects.

Chemical and physical properties (30% solution of DCOIT in xylene)

Appearance	Yellow fluid
Boiling point	157–174°C
Solubility	Highly soluble in organic solvents; water solubility of the a.i.: 2 mg/litre
Stability	The reactive a.i. is unstable in alkaline media and sensitive to amines

Toxicity data

LD_{50} oral	1·89 g/kg rat
LC_{50} on inhalation	0·72 mg/litre for rats (exposure: 4 h)

Strong skin sensitizer.

Irritant to skin and mucous membranes.

Antimicrobial effectiveness/applications

DCOIT is a broad spectrum biostat. Its very low water-solubility is in favour of the applications which are preferably recommended: a.i. in coatings, especially antifouling coatings and in wood preservatives.

Table 105. Minimum Inhibition Concentrations (MIC) of a 30% DCOIT Solution in Nutrient Agar

Test organism	*MIC (mg/litre)*
Alternaria alternata	35
Aspergillus niger	20
Aureobasidium pullulans	150
Chaetomium globosum	20
Cladosporium cladosporioides	20
Lentinus tigrinus	75
Penicillium glaucum	50
Sclerophoma pityophila	100
Trichoderma viride	350
Slime organisms[a]	0·75
Algae	~5

[a] According to Kato & Fukumura (1962).

13.6 1,2-Benzisothiazolin-3-one (BIT)

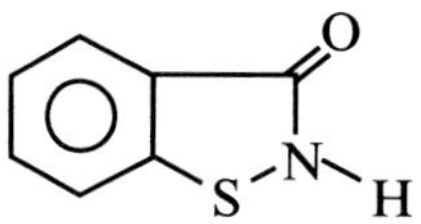

C_7H_5NOS M_r 151·19
CAS-No. 2634-33-5

Chemical and physical properties (33% aqueous dispersion)

Appearance	Brown, pourable dispersion
Viscosity at 25°C	2·5 mPa s
Density at 25°C	1·15 g/ml
pH at 25°C	7·0
Boiling point	100°C
Melting pointy of the a.i.	157–158°C
Solubility	The a.i. (BIT) is soluble in hot water; it forms water soluble salts with alkalis and amines and is highly soluble in organic solvents, preferably in alcohols and glycols
Stability	Comparatively heat resistant, stable over a wide pH range (3–11)

Toxicity data

LD_{50} oral	Approx. 3000 mg/kg rat

Skin sensitizer.
Corrosive to the skin and mucous membranes.

Ecotoxicity

LC_{50} for Rainbow trout	4·2 mg/litre (exposure: 96 h)
for Bluegill sunfish	15·3 mg/litre (exposure: 96 h)

2–10 ppm BIT do not significantly disturb the activity of biological sewage plants.

Antimicrobial effectiveness/applications

Preservatives based on BIT may be used for the in-tank/in-can protection of polymer emulsions, latex paints, pigment and filler slurries, paper coatings, adhesives and other aqueous functional fluids. Because of the heat stability and non-volatility of BIT the preservatives can be incorporated in fluids which are still hot. BIT exhibits no headspace activity. It is compatible with non-ionic and anionic compounds and active in acid and alkaline media.

Table 106. Minimum Inhibition Concentrations (MIC) of a 20% Alkaline Aqueous BIT Solution

Test organism	*MIC (mg/litre)*
Alternaria alternata	200
Aspergillus niger	1000
Aureobasidium pullulans	500
Chaetomium globosum	500
Cladosporium cladosporioides	100
Lentinus tigrinus	500
Penicillium glaucum	500
Sclerophoma pityophila	350
Trichoderma viride	>1000
Aerobacter aerogenes	100
Aeromonas punctata	100
Bacillus mycoides	75
Bacillus subtilis	100
Escherichia coli	500
Leuconostoc mesenterioides	100
Proteus mirabilis	500
Pseudomonas aeruginosa	>1000
Pseudomonas fluorescens	500
Staphylococcus aureus	500

Because of the sensitization properties of BIT one should not use it for the protection of functional fluids coming in contact with the human skin, e.g. cosmetics, metal working fluids. Using BIT based preservatives one has to bear in mind that for the control of Pseudomonades and some species of fungi (e.g. *Aspergillus niger*) one needs higher addition rates than needed for other microbe species.

BIT based preservatives:

9·5% BIT in an aqueous propylene glycol solution
20·0% BIT in an aqueous dipropylene glycol solution
33·0% BIT in an aqueous ethylene diamine solution.

13.7 2-Methyl-4,5-trimethylene-4-isothiazolin-3-one

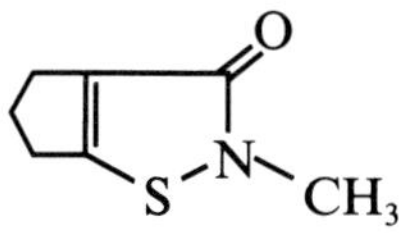

C_7H_9NOS M_r 155·22
CAS-No. 82633-79-2

Chemical and physical properties (5% aqueous solution)

Appearance	Pale yellow liquid
Freezing point	−2°C
Flash point	Boils without flashing
Stability	The solution is stable under normal storage conditions; the a.i. starts to decompose at 216°C; it is relatively stable in aqueous solutions of ammonium hydroxide of pH 9, that means significantly more stable than CMIT (Section 13.2) but less stable than BIT (Section 13.6)

Toxicity data

Causes irritation following eye contact. Potent skin sensitizer, however, according to assessments of Botham *et al.* (1990) the sensitizing potency of the a.i. is lower than that of CMIT (Section 13.2).

Antimicrobial effectiveness/applications

Table 107. Minimum Inhibition Concentrations (MIC) of the Active Ingredient in Nutrient Agar

Test organism	*MIC (mg/litre)*
Escherichia coli	5
Pseudomonas aeruginosa	25
Pseudomonas putida	25
Staphylococcus aureus	25

The a.m. solution offers a broad spectrum of antibacterial activity which includes Pseudomonades. Fungal growth is controlled by comparatively high concentrations of 2-methyl-4,5-trimethylene-4-isothiazolin-3-one.

As a liquid product the 5% aqueous formulation of 2-methyl-4,5-trimethylene-4-isothiazolin-3-one offers a number of processing advantages when the product is used for the in-can/in-tank protection of aqueous functional fluids, preferably paints, adhesives and polymer emulsions. In these applications the remarkable water-solubility of the a.i. is a special advantage with an addition rate between 0·05 and 0·2% of the 5% preparation.

13.8 2-(4-Thiazolyl)-benzimidazole — Thiabendazole (TBZ)

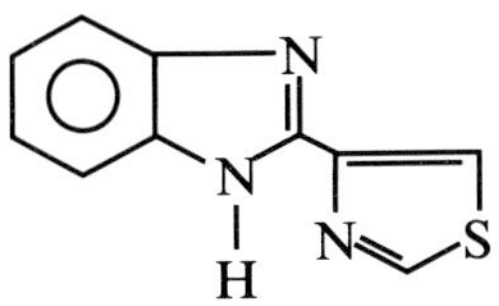

$C_{10}H_7N_3S$ M_r 201·25
CAS-No. 148-79-8

Chemical and physical properties

Appearance	Off-white, free flowing powder	
Melting point	304–305°C (sublimes)	
Density	1·44 g/ml	
Solubility (g/litre)	In water	0·03
	water of pH 2·2	38·4
	acetone	2·8
	benzene	2·3
	DMF	39·0
	DMSO	80·0
	dioxane	4·5
	ethanol	6·8
	ethylene glycol	7·7
	methanol	9·3
Stability	Heat resistant, stable in acid and alkaline media	

The hypophosphite salt of TBZ is highly soluble in water and may be used for the formulation of TBZ concentrates. Typical concentrate: 26·6% TBZ-hypophosphite salt (corresponds to 20% TBZ) in water (pH 1·7).

Toxicity data

LD_{50} oral 3·1–3·8 g/kg rat, rabbit or mouse

Good skin compatibility.

Further studies on the toxicologic and pharmacologic properties of TBZ were performed by Robinson *et al.* (1965), Franz (1963) and Mizutani *et al.* (1990).

Antimicrobial effectiveness/applications

TBZ is primarily fungistatic and of special interest because of its effectiveness against *Penicillium digitatum, Penicillium italicum, Diplodia natalensis* and *Gloeosporium* species which led to its use for the protection of citrus fruits and bananas. It has also been proposed to use TBZ as a fungicide in paint compositions, in adhesives, sealants, paper (e.g. soap wrappers), cardboard, fibreboard, fabric, leather goods, etc., where one can profit from the advantages of TBZ: poor water-solubility, non-volatility, stability, heat resistance and low toxicity. However, there are deep gaps in the activity spectrum of TBZ (see Table 108) similar to those of BCM (Section 9.4). It is therefore recommendable to combine TBZ with other fungicides, e.g. Ziram (Section 9.10.2), Zinc Pyrithione (Section 11.1.4), Dichlofluanide (Section 14.5), Fluorfolpet (Section 14.2) and others. Specific mixtures were proposed by Brake (1974).

Table 108. Minimum Inhibition Concentrations (MIC) of TBZ in Nutrient Agar

Test organism	*MIC (mg/litre)*
Alternaria alternata	>5000
Aspergillus niger	40
Aureobasidium pullulans	1
Chaetomium globosum	1
Cladosporium herbarum	<1
Coniophora puteana	100
Diplodia natalensis	3
Lentinus tigrinus	>5000
Penicillium digitatum	50
Penicillium glaucum	<1
Penicillium italicum	50
Polyporus versicolor	20
Trichoderma viride	10
Candida albicans	>1000
Candida krusei	>1000
Rhodotorula mucilaginosa	35
Sporobolomyces roseus	75
Torula rubra	150

13.9 2-Mercaptobenzothiazole (MBT) — 2-Benzothiazolthiol (A) ⇌ Benzothiazolin-2-thione (B)

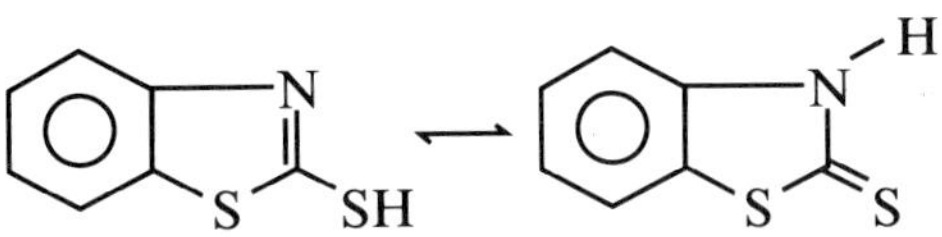

A B

$C_7H_5NS_2$ M_r 167·26
CAS-No. 149-30-4

Chemical and physical properties

Appearance	Slightly yellow, crystalline powder
Melting point	177–181°C
Density	1·42–1·52 g/ml
Solubility	Sparingly soluble in water and acids; soluble in alkalis, highly soluble in acetone, moderate solubility in alcohol, ether, benzene

Characteristics (50% sodium MBT solution) (MBT-Na CAS-No. 2492-26-4)

Appearance	Yellow-brown fluid
Density at 20°C	Approx. 1·26 g/ml
Viscosity at 20°C	<90 mPa s
Vapour pressure (hPa)	Approx. 20 at 20°C Approx. 107 at 50°C
Solidification point	Approx. –14°C
Boiling point	Approx. 108°C
Flash point	>108°C
Ignition temperature	Approx. 480°C
Solubility	Miscible with water, alkalis, ethylene glycol, propylene glycol
pH value (1% solution)	Approx. 9·5
Stability	Stable between pH 7 and 14

Toxicity data (50% Na-MBT solution)

LD_{50} oral	Approx. 3200 mg/kg rat
dermal	>5000 mg/kg rabbit (exposure: 24 h)

Strongly irritant to skin and eyes.

Fish toxicity — LC_0 for *Leuciscus idus*: 5 mg/litre (exposure: 48 h)

Antimicrobial effectiveness/applications

As can be seen from the structural formula the dithiocarbamate configuration
-N-C-S-
|| is a structural element of the MBT molecule. Its antimicrobial effective-
S
ness is due to its chelation qualities (Albert *et al.*, 1947). The minimum inhibition concentrations in Table 38 inform about the activity spectrum of MBT in comparison to the corresponding *N*-hydroxymethyl derivative (Section 3.4.10.2) which is a formaldehyde releasing substance. As a fungicide MBT is especially toxic to superficial moulds and cellulose-decomposing fungi. The antibacterial activity of MBT is marked by a lack of efficacy for Pseudomonades. In practical application, mainly as slimicides, MBT sodium salt solutions therefore contain additionally other dithiocarbamates (e.g. Sections 9.9 or 9.10.1). Moreover MBT/sodium MBT is characterized by dual utility, as it serves also as a popular corrosion inhibitor for non-ferrous metals.

13.10 2-(Thiocyanomethylthio)benzthiazole (TCMBT)

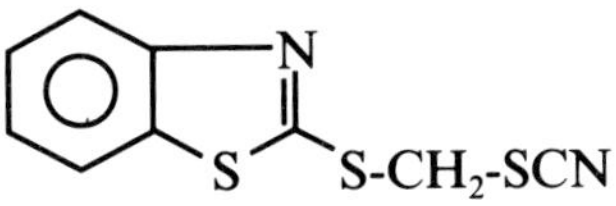

$C_9H_6N_2S_3$ M_r 238·36
CAS-No. 21564-17-0

Chemical and physical properties

Appearance	Yellow crystals or oil
Density at 22°C	1·38 g/ml
Vapour pressure at 25°C	$4{\cdot}75 \times 10^{-6}$ hPa
Solubility	0·033 g/litre H_2O; sparingly soluble (<0·5 g/litre) in isopropanol, ethylene glycol, propylene glycol, highly soluble in acetone, DMF, ethyl glycolacetate, toluene
Stability	Hydrolysis/decomposition in alkaline media (pH > 8) and on heating (>60°C)

TCMBT is available in formulations containing 80%, 60%, 30% or 20% active ingredient.

Toxicity data (80% concentrate)

LD_{50}	oral	Approx. 500 mg/kg rat
	dermal	1414 mg/kg rat

Irritant to skin and mucous membranes.

Sensitization possible; guinea pig test according to Magnusson–Kligman positive.
Risk of methaemoglobuline formation on skin contact.
No evidence of mutagenic potential.

Ecotoxicity (tests performed with a 20% a.i. formulation)

Activated sludge organisms tolerate:
2·5–5 mg/litre (0·8–1·25 mg a.i./ml)

LC_0	for *Leuciscus idus*	0·5 m/litre (exposure: 48 h)
	for *Brachydanio rerio*	0·24 mg/litre (exposure: 96 h)

Antimicrobial effectiveness/applications

As is demonstrated by the MIC in Table 109 TCMBT is a broad spectrum microbistat, active against fungi, algae, yeasts and bacteria. It is mainly used as a fungicide; important application fields; leather industry (protection of wet blues); wood protection (including particle board), especially sapstain prevention; mould and algae resistant coatings.

TCMBT has also gained importance as an active ingredient in slimicides, algicides and preservatives (although the activity spectrum for bacteria is not very equalized). TCMBT inhibits algal growth in concentrations of 0·5–1 mg/litre.

Table 109. Minimum Inhibition Concentrations (MIC) of TCMBT in Nutrient Agar

Test organism	*MIC (mg/litre) (20% a.i. formulation)*
Alternaria alternata	25
Aspergillus niger	25
Aureobasidium pullulans	75
Cephaloascus fragrans hanava	35
Ceratocystis pilifera	15
Chaetomium globosum	10
Cladosporium cladosporioides	7·5
Lentinus tigrinus	10
Penicillium glaucum	20
Phialophora fastigiata	25
Rhizopus stolonifer	200
Sclerophoma pityophila	20
Trichoderma viride	200
Candida albicans	50
Candida krusei	50
Rhodotorula mucilaginosa	20
Saccharomyces bailii	20
Saccharomyces cerevisiae	50
Torula rubra	20
Torula utilis	50
Aerobacter aerogenes	50
Aeromonas punctata	75
Bacillus mycoides	75
Bacillus subtilis	100
Escherichia coli	500
Leuconostoc mesenterioides	100
Proteus mirabilis	1000
Pseudomonas aeruginosa	1000
Pseudomonas fluorescens	500
Staphylococcus aureus	150

14 *N*-HALOALKYLTHIO COMPOUNDS

N-Haloalkylthio compounds with antimicrobial activity are obtained in almost any desired number according to the following reaction scheme, provided that the H atom has acidic character:

$$R_1R_2N\text{-}H + Cl\text{-}S\text{-}alkylX - R_1R_2N\text{-}S\text{-}alkylX + HCl \qquad X = \text{halogen}$$

The most important microbicides within this substance class are the compounds bearing a trihalomethylthio group (S-CX_3) as toxophor. The antimicrobial efficacy of *N*-haloalkylthio compounds is based on the capacity of the N-S bond to open and react with nucleophilic components of the microbial cell (Paulus & Kühle, 1986). The most effective compounds are those containing the S-CCl_2F group. These also contrary to the trichloromethylthio derivatives do not show mutagenic activities (Schuphan *et al.*, 1981). Optimum antimicrobial activity is achieved with compounds whose N-S bond has a medium stability. Taking account of the facts mentioned, it is possible to select from the wide range of *N*-haloalkylthio compounds already available, microbicides with such chemico-physical and toxicological properties as are suited to defined uses, e.g. as fungicides and algicides in plastics, paint films, wood preservatives and antifouling coatings.

In Table 110 the minimum inhibition concentrations (MIC) of two N-S-CCl_2F compounds (Dichlofluanid and Fluorfolpet) are compared with the MIC of the

Table 110. Minimum Inhibition Concentrations (MIC) (mg/litre) of *N*-trihalomethylthio Microbicides — Comparison of N-S-CCl_2F Compounds with N-S-CCl_3 Compounds

Test organism	*Dichlofluanide* N,N-*dimethyl-* N'-*phenyl-*N'-*fluorodichloro-methylthio-sulphamide*	N,N-*dimethyl-* N'-*phenyl-*N'-*trichloro-methyl-thio-sulphamide*	*Fluorofolpet* N-*fluorodichloro-methylthio-phthalimide*	*Folpet* N'-*trichloromethyl-thio-phthalimide*
Alternaria alternata	10	1000	10	750
Aspergillus niger	50	>4000	200	350
Aureobasidium pullulans	10	200	50	50
Chaetomium globosum	20	1000	100	50
Coniophora puteana	10	500	2	75
Lentinus tigrinus	2	200	2	35
Penicillium glaucum	20	4000	35	50
Polyporus versicolor	10	>1000	50	100
Sclerophoma pityophila	10	500	10	150
Trichoderma viride	≥5000	>5000	1000	>5000
Trichophyton pedis	10		20	
Candida albicans	20	>5000	5	150
Candida krusei	20	>5000	20	500
Rhodotorula mucilaginosa	20	>5000	10	100
Saccharomyces cerevisiae	10	750	5	50
Torula rubra	20	>5000	15	50
Torula utilis	10	350	10	50
Algae	10	50	20	20

corresponding N-S-CCl_3 compounds. The higher efficacy of the N-S-CCl_2F compounds is demonstrated. Remarkable is the resistance of *Trichoderma viride* against the *N*-trihalomethylthio compounds listed in Table 110.

14.1 *N*-Trichloromethylthiophthalimide — Folpet

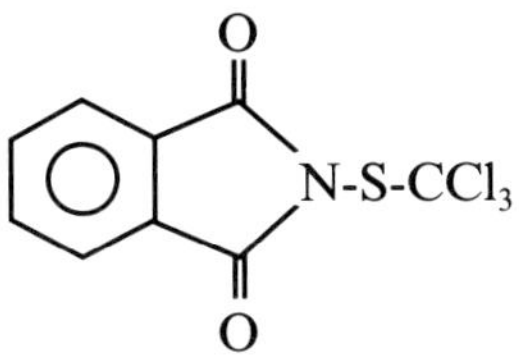

$C_9H_9Cl_3NO_2S$ M_r 296·56
CAS-No. 719-96-0

Chemical and physical properties

Appearance	White, crystalline powder with a faint mercaptaneous odour
Melting point	180°C (decomposition)
Density at 20°C	1·75 g/ml
Vapour pressure at 20°C	$>10^{-5}$ hPa
Solubility	In water $<1{\cdot}4$ mg/litre ketones 30–40 g/litre non-polar solvents 1–10 g/litre
Stability	Decomposition starts at 180°C. Hydrolyses in aqueous media increasing with pH and temperature (at pH 7·8 in solution 100% decomposition at 20°C within 2 h). Reacts under decomposition with sulphides and mercapto compounds, e.g. cysteine

Toxicity data

LD_{50} oral		>5000 mg/kg rat
		1546 mg/kg mouse
		1115 mg/kg rabbit
	dermal	>22 600 mg/kg rabbit
	intraperitoneal	68 mg/kg rat

Mutagenic and tumorigenic in a number of different essays; not a carcinogen; teratogenicity of Folpet is under discussion (McCann & Ames, 1975; Moriya *et al.*, 1983; Gold *et al.*, 1984).

Only moderately irritant to the skin, very irritant to mucous membranes.

Ecotoxicity

Hydrolyses in water to H_2S, HX, CO_2 and phthalimide. LC_0 of phthalimide for fish: approx. 800 mg/litre. Activated sludge organisms tolerate up to 1000 mg phthalimide/litre without losing activity.

Antimicrobial effectiveness/applications

Folpet's antimicrobial activity is particularly directed against fungi (Table 110). The application of Folpet as a fungicide for the protection of materials is, however, limited by its poor solubility properties; another limitation is its tendency to hydrolyse in aqueous media. Folpet is usually applied as a fungicide in pigmented coatings or wood stains with a relatively high resin content. It may be used for the fungicidal treatment of adhesives and sealants and also of plastic material, if the processing temperature does not exceed 180°C.

14.2 *N*-dichlorofluoromethylthiophthalimide — Fluorfolpet

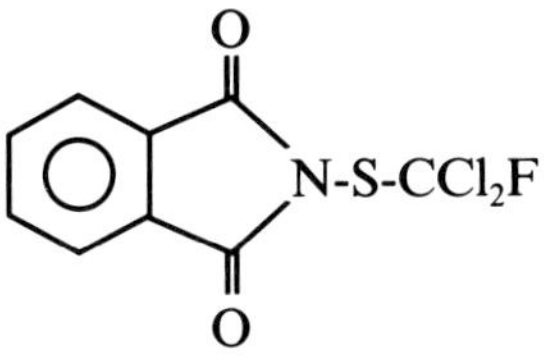

$C_9H_4Cl_2FNO_2S$ M_r 280·11
CAS-No. 719-96-0

Chemical and physical properties

Appearance	White crystalline powder with a faint musty odour	
Melting point	142°C	
Density at 20°C	Approx. 1·88 g/ml	
Vapour pressure	10^{-6} hPa at 20°C; 10^{-5} hPa at 80°C	
Solubility at 20°C	In water	Approx. 0·015 g/litre
	organic solvents (percentage by wt)	
	acetone	12·0
	butyl acetate	6·4
	methanol	1·2
	methylene chloride	14·2
	xylene	3·6
	white spirit	0·3
	Solvesso 100	4·2
	alkyd resins	2–4
	dioctyl phthalate	approx. 2
	dibenzyl phthalate	approx. 4
Stability	Sublimes when heated above the melting point; exothermal decomposition starts at 285°C. Hydrolyses in aqueous solution (100% decomposition at pH 7·8 within 0·3 h at 20°C), incompatible with alkalis, amines, sulphides and R-SH compounds	

Toxicity data

LD_{50} oral	2900 mg/kg rat
	1000 mg/kg rabbit
	500–1000 mg/kg guinea pig
	2500 mg/kg cat
dermal	>1000 mg/kg rat (exposure: 24 h)
intraperitoneal	37·5 mg/kg rat

On inhalation of the LD_{50} for rats is 1000 mg/m^3 air (exposure: 4 h). Fluorfolpet causes moderate irritation of the skin and mucous membranes; sensitization may occur. It is not mutagenic in the Ames test.

Ecotoxicity

LC_0 for *Leuciscus idus*: 2 mg/litre (exposure: 48 h)

Hydrolysis in aqueous media under generation of phthalimide: LC_0 for *Leuciscus idus*: 800 mg/litre (exposure: 48 h). Activated sludge organisms tolerate approx. 1000 mg phthalimide/ml without decrease in biological activity.

Antimicrobial effectiveness/applications

The spectrum of activity of Fluorfolpet (see Table 110) covers all types of mould fungus, including mercury-resistant *Penicillium* species, which may infest a painted surface. The product is particularly effective against wood-staining fungi (blue stain mould). As it is insoluble in water, non-volatile and light-stable, coatings containing Fluorfolpet retain their fungicidal effect for years.

Fluorfolpet also has an algicidal effect and eliminates both freshwater and seawater algae. For this reason it is regarded as a possible active ingredient in marine antifouling coatings, able to substitute for the ecotoxic organotin compounds (section 17.6). A preferred application of Fluorfolpet is the fungicidal treatment of plastic material where it performed well even under extreme severe exposure conditions (tropical climate for 3 years; Upsher & Roseblade, 1984).

14.3 *N*-trichloromethylthio-4-cyclohexene-1,2-dicarboximide — *N*-trichloromethylthio-tetrahydrophthalimide — Captan

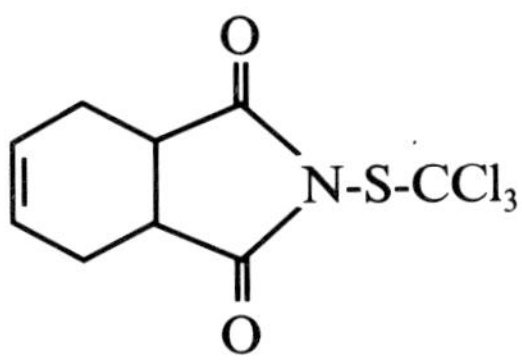

$C_9H_8Cl_3NO_2S$ M_r 300·60
CAS-No. 133-06-2

Chemical and physical properties

Appearance	White to yellow crystalline powder
Melting point	168–174°C
Density at 20°C	1·54 g/ml
Solubility	Practically insoluble in water (< 100 mg/litre) and white spirit; moderately soluble in alcohols/glycols (approx. 2–5 g/litre); soluble in acetone, cyclohexanone, tetrachloroethane, toluene
Stability	Most stable in non-aqueous, non-alkaline systems; hydrolyses in alkaline systems; heat resistant under PVC processing temperatures

Toxicity data

LD_{50}	oral	9000 mg/kg rat
		2000 mg/kg rabbit
	intraperitoneal	50–100 mg/kg rat

Only moderately irritant to skin, irritant to mucous membranes; disposes some sensitization potential.
Mutagenic (Schupan *et al.*, 1981; Xu & Schurr, 1990).
Carcinogenic (Heil *et al.*, 1991).

Antimicrobial effectiveness/applications

The activity spectrum of Captan covers Gram-positive and Gram-negative bacteria, fungi and yeasts; the minimum inhibition concentrations are in the range of 25–100 mg Captan/ml nutrient agar. Captan has been recommended for the microbicidal treatment of coatings and plastic materials, especially PVC mixtures.

14.4 *N*-1,1,2,2-tetrachloroethylthio-4-cyclohexene-1,2-dicarboximide — *N*-1,1,2,2-tetrachloroethylthio-tetrahydrophthalimide — Captafol

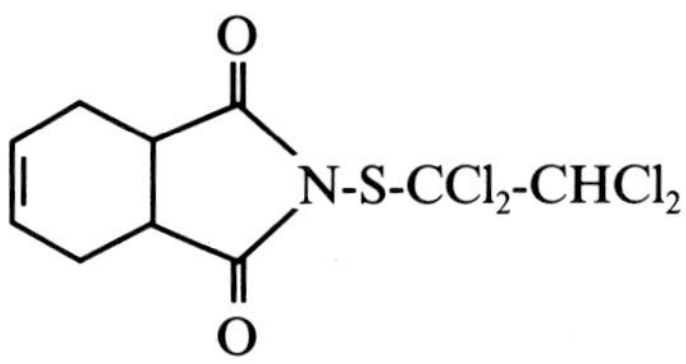

$C_{10}H_9Cl_4NO_2S$ M_r 349·07
CAS-No. 2425-06-1

Chemical and physical properties

Appearance	Crystalline powder
Solubility	Sparingly soluble in water, soluble in organic solvents
Stability	Hydrolyses in alkaline media

Toxicity data

LD_{50} oral	4600 mg/kg rat

Antimicrobial effectiveness/applications

Captafol exhibits significant antifungal activity; it has been developed as an agricultural fungicide. Attempts to use Captafol as a fungicide for the protection of materials, especially wood, were not very successful.

14.5 *N,N*-dimethyl-*N'*-phenyl-*N'*-dichlorofluoromethylthiosulphamide — *N*-dimethylaminosulphonyl-*N*-phenyl-dichlorofluoromethanesulphenamide — Dichlofluanide

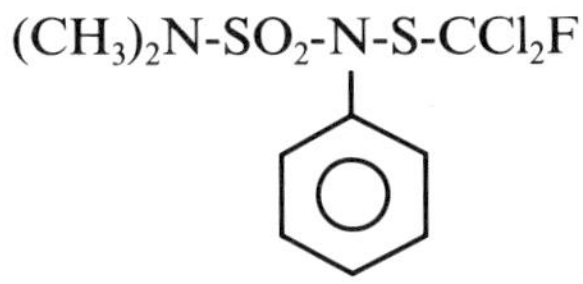

$C_9H_{11}Cl_2FN_2O_2S_2$ M_r 333·24
CAS-No. 1085-98-9

Chemical and physical properties

Appearance	White to sightly yellowish, crystalline powder with a faint musty odour	
Melting point	Approx. 105°C	
Density at 20°C	1·6 g/ml	
Vapour pressure	Approx. 10^{-6} hPa at 20°C; approx. 10^{-5} hPa at 45°C	
Solubility at 20°C	In water	approx. 1·3 mg/l
	organic solvents (percentage by wt)	
	acetone	30·0
	methyl glycol acetate	9·5
	ethyl acetate	11·7
	ethyl glycol acetate	13·0
	butyl acetate	10·3
	isopropanol	2·0
	isooctyl alcohol	6·1
	glycolic acid butyl ester	6·1
	methylene chloride	32·5
	chloroform	31·5
	dimethyl formamide	31·0
	xylene	6·5
	cyclohexanone	26·0
	linseed oil	2·0
	Solvesso 100	7·5
	stand oil CP	1·0
	white spirit	1·2
Stability	Hydrolyses in aqueous solutions (100% decomposition at pH 7·8 within 55 h at 20°C); incompatible with alkalis and amines, sulphides and mercapto compounds	

Toxicity data

LD_{50} oral	>5000 mg/kg rat and mouse
dermal	>5000 mg/kg rat (exposure: 24 h)
intraperitoneal	100–150 mg/kg rat
LC_{50} inhalation	1300 mg/m^3 air for rats (exposure: 24 h)

Moderately irritant to skin and eyes. Sensitization may occur. Investigations on mutagenic or carcinogenic effects show controversial results (e.g. Moriya *et al.*, 1983; Barrueco & de la Pena, 1988; Heil *et al.*, 1991).

Ecotoxicity

LC_0 for *Leuciscus idus*	0·2 mg/l (exposure: 48 h)
Half-life in aqueous solution at 22°C	15·3 days at pH 4; 18·8 h at pH 7; <10 min at pH 9

Hydrolysis leads to the generation of *N,N*-dimethyl-*N*'-phenylsulphamide (DMSA):

LC_0 for *Leuciscus idus*	125 mg/l (exposure: 48 h)

Activate sludge organisms tolerate approx. 200 mg DMSA/litre without losing activity.

Antimicrobial effectiveness/applications

Dichlofluanide has a broad spectrum of activity (see Table 110) which also covers mercury resistant *Penicillium* species, but not *Trichoderma viride*. It is particularly effective against wood-staining fungi (blue stain mould) and therefore one of the most important fungicides in non-film forming decorative wood stains, in wood coatings and primers.

Favourable toxicity data, insolubility in water, non-volatility and light stability are advantages of Dichlofluanide, which also exhibits algicidal activity. For this reason it is considered as a non-ecotoxic alternative to organo-tin compounds (Section 17.6) in antifouling coatings. Dichlofluanide is not phytotoxic which means that coatings and sealants containing Dichlofluanide can safely be used in greenhouses.

14.6 *N,N*-dimethyl-*N'*-tolyl-*N'*-dichlorofluoromethylthiosulphamide — *N*-dimethylaminosulphonyl-*N*-tolyl-dichlorofluoromethanesulphenamide — Tolylfluanide

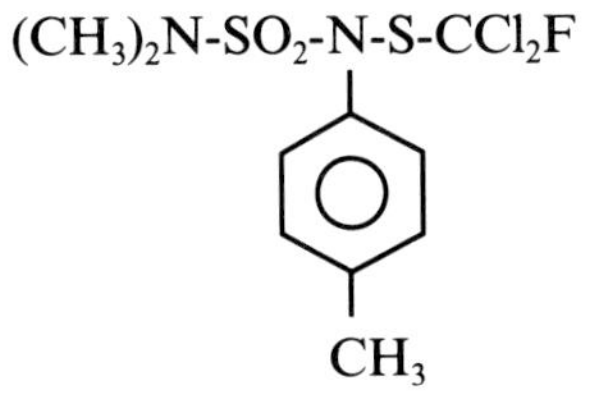

$C_{10}H_{13}Cl_2FN_2O_2S_2$ M_r 347·26
CAS-No. 731-27-1

Chemical and physical properties

Appearance	White crystalline powder
Melting point	96°C
Density at 20°C	Approx. 1·5 g/ml
Vapour pressure	Approx. 10^{-7} hPa at 20°C; approx. 10^{-4} hPa at 60°C
Ignition temperature	Approx. 400°C
Solubility at 20°C	In water approx. 2 mg/litre

organic solvents (percentage by wt)

	at 10°C	at 20°C	at 40°C
acetone	40	46	
ethyl acetate	24	30	49
Shellsol AB	10	12	28
Solvesso 100	9	18	31
white spirit*	1	1·5	4
xylene	20	22	38

*Content of aromatic hydrocarbons — 30% (calculated as xylene)

Stability: Hydrolyses in aqueous solutions: half-life at 22°C:
>15 days at pH 4
>19 h at pH 7
<10 min at pH 9

Incompatible with alkalis and amines, sulphides and mercapto compounds

Toxicity data

LD_{50} oral	>5000 mg/kg rat
dermal	>5000 mg/kg rat
intraperitoneal	20–25 mg/kg rat
LC_{50} on inhalation	265 mg/m³ air for rats (exposure: 4 h)

Not irritant to the skin, but severely irritant to the eyes.
Sensitization may occur.
Neither carcinogenic nor mutagenic nor teratogenic.

Ecotoxicity

Tolylfluanide is unstable in water; the active ingredient is toxic for fish.

LC_{50} for *Leuciscus idus*: 0·061 mg/litre; exposure: 96 h

Antimicrobial effectiveness/applications

Table 111. Minimum Inhibition Concentrations (MIC) of Tolylfluanide in Nutrient Agar

Test organism	*MIC (mg/litre)*
Alternaria alternata	10
Aspergillus flavus	75
Aspergillus niger	100
Aspergillus terreus	20
Aspergillus ustus	50
Aureobasidium pullulans	10
Cephaloascus fragrans hanava	10
Ceratocystis pilifera	10
Chaetomium globosum	20
Cladosporium cladosporioides	35
Cladosporium herbarum	10
Coniophora puteana	10
Fusarium culmorum	20
Fusarium moniliforme	20
Gliocladium virens	>1000
Lentinus tigrinus	5
Mucor racemosus	10
Penicillium glaucum	20
Polyporus versicolor	10
Poria vaporaria	2
Sclerophoma pityophila	10
Trichoderma viride	>1000
Candida albicans	20
Candida krusei	75
Rhodotorula mucilaginosa	150
Saccharomyces cerevisiae	5
Torula rubra	10
Torula utilis	10
Slime bacteria[a]	2·5
Algae	5–10

[a]Described by Kato & Fukumura (1962).

In antimicrobial effectiveness Tolylfluanide is very similar to Dichlofluanide (Section 14.5). It is also, as is Dichlofluanide, light-stable, non-volatile, colourless, virtually insoluble in water and of low toxicity. But the better solubility of Tolylfluanide in organic solvents eases its incorporation into coating formulations and impregnation agents (preferably for wood) and makes this microbicide increasingly popular as an alternative to Dichlofluanide. Its suitability for combating wood rotting fungi is reported by Kühle *et al.* (1981).

14.7 *N*-methylsulphonyl-*N*-phenyl-(1,1-dichloro-2,2-dichloro-2-fluoro-) ethanesulphenamide

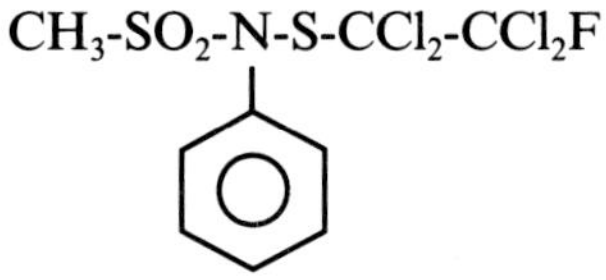

$C_9H_8Cl_4FNO_2S_2$ M_r 387·10
CAS-No. 22729-75-5

Chemical and physical properties

Appearance	White crystalline powder
Melting point	69°C
Solubility	Virtually insoluble in water, moderately soluble in white spirit, highly soluble in xylene

Toxicity

LD_{50} oral	3000 mg/kg rat

The microbicide has not gained importance and is mentioned here for completeness only. In behaviour and effectiveness it corresponds to the *N*-haloalkylthio compounds described above.

14.8 *N*-methyl-*N'*-3,4-dichlorophenyl-*N'*-dichlorofluoromethylthiourea

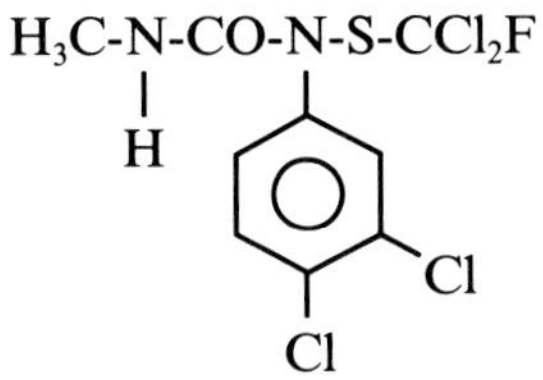

$C_9H_7Cl_4FN_2OS$ M_r 352·05
CAS-No. 88308-77-4

Chemical and physical properties

Appearance	White crystalline	
Melting point	1·25–128°C	
Solubility (% by wt)	In acetone	50
	ethylacetate	30
	xylene	6
	white spirit	0·1
	practically insoluble in water	
Stability	Stable in aqueous dispersion, e.g. in water based paint formulations at pH 8·5–9 for longer than 1 year at room temperature; very slow hydrolysis in aqueous solution (half-life at pH 7·8 (20°C): >208 days)	

Toxicity data

LD_{50} oral Approx. 650–1650 mg/kg rat

Irritant to skin and mucosa.

Antimicrobial effectiveness/applications

This sulphenylated urea derivative, which was not detected before 1985 (Kühle *et al.*) occupies a very special position within the range of *N*-haloalkylthio compounds because of its stability coupled with high antimicrobial activity and a broad spectrum of effectiveness (Table 112). The starting material in the synthesis of the sulphenylated urea derivative is *N*-methyl-*N'*-3,4-dichlorophenylurea, a potent photosynthesis inhibitor (similar to Diuron, Section 8.14) and consequently active as an algicide. But the transfer of the S-CCl_2F group to the urea derivative leads to a considerable increase in algicidal activity and additionally equips the molecule with strong fungicidal efficacy. With regard to the use of *N*-methyl-*N'*-3,4-dichlorophenyl-*N'*-dichlorofluoromethylthiourea as an algicide in antifouling coatings it is remarkable that even after a hydrolytic cleavage of the S-CCl_2F

group a degradation product, namely *N*-methyl-3,4-dichlorophenylurea is left which still exhibits strong antialgal activity.

The extraordinary stability of the sulphenylated urea derivative opens many fields of application to the microbicide. It can act as a fungicide and algicide in aqueous and solvent based coatings for exterior and interior application; its suitability as a substitute for organotin compounds (Section 17.6) in anti-fouling paints is under evaluation. Plastic material and sealants are other examples of materials which may be protected against fungal attack by the incorporation of *N*-methyl-*N*'-3,4-dichlorophenyl-*N*'-dichlorofluoromethylthiourea.

Table 112. Minimum Inhibition Concentrations (MIC) of *N*-methyl-*N*'-3,4-dichlorophenyl-*N*'-dichlorofluoromethylthiourea in Nutrient Agar

Test organism	*MIC (mg/litre)*
Alternaria alternata	5
Aspergillus niger	10
Aureobasidium pullulans	10
Chaetomium globosum	5
Cladosporium cladosporioides	5
Coniophora puteana	0·5
Lentinus tigrinus	1
Penicillium glaucum	50
Polyporus versicolor	10
Sclerophoma pityophila	5
Trichoderma viride	100
Candida albicans	50
Candida krusei	35
Rhodotorula mucilaginosa	10
Sporobolomyces roseus	5
Saccharomyces cerevisiae	50
Torula rubra	5
Torula utilis	50
Algae	0·1–1

15 COMPOUNDS WITH ACTIVATED HALOGEN GROUPS

Microbicides of this kind are electrophilic active substances having at their disposal an activated halogen group in the α-position and/or in the vinyl position to an electronegative group E (Fig. 34). The antimicrobial activity of these substances arises from the fact that nucleophilic entities (H-Nu|) of the microbial cell react with the positive carbon atom of the substances. It is apparent that a variety of modifications of both the electronegative group E and the portion of the molecule bearing the halogen group are possible. Thus microbicides having a wide range of chemical and physical properties have been synthesized within this group of active ingredients.

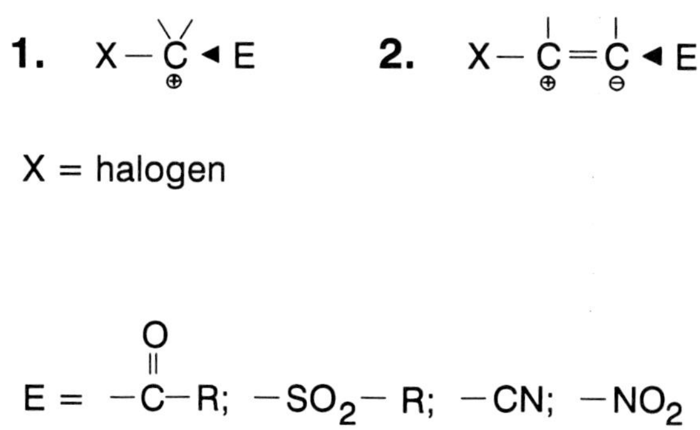

Fig. 34. Activated halogen groups.

The following microbicides with activated halogen groups have already been described within other substance classes:

Section	3.2	C-methylols
	3.2.2	2-Bromo-2-nitro-propan-1,3-diol (Bronopol)
	3.2.3	2-Bromo-2-nitro-propan-1-ol (BNP)
	3.2.5	5-Bromo-5-nitro-1,3-dioxane (Bronidox)
Section	3.4	Reaction products of amides and formaldehyde
	3.4.1	*N*-Hydroxymethyl-chloracetamide
	3.4.2	2,2,3-Trichloro-*N*-hydroxymethyl-propionamide
Section	7	Carbonic acid esters
	7.1	2-Bromo-ethylacetate
	7.2	2-Bromo-benzylacetate
	7.3	2,3-Dibromo-(2-hydroxy-)ethylpropionate
	7.4	Bis-1,2-(bromoacetoxy)ethane
	7.5	Bis-1,4-(bromoacetoxy)-2-butene
	7.6	(2-Bromo-1,2-diiodoacryl-)ethylcarbonate

Section	8	Amides
	8.1	2-Chloracetamide
	8.2	2-Bromacetamide
	8.3	2-Iodacetamide
	8.4	*N*-(4-Bromo-2-methylphenyl)-2-chloracetamide (BMPCA)
	8.5	2,2-Dibromo-3-nitrilopropionamide (DBNPA)
Section	11	Pyridine derivatives
	11.3	2,3,5,6-Tetrachloro-4-(methylsulphonyl)pyridine
Section	13	Heterocyclic *N,S* compounds
	13.2	5-Chloro-2-methyl-4-isothiazolin-3-one (CMIT)
	13.5	4,5-Dichloro-2-(*n*-octyl)-4-isothiazolin-3-one (DCOIT)

15.1 Hexachlorodimethylsulphone

$Cl_3C\text{-}SO_2\text{-}CCl_3$ $C_2Cl_6O_2S$ M_r 300·81
CAS-No. 3064-70-8

Chemical and physical properties

Appearance	White crystalline powder with a pungent odour (lachrymatory)
Melting point	35–36°C
Solubility	In water 70 mg/litre; soluble in organic solvents
Stability	Decomposition starts at 100–140°C; hydrolyses in alkaline media

Toxicity data

LD_{50}	oral	708 mg/kg rat
	dermal	5620 mg/kg rabbit

Irritant and corrosive to the skin and mucosa.

Antimicrobial effectiveness/applications

Hexachlorodimethylsulphone is a reactive microbicide which exhibits a broad spectrum of activity. The minimum inhibition concentrations for bacteria, fungi, yeasts and algae range between 10 and 100 mg/litre. The slime forming bacteria described by Kato & Fukumura (1962) are especially sensitive to hexachlorodimethylsulphone; MIC: 0·5 mg/litre. In consequence of its limited stability in water the compound is mainly used as a non-persistent slimicide and algicide in process water circuits, e.g. in the paper industry, and not so much as a preservative.

15.2 Diiodomethyl-*p*-tolylsulphone

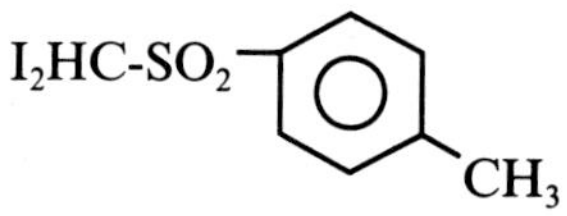

$C_8H_8I_2O_2S$

M_r 422·03
CAS-No. 20018-09-1

Chemical and physical properties

Appearance	Fine tan powder	
Melting point	157°C	
Density	2·20 g/ml	
Solubility (g/litre) at 25°C	In water	0·0001
	ethanol	20
	isopropanol	10
	ethylene glycol	10
	acetone	350
	hexane	2
	mineral spirits	4
	xylene	33
Stability	Stable through the pH range 4–10	

Toxicity data

LD_{50} oral	>10·0 g/kg rat	
	9·4 g/kg mouse	
dermal	>10·0 g/kg rabbit	

Not irritating to the skin, can cause slight, temporary irritation of the eyes.
Is not a sensitizing agent according to the guinea pig test.

Is not considered to be a toxic substance when inhaled (1·9 mg/litre air; exposure of rats for a 4 h period).

Diiodomethyl-*p*-tolylsulphone did not demonstrate mutagenic activity in a series of in-vitro and in-vivo assays.

Ecotoxicity

LC_{50} for Rainbow Trout	0·13 mg/litre
for Bluegill Sunfish	0·75 mg/litre

Antimicrobial effectiveness/applications

Diiodomethyl-*p*-tolylsulphone can be regarded as a broad spectrum microbicide which is more active against fungi, yeasts and algae than against bacteria. Being not volatile and practically insoluble in water it is recommended for the fungicidal treatment of aqueous or solvent based coatings, adhesives, sealants, etc., for interior or exterior applications. As an aliphatic iodo compound it can cause some yellow coloration, especially in white coatings. The addition of colour suppressants to the active ingredient can alleviate the coloration problem. It is also recommended to apply diiodomethyl-*p*-tolylsulphone in the tanning process for the protection of wet blues against mould infestation. It has been found effective as a preservative for wood, particularly for the prevention of sap stain. Another proposal is the application of diiodomethyl-*p*-tolylsulphone in metal working fluids to overcome problems caused by fungal growth.

Table 113. Minimum Inhibition Concentrations (MIC) of Diiodomethyl-*p*-tolyl-sulphone in Nutrient Agar

Test organism	*MIC (mg/litre)*
Alternaria alternata	3·5
Aspergillus niger	5
Aureobasidium pullulans	7·5
Chaetomium globosum	5
Cladosporium cladosporioides	1
Lentinus tigrinus	5
Penicillium glaucum	5
Rhizopus nigricans	100
Sclerophoma pityophila	1
Trichoderma viride	10
Candida albicans	20
Candida krusei	50
Rhodotorula mucilaginosa	10
Saccharomyces bailii	5
Saccharomyces cerevisiae	10
Torula rubra	10
Torula utilis	10
Escherichia coli	200
Proteus vulgaris	>1000
Pseudomonas aeruginosa	>1000
Staphylococcus aureus	50

15.3 4-Chlorophenyl-diiodomethylsulphone

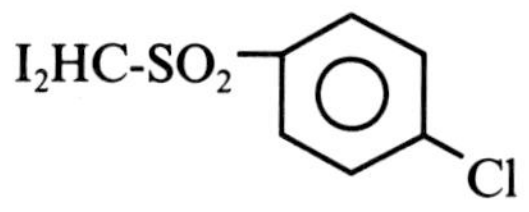

$C_7H_5ClI_2O_2S$

M_r 442·45
CAS-No. 20018-12-6

Chemical and physical properties

Appearance	Fine tan powder	
Melting point	134–138°C	
Solubility (g/litre) at 25°C	In water	0·0002
	ethanol	40
	isopropanol	20
	ethylene glycol	20
	acetone	350
	toluene	95
Stability	Stable between pH 4 and 10	

Toxicity data

LD_{50} oral	3·6 g/kg mouse
	0·6 g/kg rat

Not irritant to the skin, slightly irritant to the eyes (according to tests with rabbits).

Ecotoxicity

LC_{50} for Rainbow Trouts	0·14 mg/litre
for Bluegills	0·24 mg/litre

Antimicrobial effectiveness/applications

4-Chlorophenyl-diiodomethylsulphone may be used as is reported for diiodomethyl-*p*-tolylsulphone (Section 18.2), which is, however, the preferred microbicide.

15.4 3,3,4,4-Tetrachloro-tetrahydro-thiophene-1,1-dioxo

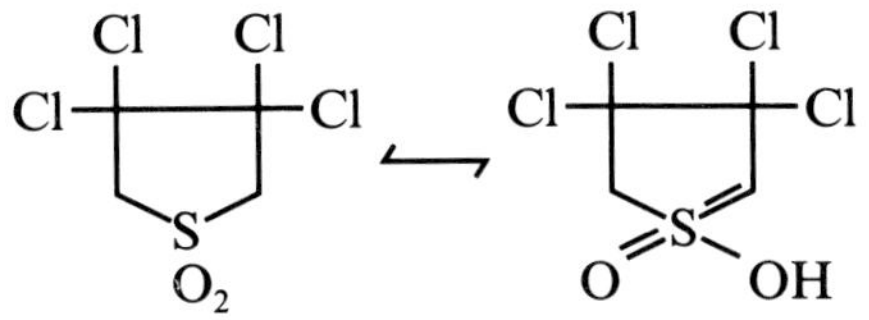

$C_4H_4Cl_4O_2S$ M_r 257·95
CAS-No. 3737-41-5

Chemical and physical properties

Appearance	White powder with a pungent odour	
Melting point	67–70°C	
Solubility (g/litre)	In methanol	100
	ethanol	50
	propylene glycol	20
	acetone	330
	dichloromethane	150
	toluene	80
	Sparingly soluble in water	
Stability	Acid stable; hydrolysis/decomposition at pH > 8	

Toxicity data

LD_{50} oral	250 mg/kg rat
dermal	8700 mg/kg rabbit

Irritant to the skin, corrosive to mucous membranes (5% in ethanol).

Antimicrobial effectiveness/applications

The reactive compound is effective against all kinds of bacteria, aerobic and anaerobic, and against fungi and yeasts.

Minimum inhibition concentrations

For bacteria	25–100 mg/litre
For fungi	5– 10 mg/litre

The microbicide is used as an active ingredient for the formulation of non-persistent slimicides to be applied in paper mills and cooling water systems. In neutral to acid aqueous functional fluids it may also be used as a preservative.

15.5 1-Chloro-1-cyano-2-phenylsulphonylethylene — α-Chloro-β-phenyl-sulphonylacrylnitril — (2-Chloro-2-cyanovinyl)-phenylsulphone

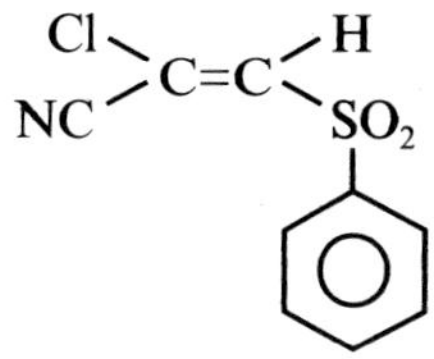

$C_9H_6ClNO_2S$

M_r 227·67
CAS-No. 60736-58-5

Chemical and physical properties (mixture containing approx. 35% of the above compound in 2-methoxyethylacetate in association with a small amount of (1,2-dichloro-2-cyanovinyl)-phenylsulphone)

Appearance	Yellow to brown fluid with aromatic odour	
Density at 20°C	Approx. 1·13 g/ml	
Vapour pressure at 20°C	38 hPa	
Boiling point	Approx. 140°C	
Flash point	44°C	
pH (10% solution in water)	Approx. 2·5	
Solubility (g/100 g solvent) at 20°C	In water	0·045
	ethanol	10
	triethylene glycol	21
	tetraethylene glycol	31
	polyglycol 300	35
	acetone	210
	methylenechloride	148
Stability	Hydrolysis and inactivation in aqueous media increasingly with increasing pH and temperature	

Toxicity data

LD_{50} oral — 521 mg/kg male rat; 237 mg/kg female rat

Irritant to skin and mucosa.
Non-mutagenic according to Salmonella microsome test.

Ecotoxicity

LC_0 for *Brachydanio rerio*	0·1 mg/litre (exposure 96 h)
LC_{50} after hydrolysis (24 h in water) for *Leuciscus idus*	≥10 mg/litre

Antimicrobial effectiveness/applications

α-Chloro-β-phenylsulphonylacrylnitril is active against bacteria, fungi and yeasts. The minimum inhibition concentrations of the 35% solution described above range between 10 and 100 mg/litre nutrient agar. But the solution is especially toxic for slime forming micro-organisms such as characterized by Kato & Fukumura (1962); MIC: 1 mg/litre nutrient solution. Accordingly the microbicide is used as an active ingredient for the formulation of non-persistent slimicides to be used in paper mills and cooling water circuits.

15.6 4,5-Dichloro-1,2-dithiolone — 5-Oxo-3,4-dichloro-1,2-dithiolan — 4,5-Dichloro-1,2-dithiacyclopenten-3-one

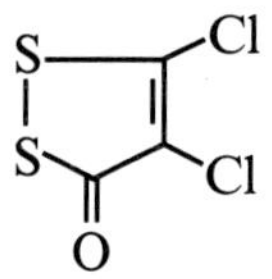

$C_3Cl_2OS_2$ M_r 187·07
CAS-No. 1192-52-5

Chemical and physical properties (30% emulsion)

Appearance	Yellow liquid with a characteristic odour
Solubility	Emulsifiable in water, soluble in organic solvents
Stability	Hydrolyses in alkaline media

Toxicity data

Corrosive to skin and eyes.

Antimicrobial effectiveness/applications

The chlorinated 1,2-dithiolone is especially effective against slime forming microbe species such as described by Kato & Fukumura (1962); MIC approx. 1 mg/litre nutrient solution. However the very reactive compound is stable and active only in acid media; reduction in activity is observed in neutral media and decomposition under alkaline conditions. The active ingredient may be used for the preparation of slimicides to be applied in pulp and paper mills and in cooling water systems (Clarkson & Clifford, 1991).

15.7 Tetrachloroisophthalodinitril — 1,3-Dicyano-2,4,5,6-tetrachlorobenzene — Chlorthalonil

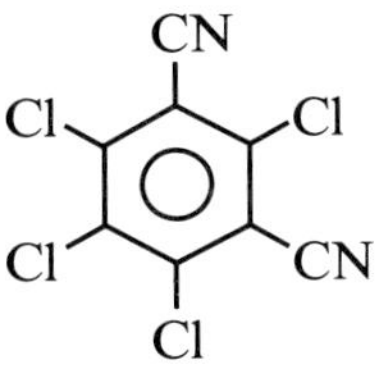

$C_8Cl_4N_2$ M_r 265·91
CAS-No. 19897-45-6

Chemical and physical properties

Appearance	White, crystalline powder with slight pungent odour	
Melting point	Approx. 250°C	
Boiling point	Approx. 350°C (sublimation)	
Vapour pressure	<1·3 Pa at 40°C	
Flash point	Non-flammable	
Solubility	In water	0·6 mg/litre
	organic solvents (g/kg)	
	acetone	20
	butanone	20
	cyclohexanone	30
	dimethylformamide	30
	dimethylsulphoxide	20
	xylene	80
Stability	Stable in neutral and acid aqueous solutions, hydrolyses in alkaline solutions, increasing with pH and temperature; reacts with thiol compounds; stable to UV light; sublimes when heated above 100°C	

Toxicity data

LD_{50}	oral	5000 mg/kg rat
	dermal	5000 mg/kg mouse
		2000 mg/kg rabbit

Irritant to skin and mucosa.

LC_{50} for fish 250–230 ppb

Antimicrobial effectiveness/applications

The reactivity of Chlorthalonil, for example, with thiol groups of the microbial cell, which is the key to Chlorthalonil's antimicrobial activity, involves also the detoxification of the compound. As can be seen from the MIC in Table 114

Chlorthalonil is especially effective against fungi, yeasts and algae; however, there are some gaps in the spectrum of effectiveness. Nevertheless Chlorthalonil is used as a non-leachable fungicide in paints, adhesives, sealants, putty, plastic materials, etc. Because of its extremely low solubility in water, it is relatively stable also in aqueous paint formulations which in general react alkaline. The anti-algal activity of Chlorthalonil (MIC approx. 1–2 mg/litre) has led to proposals to use the compound as a substitute for organotin compounds (Section 17.6) in antifouling coatings.

Table 114. Minimum Inhibition Concentrations (MIC) of Chlorthalonil in Nutrient Agar

Test organism	*MIC (mg/litre)*
Alternaria alternata	0·75
Aspergillus niger	1000
Aspergillus terreus	>1000
Aureobasidium pullulans	1
Chaetomium globosum	5
Coniophora puteana	0·75
Lentinus tigrinus	2
Penicillium expansum	500
Penicillium funicolosum	5000
Penicillium glaucum	2
Polyporus versicolor	>1000
Rhizopus stolonifer	>5000
Trichoderma viride	5000
Candida albicans	1000
Candida krusei	350
Rhodotorula mucilaginosa	750
Saccharomyces cerevisiae	350
Torula rubra	1000
Torula utilis	500

15.8 2,4-Dichloro-6-(2-chloranilino)s-triazine — Dyrene

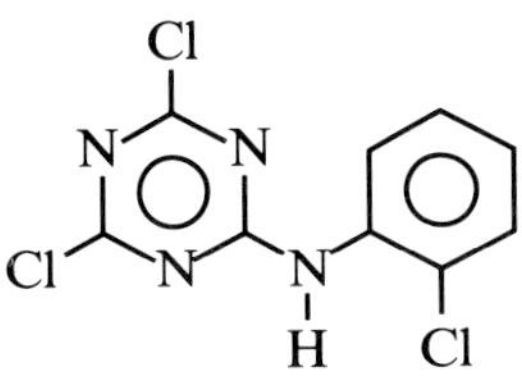

$C_9H_5Cl_3N_4$

M_r 275·53
CAS-No. 101-05-3

Chemical and physical properties

Appearance	White, crystalline powder
Melting point	159°C
Solubility	Sparingly soluble in water, soluble in organic solvents
Stability	Hydrolyses in alkaline media (separation of 1 chloro atom from the triazine ring at 40°C within 3·5 h at pH 10·4, 1·3 h at pH 11, 0·7 h at pH 12·5; 2 chloro atoms are separated by hydrolysis at pH 12·5 within 3·5 h)

Toxicity data

LD_{50} oral 2710 mg/kg rat

Irritant to skin and eyes.

Antimicrobial effectiveness/applications

As is demonstrated by the hydrolysis data, Dyrene is a reactive substance which owes its antimicrobial activity to its reactive chloro atoms enabling the substance to react with nucleophilic cell entities; preferred are those bearing amino and thiol groups. Substituted *s*-triazines were described as highly fungitoxic by Schuldt & Wolf (1958). Because of being virtually non-phytotoxic Dyrene has become an important fungicide for plant protection. Attempts to introduce Dyrene also for the fungicidal treatment of materials failed more or less, as the activity and the efficacy spectrum could not successfully compete with that of other appropriate fungicides. The antimicrobial treatment of textile material by reacting cotton with Dyrene without affecting fibre properties is reported by Paulus & Pauli (1970); rot-proofing effects can be produced on cotton at substitution levels of around 0·01 only. Such finishes are much faster to washing or bleaching and to solvents than those achieved by simple impregnation procedures. According to the findings of Paulus & Pauli (1970) the chloro atom at the 2-position of the s-triazine ring is responsible for the reactivity of Dyrene and essential for its antimicrobial effectiveness. The reaction of Dyrene with cellulose apparently blocks parts of the cellulose molecule against the attack of cellulose degrading enzymes (cellulases).

Table 115. Minimum Inhibition Concentrations (MIC) of Dyrene in Nutrient Agar

Test organism	*MIC (mg/litre)*
Alternaria alternata	200
Aspergillus niger	>1000
Aureobasidium pullulans	200
Chaetomium globosum	200
Cladosporium cadosporioides	100
Lentinus tigrinus	100
Penicillium glaucum	>1000
Rhizopus nigricans	>1000
Sclerophoma pityophila	200
Trichoderma viride	>1000

15.9 2-Bromo-4'-hydroxyacetophenone — 4-(Bromacetyl)phenol

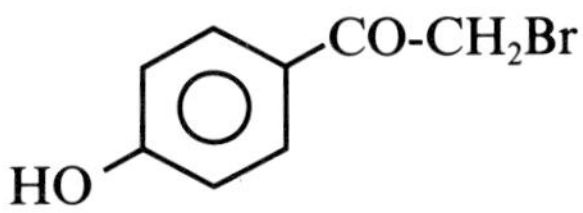

$C_8H_7BrO_2$

M_r 215·05
CAS-No. 2491-38-5

Chemical and physical properties

Appearance	Violet coloured crystals
Melting point	127–129°C
Solubility	Sparingly soluble in water, soluble in organic solvents
Stability	Hydrolyses slowly in aqueous media (increasingly with increasing pH and temperature)

Toxicity data

Strongly irritant and corrosive to the skin and eyes.

Antimicrobial effectiveness/applications

4-(Bromoacetyl)phenol has been used in the past as an active ingredient in slimicides for the application in process water systems, predominantly in pulp and paper mills. In the meantime the activity has been widely substituted by microbicides which are safer with regard to their toxicity.

Table 116. Minimum Inhibition Concentrations (MIC) of 4-(Bromoacetyl)phenol

Test organism	*MIC (mg/litre)*
Aspergillus niger	100
Chaetomium globosum	500
Penicillium glaucum	200
Escherichia coli	100
Pseudomonas aeruginosa	200
Slime bacteria[a]	1–2

[a] Described by Kato & Fukumura (1962).

15.10 2-Bromo-2-nitrostyrene (BNS)

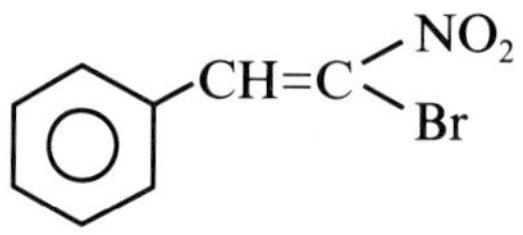

$C_8H_6BrNO_2$ M_r 228·05
CAS-No. 7166-19-0

Chemical and physical properties

Appearance	Yellow, crystalline powder
Melting point	59–62°C
Solubility	Sparingly soluble in water, soluble in organic solvents
Stability	Hydrolyses quickly in aqueous media; reacts under inactivation with inorganic or organic sulphite compounds and oxidizing agents such as hydrogen peroxide and potassium permanganate

Toxicity data

LC_{50} for fish (Rainbow trout, Bluegill sunfish) 1–2 mg/litre (exposure 96 h)

Antimicrobial effectiveness/applications

Table 117. Minimum Inhibition Concentrations (MIC) of BNS in Nutrient Agar

Test organism	*MIC (mg/litre)*
Aspergillus niger	10
Chaetomium globosum	35
Penicillium glaucum	5
Escherichia coli	20
Pseudomonas aeruginosa	10
Slime bacteria[a]	0·25

[a] Described by Kato & Fukumura (1962).

The cis, trans or mixed forms of BNS are highly effective against a broad spectrum of microbe species. The extraordinarily good activity of BNS against slime bacteria has resulted in the utilization of the compound in compositions useful for inhibiting slime formation in cooling water and pulp and paper mill systems. Due to the natural property of BNS to hydrolyse quickly to less toxic products and the susceptibility of the active ingredient to detoxification by various oxidizing agents it is recommended to use BNS for controlling the fouling potential of molluscs in once-through cooling water systems (Davis & Doherty, 1985).

15.11 1,2-Dibromo-2,4-dicyanobutane — 2-Bromo-2-bromomethylglutaronitril

CH_2-Br | NC-C-CH_2-CH_2-CN | Br	$C_6H_6Br_2N_2$	M_r 265·94 CAS-No. 35691-67-7

Chemical and physical properties

Appearance	White to yellow crystalline powder
Melting point	51–53°C
Solubility	3·8 g/litre H_2O; soluble in organic solvents, preferably in lower alcohols
Stability	Sufficiently stable in aqueous formulations up to pH 9 and 60°C

Toxicity data

LD_{50} oral	541 mg/kg rat
dermal	>5000 mg/kg rabbit

Inhalation toxicity

LC_{50} > 200 mg/litre of air when administered as a powder for 1 h at an air flow of 5 litres/min.

Irritant to the skin, corrosive to the eyes (test with rabbits).
No sensitization or mutagenic effects have been observed.

Ecotoxicity

LC_{50} (mg/litre) after 96 h exposure	Bluegill sunfish	4·09
	Rainbow Trout	1·75
	Daphnia Magna	2·20

Antimicrobial effectiveness/applications

According to its chemical structure 1,2-dibromo-2,4-dicyanobutane contains several highly electron-withdrawing centres making the compound strongly reactive with nucleophilic groups in the microbial cell. The consequence is a broad activity spectrum covering bacteria, yeasts, fungi and algae (Table 118).

Because of its equalized spectrum of effectiveness and its favourable toxicological properties the microbicide is gaining more and more importance as a preservative for a wide variety of aqueous functional fluids, such as polymer emulsions, paints, adhesives (including protein based adhesives), mineral slurries, concrete additives, metal working fluids, detergent solutions, cosmetic products, etc. The EC list of preservatives permitted for use in cosmetics mentions 1,2-dibromo-2,4-dicyanobutane with a maximum authorized concentration of 0·1% and not to be used in cosmetic sunscreen products.

Table 118. Minimum Inhibition Concentrations (MIC) of 1,2-Dibromo-2,4-dicyanobutane in Nutrient Agar (Lederer *et al.*, 1982)

Test organism	*MIC (mg/litre)*
Aerobacter aerogenes ATCC 7356	5
Bacillus mycoides IPC 509	10
Escherichia coli ATCC 4352	10
Proteus mirabilis ATCC 7002	10
Pseudomonas aeruginosa ATCC 10145	100
Pseudomonas sp. (adhesive)	150
Aspergillus niger ATCC 6275	100
Chaetomium globosum ATCC 6205	10
Geotrichum sp. (latex emulsion)	100
Penicillium luteum ATCC 10466	50
Pullularia pullulans ATCC 9348	25
Trichoderma viride ATCC 9678	100
Saccharomyces cerevisiae ATCC 4111	5
Rhodotorula sp. latex emulsion isolate	100
Chlorella pyrenoidosa Wisc 2005	5
Phormidium innundatum Wisc 1093	3
Phormidium retzii Wisc 1094	2

16 SURFACE ACTIVE AGENTS

Surfactants are characterized by their ability to reduce the surface tension of aqueous fluids; this enables them to act as detergents, wetting agents, emulsifiers. James (1965) has defined them as molecules with two different structural elements, one being a hydrophobic hydrocarbon (water-repellent) group, and the other a hydrophilic polar (water attracting) group. Depending on the charge of the hydrophilic structural element surface active agents are classified as anionic, cationic, amphoteric and non-ionic compounds.

Anionic surfactants exhibit some bactericidal effect only in acid media (pH 2–3), that means in their undissociated state. They present themselves as alkali or amine salts of long-chain fatty acids or alkane sulphonates (e.g. $R\text{-}COO^-Na^+$, $R\text{-}SO_3^-Na^+$; $R = C_{10}–C_{12}$ alkyl); in aqueous solution they dissociate to a large anion, responsible for the strong detergent properties and a small cation. The antimicrobial effect of anionic surfactants is restricted mainly to Gram-positive bacteria. Their point of attack is apparently the microbial cell membrane. A survey on the mechanism of the bactericidal action is given by Newton (1960). Acid formulations of anionic surfactants are used as sanitizers in the dairy, beverage and food processing industries, in institutions and homes.

In non-ionic surface active compounds the hydrophilic group usually consists of a chain of ethylene oxide units (ethoxylated compounds). Sorbitan derivatives, such as polysorbates (Tweens) are other examples of non-ionic agents. They do not exhibit significant antimicrobial activity. At low concentrations non-ionic surfactants, and anionic surfactants, too, may potentiate the antimicrobial action of microbicides, such as *p*-hydroxy-benzoates (Section 6.1.10) and phenol derivatives (Section 5), by increasing the cellular permeability to the microbicides. At higher concentrations of surface active agents one then observes antagonistic effects caused by the inclusion of active ingredients in the micellar phases of surface active compounds or by complex formation. The antagonistic effect starts at the so-called critical micell concentration. The following relationship is valid and enlightened by Fig. 35 (Block, 1983).

$$R = SC + 1$$

R = ratio of total to free (not bound) microbicide concentration
S = surfactant concentration
C = constant, characteristic for each surfactant microbicide mixture

The neutralizing effect of a microbicide binding to surface active agents above the critical micell concentration (CMC) can be overcome by diluting the mixture to or below the CMC or by increasing the microbicide concentration in the mixture to an amount such that the concentration of free microbicide reaches the effective level.

Anionic and non-ionic surface active compounds are not important as microbicides for the protection of materials. On the contrary, aqueous anionic or non-ionic detergent solutions need the addition of in-tank/in-can preservatives for protection against contamination and proliferation of micro-organisms.

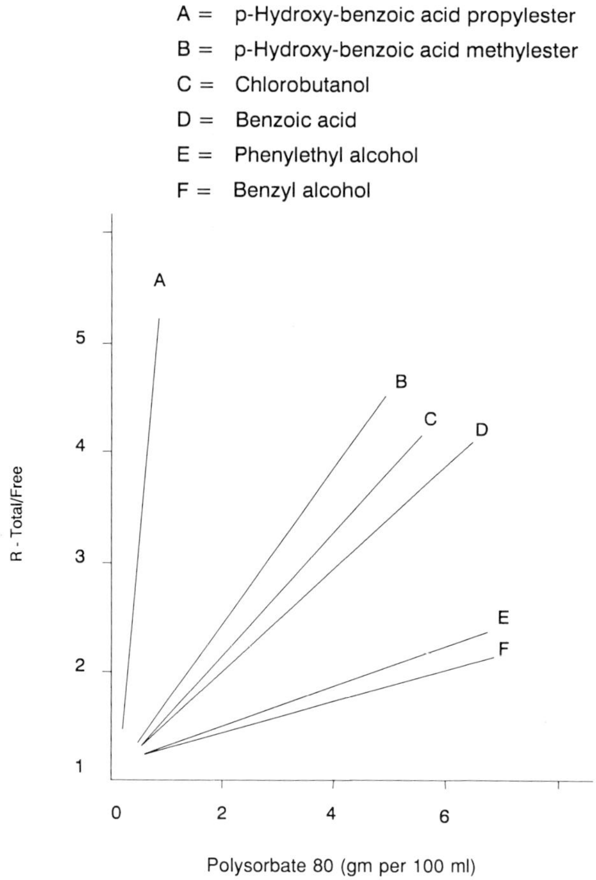

Fig. 35. Binding of representative preservatives by a non-ionic surface active agent, polysorbate 80, in aqueous solution at 30°C.

Long-chain quarternary ammonium compounds (QACs) are cationic surface active agents according to the general formula

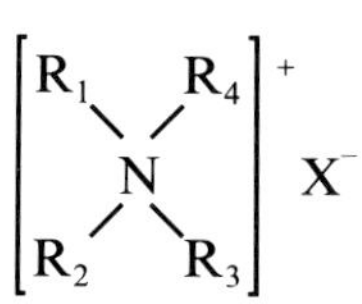

X is usually chloride or bromide.

The most important QACs may be characterized as follows:

Monoalkyltrimethylammonium salts, e.g. cetyltrimethylammonium bromide.
Monoalkyldimethylbenzylammonium chloride, e.g. benzalkonium chloride.
Dialkyldimethylammonium salts, e.g. didecyldimethylammonium halides.

Heteroaromatic ammonium salts (one R stands for a long-chain alkyl group, and the remaining three R are components of an aromatic system such as pyridine, quinoline or isoquinoline), e.g. cetylpyridinium halide and alkylisoquinolinium bromide.

Long-chain alkylamines and the corresponding ammonium salts can also be characterized as cationic surface agents; they exhibit considerable antimicrobial activity if the alkyl chain consists of 12–15 carbon atoms. This is also valid for aliphatic diamines, guanidines and biguanides (Section 16.2).

Examples are:

Dodecylammonium benzoate, -salicylate, -lactate
Laurylaminopropylamine
Dodecylguanidine

Disinfectants based on long-chain alkyl amines as active ingredients to be used in the food industry, e.g. in bottling plants, are described by Falter *et al.* (1991).

Instead of the names of many researchers which were involved in the synthesis of QACs and the exploration of their antimicrobial effectiveness, Domagk (1935) is cited, who disclosed the antibacterial activity of the long-chain quaternary ammonium salts. QACs act as algicides, bactericides and fungicides; they are virucidal against lipophilic viruses, but not against hydrophilic viruses; and they are not tuberculocidal or sporicidal. Concentrations of 5–10 mg QAC/litre are sufficient to kill Gram-positive bacteria; whereas one needs approximately 50–100 mg QAC/litre to kill Gram-negative bacteria; however, resistant species of Pseudomonades are also known. The antimicrobial activity of QACs depends on their structure and size, but especially on the length of the long-chain alkyl group; QACs bearing the C_{14} alkyl group exhibit maximum activity (Cutler *et al*, 1966). The efficacy of QACs increases with temperature and pH. Alkaline media are most favourable. At pH < 3 QACs are widely ineffective. Aralkyl alcohols, especially 3-phenylpropanol potentiate the action of QACs on *Pseudomonas aeruginosa* (Richards & McBridge, 1973).

The QACs' mode of action is rather complex. Of course they belong to the membrane-active microbicides which damage the cytoplasmic membrane controlling the cell permeability (Hugo, 1965). QACs can be characterized as microbicides of low toxicity which with normal precautions may be handled safely especially with regard to the very low use dilutions. They find many and varied uses. However, their usefulness in practice is severely limited by the fact that they can cause foam problems and are incompatible with a wide variety of compounds, especially anionic surfactants, organic matter, including milk, serum and faeces. If one pays attention to these properties of the QACs, they can successfully be used as active ingredients in disinfectants for application in hospitals, households and the food and beverage industry; QACs are used as microbicides in the sugar refining industry; they are added to process water to inhibit the proliferation of slime forming organisms and algae. 5–20 mg QAC/ml injection water or brine used in secondary oil recovery prevent the growth of microorganisms which cause the plugging of the subterranean sand formations thus stopping the operation. In outdoor swimming pools QACs may act as powerful algicides. Algae and lichens on stone surfaces are killed by a wash with an aqueous solution containing 0·5–1% QAC. In combination with other fungicides QACs are applied for the temporary protection of freshly cut and sawn timber against stain and mould fungi.

Due to the positively charged structural element of the quaternary molecule it has an affinity for negatively charged fabric. The antimicrobial treatment of textile material with QACs takes advantage of this phenomenon. The substantivity of QACs is extraordinarily strong to cotton which has gained anionic properties by chemical modification (Paulus, 1971).

Another possibility for surface-bonded antimicrobial activity is the use of organosilicon quaternary ammonium salts (Si-QAC). Isquith *et al.* (1972) has demonstrated that the hydrolysis product of 3-(trimethoxysilyl)-propyldimethyl-octadecyl ammonium chloride exhibits antimicrobial activity against a broad range of micro-organisms while chemically bonded to a variety of surfaces, such as siliceous surfaces (glass, stone), natural fibres, man-made fibres, metals, leather, wood and rubber. Hydrolysis and condensation of Si-QACs on surfaces containing reactive functional groups occurs according to the scheme in Fig. 36.

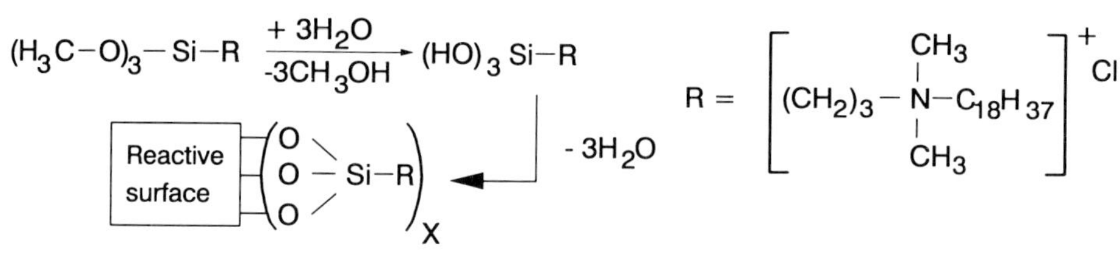

Fig. 36. Hydrolysis and condensation of 3-(trimethoxysilyl)-propyldimethyloctadecyl ammonium chloride.

Isquith used ^{14}C-Si-QAC-treated cellulose acetate sheet to gain evidence of substantial antimicrobial activity of Si-QAC. The fact that the efficacy of Si-QAC is not dependent on the slow release of the QAC lends support to the theory that the QACs act on the membrane or cell wall but probably not on intracellular organelles.

The microbicidal activity of quaternary surface active compounds is not restricted to ammonium salts; long-chain quaternary phosphonium salts are similar in effectiveness and properties, e.g.

Tetra-alkyl phosphonium chloride (TAPC)
Dodecyl-triphenyl phosphonium bromide

TAPC is available as a 50% aqueous solution which is completely miscible with water in all proportions and highly soluble in methanol, isopropanol and ethylene glycol. The active ingredient is stable in neutral, alkaline and acidic solutions and has a high activity against aerobic and anaerobic bacteria, algae and fungi.

The excellent microbicidal effectiveness of TAPC, its high surface activity and low foaming properties make it especially suitable for cleaning cooling systems which are fouled with biological slimes. It is recommended to use TAPC within the pH range 2–12 at addition rates between 10 and 30 mg a.i./litre.

Amphoteric surface active agents consist of an amino acid, usually glycine, substituted with a long-chain alkyl group. Their dissociation in water leads to zwitterions (ions with a positive and negative charge in the same molecule)

$$\text{R-NH-CH}_2\text{-C}\overset{\text{O}}{\underset{\text{OH}}{\lessgtr}} \rightleftharpoons \text{R-NH}_2^{+}\text{-CH}_2\text{-C}\overset{\text{O}}{\underset{\text{O}^-}{\lessgtr}}$$

Contrary to anionic and non-ionic agents the ampholyte soaps exhibit distinct and broad antimicrobial effectiveness; they are bactericidal to Gram-positive and Gram-negative bacteria, including myobacteria which are not safely killed by other surface active agents such as quaternary ammonium compounds. Ampholytes are also effective on lipophilic viruses but not on hydrophilic ones. Amphoteric microbicides are applied in disinfecting cleansing agents to be used in hospitals and medicine and in the food, beverage and cosmetic processing industry. In these applications they profit from having low toxicity, good skin compatibility and from being non-corrosive, odourless and relatively resistant to inactivation by proteins. The antimicrobial activity of the ampholytes remains virtually constant over a wide pH range.

16.1 Quaternary ammonium compounds (QACs)

The series of QACs described in this section does not claim to be complete. As can be seen from the general formula cited above it is apparent that a wide scope exists for synthesis of antimicrobial QACs; in consequence numerous QACs are offered for practical application as microbicides. The QACs listed in the following may be regarded as prototypes illustrating the wide variety of modifications which are possible within this class of microbicides.

16.1.1 Cetyltrimethylammonium bromide (CTAB) — Hexadecyltrimethylammonium bromide — Cetrimoniumbromide

$$\left[\begin{array}{c} CH_3 \\ | \\ CH_3\text{-}(CH_2)_{15}\text{-}N\text{-}CH_3 \\ | \\ CH_3 \end{array}\right]^+ Br^- \qquad C_{19}H_{42}BrN$$

M_r 346·46
CAS-No. 57-09-0

Chemical and physical properties

Appearance	Almost colourless and odourless crystals
Melting point	250–256°C
Solubility	Unlimited in water, reacting slightly acidic (pH 5–7) and promoting foam; highly soluble in lower alcohols, poorly soluble in acetone, virtually insoluble in non-polar solvents
Stability	Stable in acid solutions; incompatible with anionic detergents, peptides, zinc salts, polymeric phosphates, pectins, strong oxidizing and reducing agents

Toxicity data

Chronic toxicity in rats: application of 10 and 20 mg/kg/day for 1 year with the drinking water caused no effect, 45 mg/kg/day reduced the body weight. Intra-peritoneal application produced embryotoxic and teratogenic effects in pregnant mice.

Ames test negative.

Antimicrobial effectiveness/applications

The microbial efficacy and activity spectrum of CTAB is demonstrated by the microbicidal concentrations of CTAB for different species of bacteria, yeasts and fungi (Wallhäusser, 1984) in Table 119.

Table 119. Minimum Microbicidal Concentrations of CTAB (MMC) after 24 h at 22°C

Test organism	*MMC (mg/litre)*
Staphylococcus aureus	1·5
Streptococcus lactis	0·6
Bacillus subtilis	6
Escherichia coli	25
Pseudomonas aeruginosa	>50
Serratia marcescens	25
Desulfovibrio desulfuricans	50
Mycobacterium 607	6
Streptomyces griseus	25
Candida albicans	25
Penicillium chrysogenum	>50

Apparently CTAB is highly effective against Gram-positive bacteria, but by far not that active against Gram-negative bacteria and fungi. The complex Gram-negative cell wall is in comparison to the Gram-positive cell membrane the more solid barrier for CTAB and many other microbicides. Concentration dependent effects of CTAB on the structure of cell walls and the permeation of cell membranes especially those of Pseudomonades were examined by Wölfel *et al.* (1985). The survival of *Pseudomonas aeruginosa* cells even in the presence of concentrations of CTAB as high as 160 mg/litre is explained by the formation of cell aggregates which is induced by the surface active compound and protect the cells within the conglomerates.

CTAB is mainly used as an active ingredient in disinfectants. As a preservative permitted for use in cosmetics it is listed in the corresponding EC positive list with a maximum authorized concentration of 0·1%.

16.1.2 *N*-Alkyl-*N*,*N*-dimethyl-*N*-benzylammonium chloride — Benzalkonium chloride

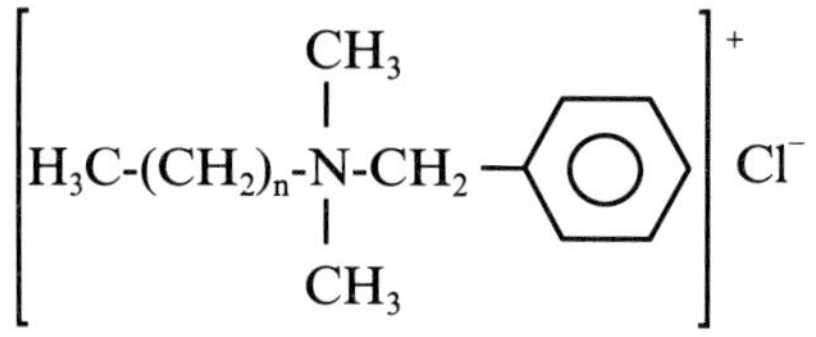

M_r approx. 354
CAS-No. 68391-01-5

Carbon chain distribution (example):

C_{12} chain	Approx. 60–80%
C_{14} chain	Approx. 20–40%
$C_8 + C_{10} + C_{16} + C_{18}$	Max. 15%
C_8, C_{10}, C_{16}, C_{18} each	Max. 10%

Chemical and physical properties (80% concentrate (carbon chain distribution as cited above) containing approx. 10% isopropanol and 10% water)

Appearance	Colourless to pale yellow, viscous liquid
Density at 20°C	Approx. 0·95 g/cm^3
Viscosity at 20°C	Approx. 860 mPa s
Vapour pressure at 50°C	Approx. 168 hPa
Solidification point	Approx. +5 to −12°C
Boiling point	Approx. 80°C
Flash point	Approx. 34°C
Ignition temperature	Approx. 362°C
Solubility	Miscible in any ratio with water, alkalis, lower alcohols, chlorohydrocarbons and ketones
pH (1% aqueous solution)	7–8 at 20°C
Surface tension (0·1% aqueous solution)	42 dyn/cm
Stability	Stable between pH 1 and 12, not compatible with anionic compounds, inactivation also by organic matter (such as dirt, protein, blood and sputum) and strong oxidizing agents, tolerates 120°C for at least 0·5 h

Toxicity data

LD_{50} oral Approx. 350 mg/kg rat

Irritant to skin and mucosa.
Ames test negative.

Ecotoxicity

LC_0 for *Brachydanio rerio*	Approx. 2 mg/litre (exposure time: 96 h)
LC_{100}	Approx. 3 mg/litre

Behaviour in waste water treatment plants: tolerable concentration for micro-organisms in activated sludge: approx. 10 mg/litre. Adsorption and biodegradation.

Antimicrobial effectiveness/applications

Benzalkoniumchloride is especially effective between pH 6 and 8. Its broad spectrum of activity covers bacteria, yeasts, fungi, algae, lichens and slime-forming organisms. Diluted solutions of Benzalkoniumchloride are therefore ideal for eliminating fungi, algae and lichens for example from finished coatings, plaster and stone, concrete and wooden surfaces. As a surface active compound it greatly reduces the surface tension of water; good wetting and penetration depth therefore are guaranteed when using 1–2% dilutions of Benzalkoniumchloride as a substrate pre-treatment agent.

Table 120. Minimum Inhibition Concentrations (MIC) of Benzalkoniumchloride in a Nutrient Solution

Test organism	*MIC (mg/litre)*
Aerobacter aerogenes	10
Aeromonas punctata	10
Bacillus mycoides	2·5
Bacillus subtilis	1·75
Escherichia coli	10
Leuconostoc mesenterioides	5
Proteus mirabilis	25
Pseudomonas aeruginosa	175
Pseudomonas fluorescens	100
Staphylococcus aureus	2·5
Candida albicans	25
Candida krusei	7·5
Rhodotorula mucilaginosa	5
Saccharomyces cerevisiae	50
Torula rubra	25
Torula utilis	50
Alternaria tenuis	40
Aspergillus flavus	100
Aspergillus niger	100
Aspergillus terreus	75
Aspergillus ustus	50
Chaetomium globosum	50
Mucor racemosus	10
Penicillium glaucum	10
Trichophyton mentagrophytes	25

Benzalkoniumchloride is also used as an active ingredient in disinfectants. With regard to this application one has carefully to observe the relatively weak activity of Benzalkoniumchloride against Pseudomonades (see Table 120) and its inactivation by anionic detergents and organic matter. In the food industry the active ingredient profits from being easily rinsed off and being tasteless and odourless.

Because of its distinctive algicidal and slimicidal effects Benzalkoniumchloride formulations may be used for the treatment of cooling and swimming pool water. However, the foam producing effect of Benzalkoniumchloride even at low concentrations (20–40 m/litre) sets limits to this application. Very limited is also the use of Benzalkoniumchloride as a preservative for the in-can/in-tank protection of aqueous functional fluids (including cosmetics) because of its cationic character.

16.1.3 *N,N*-Dimethyl-*N*-2-2-4-(1,1,3,3-tetramethylbutyl)-phenoxy-ethoxy-ethyl-benzylammonium chloride — Benzethoniumchloride

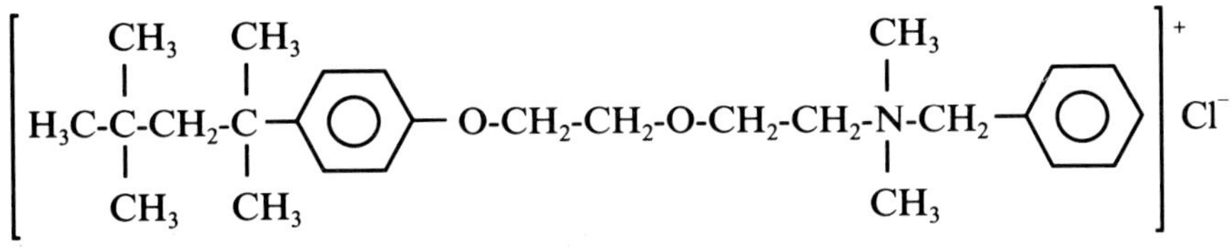

$C_{27}H_{42}ClNO_2$ M_r 448·10
CAS-No. 121-54-0

Chemical and physical properties

Appearance	White and odourless crystalline powder
Melting point	164–66°C
Solubility	Highly soluble in water, lower alcohols, glycols and acetone
pH (1% solution)	Approx. 5
Stability	Stable between pH 1 and 12, precipitation and inactivation by anionic compounds

Toxicity data

LD_{50} oral	420 mg/kg rat
	500 mg/kg mouse

Irritant to skin and mucous membranes.

Antimicrobial effectiveness/applications

As is characteristic for such quaternary ammonium compounds (e.g. Benzalkoniumchloride) they are not reliable in efficacy against Pseudomonades. Benzethoniumchloride is occasionally used in surface disinfectants. In the EC list of preservatives permitted for use in cosmetics it is mentioned with a maximum authorized concentration of 0·1%; however, Benzethoniumchloride is prohibited in products intended to come in contact with mucous membranes.

16.1.4 Di-*n*-decyl-dimethylammonium chloride

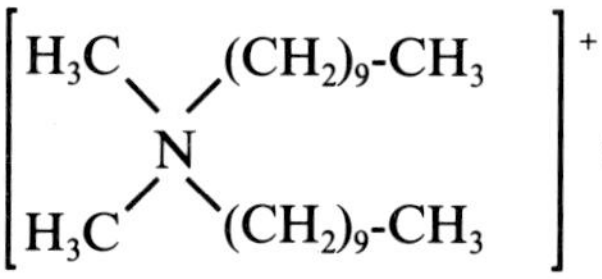

Cl^- $C_{22}H_{48}NCl$ M_r 362·09
CAS-No. 7173-51-5

Chemical and physical properties (50% solution containing 20% isopropanol and 30% water)

Appearance	Colourless to slightly yellow fluid
Density	0·91 g/ml
Viscosity	52 mPa s
Refractive index at 20°C	1·420–1·424
Flash point	29°C
Solubility	Miscible in any ratio with water and lower alcohols
pH (1% solution)	6·5–7
Decomposition	At 130–150°C (100% a.i.)
Vapour pressure	7·5 10^{-6} hPa (100% a.i.)
Stability	Not compatible with anionic compounds and strong oxidizing agents

Toxicity data

LD_{50}	oral	645 mg/kg rat
	dermal	2600 mg/kg rat

No sensitization according to the Magnusson–Kligman test.
Not mutagenic according to different in-vivo and in-vitro tests; not teratogenic in rats.

In concentrations higher than 10% skin irritation; strongly irritant to mucous membranes.

Ecotoxicity

LC_{50}	for trout	2·2 mg/litre (96 h exposure)
	for *Daphnia*	0·1 mg/litre (48 h exposure)

Biodegradability has been confirmed.

Antimicrobial effectiveness/applications

Table 121. Minimum Inhibition Concentrations (MIC) of Didecyl-dimethylammonium Chloride (100% a.i.) in a Broth Dilution

Test organism	*MIC (mg/litre)*
Escherichia coli	5
Proteus vulgaris	20
Pseudomonas aeruginosa	50
Staphylococcus aureus	0·5
Candida albicans	1·5
Aspergillus niger	5
Trichophyton mentagrophytes	5

When compared to other QACs didecyl-dimethylammonium chloride maintains an unusually high level of activity in the presence of organic matter, proteinaceous soil and hard water. This is confirmed by the fact that 400 mg a.i./ml are effective in the presence of 5% blood serum against *Staphylococcus aureus* according to the AOAC Use-Dilution Test. The compound is able to maintain its bactericidal activity even in the presence of residual amounts of anionics which normally are detrimental to the performance of QACs. Because of its favourable performance characteristics didecyl-dimethylammonium chloride is used in a great variety of application areas: as active ingredient in disinfectants, sanitizers, and cleaners for use in hospitals, homes, institutions, dairy, farm, and industrial areas; as a water treatment microbicide for cooling towers and secondary oil recovery; as a microbicide for the protection of textile material against permanent staining by the attack of mould-producing fungi and as an active ingredient in solutions for the temporary protection of freshly sawn timber against the growth of wood discolouring fungi.

16.1.5 Dioctyl-dimethylammonium chloride

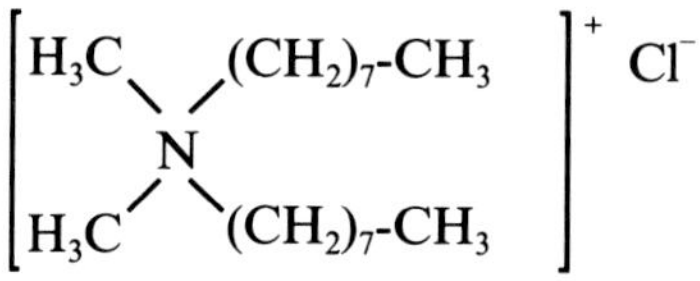

$C_{18}H_{40}NCl$

M_r 305·98
CAS-No. 5538-94-3

Chemical and physical properties (50% solution)

Appearance	Almost colourless and odourless fluid
Solubility	Miscible at any ratio with water and lower alcohols
Reaction in water	Slightly acidic; in comparison to other QACs it produces considerably less foam
Stability	Not compatible with anionic compounds, and strong oxidizing and reducing agents

Toxicity data

LD_{50}	oral	1·1 ml/kg rat
	dermal	1·51 ml/kg rabbit

Irritant to skin and mucous membranes.

Antimicrobial effectiveness/applications

The antimicrobial activity of dioctyl-dimethylammonium chloride corresponds to that of other QACs. Concentrations of 5–15 mg/litre are algicidal for fresh water algae. As the active concentrations of dioctyl-dimethylammonium chloride do not foam the microbicide is preferably used as a slimicide and algicide for water treatment.

16.1.6 Diallyl-dimethylammonium chloride (DMDAC)

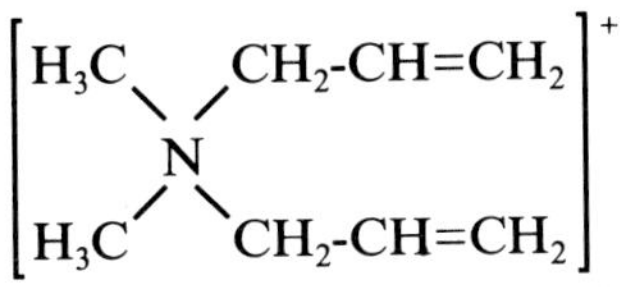

Cl^- $C_8H_{16}NCl$ M_r 161·67
CAS-No. 7398-69-8

Chemical and physical properties (polymeric 40% DMDAC solution)

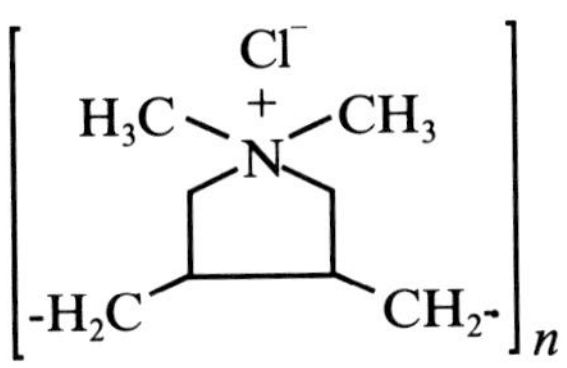

n = approx. 300

Appearance	Clear, almost colourless, viscous fluid with an odour similar to aldehydes
Density at 25°C	1·085 g/ml
Viscosity	8000–12 000 mPa s
pH	6–7
Solubility	Highly soluble in water and lower alcohols

Toxicity data

LD_{50}	oral	8·71 g/kg rat
	dermal	21·5 g/kg rabbit

Slightly irritant to skin and mucosa; no sensitization.
Not mutagenic.

Application

Preservative in cosmetics.

16.1.7 Cetylpyridinium chloride — 1-Hexadecylpyridinium chloride

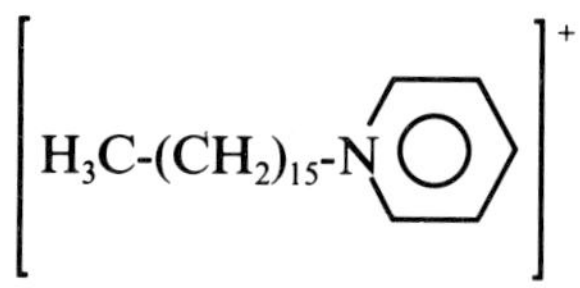

Cl^- $C_{21}H_{38}NCl$ M_r 340·10
CAS-No. 6004-24-6

Chemical and physical properties

Appearance	White, crystalline powder with a characteristic odour
Melting point	83–85°C
Solubility	Highly soluble in water, alcohol and trichloromethane, sparingly soluble in benzene and ether
pH (1% solution)	6–7
Stability	Not compatible with anionic compounds, proteins, zinc salts and polyphosphates

Toxicity data

LD_{50} oral		200 mg/kg rat
		400 mg/kg rabbit
	intravenous	20 mg/kg rat
		35 mg/kg rabbit

Irritant to skin and mucous membranes; 1:200 dilutions demonstrated no reactions (patch test on 600 human subjects); 1:5000 aqueous solution caused slight to no eye irritation on rabbits.

Antimicrobial effectiveness/applications

Cetylpyridinium chloride is used as a microbicide in disinfectants, in preservatives for cosmetics, as an active ingredient in deodorants.

16.1.8 Heteroaromatic quaternary ammonium salts — other examples

Alkyl-isoquinolinium bromide (alkyl = $C_{12}H_{25}$) M_r 378·39
CAS-No. 93-23-2

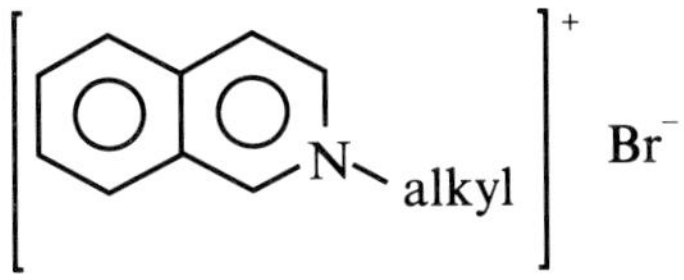

Hedaquinium chloride $C_{34}H_{46}Cl_2N_2$ M_r 553·66
CAS-No. 4310-89-8

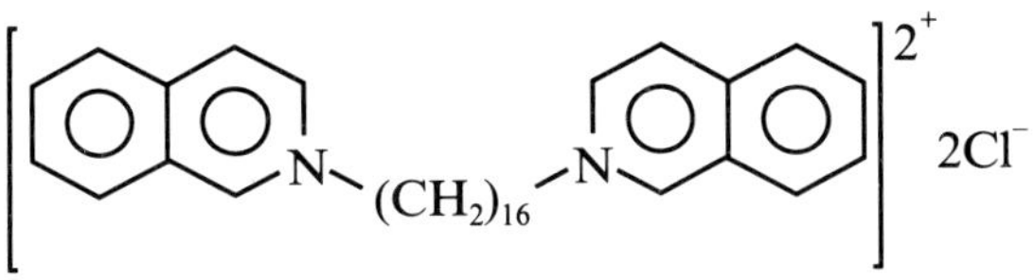

Dequalinium chloride $C_{30}H_{40}Cl_2N_4$ M_r 527·56
CAS-No. 522-51-0

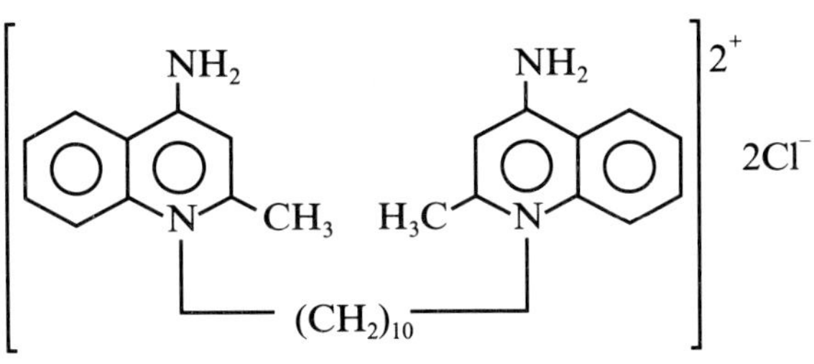

16.1.9 Polymeric quaternary ammonium salts

Polymeric quaternary ammonium salts are the result of reactions between nucleophilic monoamines or diamines and electrophilic di-halo compounds (Renbaum, 1973). They can be further cross-linked with ammonia, ethylenediamine or other reactive amines. Generally the polymeric QACs are not isolated but produced as viscous solutions containing 50–60% active ingredient. The molecular weights of linear polymeric QACs vary between 3000 and 10 000 and 30 000 and 50 000 for cross-linked compounds.

It has been found that the polymeric QACs have properties which make them uniquely different from the microbicidal QACs described heretofore. They do not produce foam, even at relatively high concentrations in water and reportedly they are of low toxicity. As ordinary microbicidal QACs they are highly effective against algae and slime forming micro-organisms and therefore used as algicides and slimicides for water treatment. But according to results presented by Hollis *et al.* (1991) commercial solutions of polymeric QACs are also capable of preserving aqueous functional fluids such as starch slurries and metal working fluids. Another possible application for polymeric QACs is the control of fouling by marine and fresh water molluscs, particularly the control of fresh water Asiatic clams of genus *Corbicula*, the most common of which is *C. fluminea* (Hollis & Lutey, 1988). Polymeric QACs can control the population of adult *C. fluminea* and inhibit the attachment of juveniles at concentrations of 2–8 mg/litre.

Examples of polymeric QACs:

- Poly[oxyethylene(dimethyliminio)ethylene(dimethyliminio)ethylene dichloride]
- Poly[hydroxyethylene(dimethyliminio)ethylene(dimethyliminio)methylene dichloride]
- Poly[hydroxyethylene(dimethyliminio)-2-hydroxypropylene(dimethyliminio)-methylene dichloride]
- [*N*-[3-(dimethylammonio)propyl]-*N'*[3-(ethyleneoxyethylenedimethylammonio) propyl]urea dichloride]
- α-4-[1-tris(2-hydroxyethyl)ammonium chloride-2-butenyl]poly[1-dimethyl-ammonium chloride-2-butenyl]-ω-tris(2-hydroxyethyl)ammonium chloride

The first example will be described in more detail showing the characteristics which are valid also for other polymeric QACs.

16.1.9.1 Poly[oxyethylene(dimethyliminio)ethylene(dimethyliminio)ethylene dichloride]

Chemical and physical properties (60% concentrate)

Appearance	Almost colourless, viscous fluid
Density at 25°C	1·15 g/ml
Flash point	>135°C
Volume/kg	870 ml
pH (0·1% solution)	6–7

Toxicity data

LD_{50} oral	3690 mg/kg male rat; 1850 mg/kg female rat

Not irritant to the skin.

Ecotoxicity

LC_0 for *Rasbora heteromorpha*	0·48 mg/litre (exposure 48 h)

Antimicrobial effectiveness/applications

Because of its high activity against algae, and slime forming micro-organisms including fungi, the polymeric QAC is used as a water treatment microbicide at concentrations between 2 and 20 mg/litre which are effective without causing foam problems. However, one has to give regard to the fact that similar to ordinary QACs the polymeric QACs, too, are easily adsorbed by organic material. It is therefore recommended to preclean the water system to be treated or to reduce the content of adsorbing materials in the system.

16.2 Guanidines and biguanides

The guanidines and biguanides exhibiting antimicrobial activity derive from guanidine and guanylguanidine = biguanide respectively. These are alkaline substances, stabilized by resonance; they form salts with acids.

$H_2N\text{-}C\text{-}NH_2$ $\Vert$ NH	$H_2N\text{-}C\text{-}NH\text{-}C\text{-}NH_2$ $\Vert \quad \Vert$ $NH \quad NH$
Guanidine	Guanylguanidine

16.2.1 Bis(guanidinooctyl)amine triacetate — Guazatine

$$\underset{\underset{NH}{\Vert}}{N_2N\text{-}C}\text{-}NH(CH_2)_8\text{-}NH(CH_2)_8\text{-}NH\text{-}\underset{\underset{NH}{\Vert}}{C}\text{-}NH_2 \cdot 3CH_3COOH$$

$C_{18}H_{41}N_7 \cdot 3CH_3COOH$ M_r 355·58 (amine)
CAS-No. 57520-17-9

Chemical and physical properties

Appearance	Colourless, crystalline solid
Melting point	Approx. 60°C
Solubility	Soluble in water and polar organic solvents

Toxicity data

LC_{50} oral 227–667 mg/kg rat

Carcinogenicity and teratogenicity were not demonstrable in 2-year feeding studies (200 mg/kg diet) with rats.

Antimicrobial effectiveness/applications

Guazatine as a membrane-active microbicide exhibits a broad spectrum of effectiveness. It is active by interfering with membrane structures of the microbial cell. Because of its antifungal activity it was developed at first as a fungicide for seed protection. But Guazatine later on also found some usage in wood preservatives for the temporary protection of freshly cut and sawn timber.

16.2.2 Oligohexamethylene biguanide hydrochloride — Poly(hexamethylene biguanide) hydrochloride (PHMB) — Polyhexanide

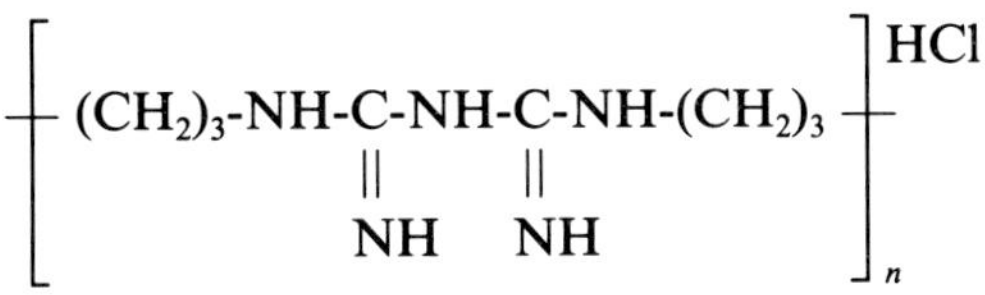

Average n = 4–7 CAS-No. 35708-79-3

Chemical and physical properties (20% aqueous formulation)

Appearance	Colourless to pale yellow liquid
Density at 25°C	1·04 g/ml
Viscosity at 25°C	5 mPa s
pH at 25°C	5·0–5·5
Flash point	Boils without flashing
Solubility	Water soluble in all portions; soluble in aliphatic alcohols, glycols and glycol ethers; insoluble in non-polar solvents
Stability	Exposure to temperatures > 80°C for prolonged periods adversely affects the microbicidal properties; in a concentration of 1 g/litre stable at pH values up to 10·0; above this precipitation may occur
Compatibility	The cationic compound gives water-insoluble precipitates with anionic surfactants, e.g. soaps, alkyl sulphates, alkylaryl sulphonates and anionic caramels; the a.i. is precipitated also by strong alkalis and complex phosphates

Toxicity data

LD_{50} oral 5000 mg/kg rat

The concentrated product can cause irritation to skin and mucosa.

Ecotoxicity

LC_0 for brown fingerling trout 10 mg/litre (exposure 96 h)
Because of its cationic character the a.i. is rapidly removed from effluents by absorption; at low concentrations biodegradation occurs reasonably rapidly.

Antimicrobial effectiveness/applications

The antimicrobial properties of PHMB, which is a membrane-active microbicide, have been described by Davis *et al.* (1968). Due to these properties and its low toxicity it has found particular use as an active ingredient in disinfecting cleansing agents for the food and beverage industry. However, the surfaces to which PHMB is applied should be widely free from organic matter and anionic detergents. The polymeric biguanide PHMB may also be used in pool sanitizers and for the treatment of raw hides against bacterial degradation in the leather industry. As can be seen from Table 122 PHMB is particularly effective against bacteria and less active against fungi. The relatively high activity of PHMB against *Pseudomonas aeruginosa* is especially interesting. In concentrations of 100–500 mg/litre PHMB exhibits bactericidal effectiveness. Quaternary ammonium compounds with their lack of activity against Pseudomonades and also phenolic microbicides may be used in combination with PHMB or bis-biguanides (Section 16.2.3.) for the formulation of disinfectants with an extraordinary broad activity spectrum (Bansemir *et al.*, 1987).

In the EC list of preservatives which cosmetic products may contain, PHMB is mentioned with a maximum authorized concentration of 0·3%.

Concentrated PHMB solutions have a marked corrosive effect on copper and are not compatible with stainless steel. Diluted PHMB solutions are not more corrosive than water and in contrast to quaternary ammonium compounds are low foaming.

Table 122. Minimum Inhibition Concentrations (MIC) of PHMB (20%) in Nutrient Agar according to a Technical Information Sheet of ICI

Test organism	*MIC (mg/litre)*
Staphylococcus aureus	20
Streptococcus faecalis	<5
Proteus vulgaris	250
Escherichia coli	20
Aerobacter aerogenes	20
Salmonella typhi	5
Pseudomonas aeruginosa	100
Saccharomyces cerevisiae	25
Saccharomyces ellipsodeus	25
Saccharomyces turbidans	25
Chaetomium globosum	125
Polystictus versicolor	1250
Trichoderma viride	250
Cladosporium resinae	1250
Alternaria tenuis	375
Cladosporium sphaerospermum	375
Aspergillus niger	375
Penicillium notatum	1250
Fusarium solani	1250
Candida albicans	250

16.2.3 1,6-Di-(4'-chlorophenyldiguanide)-hexan — Chlorhexidine

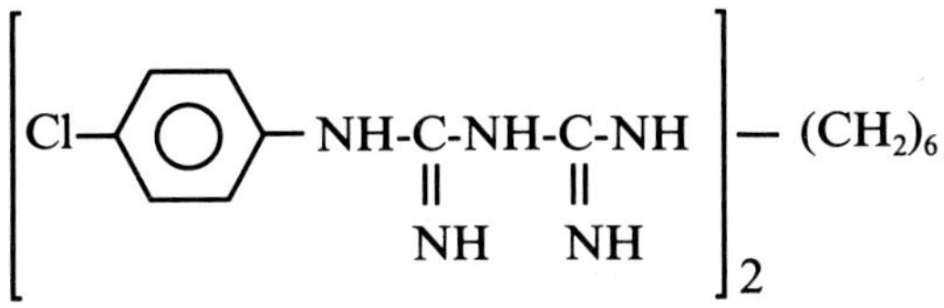

$C_{22}H_{30}Cl_2N_{10}$ M_r 505·48
CAS-No. 55-56-1

Chemical and physical properties

Appearance	Faintly cream coloured powder
Melting point	134°C
Solubility	0·08 g/litre H_2O
Reaction in water	Strongly alkaline

Salts of Chlorhexidine

Chlorhexidine dihydrochloride	
Melting point	260–262°C (decomposition)
Solubility	0·6 g/litre H_2O; highly soluble in alcohol and glycols
Chlorhexidine diacetate	
Melting point	154–155°C
Solubility	19 g/litre H_2O; soluble in ethyl alcohol, glycerine, propylene glycol, polyethylene glycols
Stability	Aqueous solutions decompose at temperatures > 70°C

LD_{50} oral	2000 mg/kg mouse
subcutaneous	320 mg/kg mouse
intravenous	25 mg/kg mouse

Irritant to skin and mucosa.
0·5% solutions do not cause skin irritation.

Toxicity data

Chlorhexidine digluconate

Solubility	Highly (approx. 70%) soluble in water; soluble in alcohols and glycols
pH (0·2% solution)	6·5–7·0

LD_{50} oral 1000 mg/kg mouse

Other Chlorhexidine salts which are especially active against Pseudomonades are described by Warner *et al.* (1980), without, however, having reached significant importance up to today:

Chlorhexidine dinalidixinate	mp 224–26°C
Chlorhexidine diphophanilate dihydrate	mp 172–74°C
Chlorhexidine disorbate monohydrate	mp 100–04°C

Antimicrobial effectiveness/applications

At low concentrations up to 200 mg/litre the bis-biguanide Chlorhexidine acts as a bacteriostat by inhibition of membrane enzymes, thus promoting the leakage of cellular constituents. Higher concentrations of Chlorhexidine (>200 mg/litre) coagulate cytoplasmic constituents accompanied by a bactericidal effect (Hugo & Longworth, 1964, 1966). The spectrum of activity covers Gram-positive and Gram-negative bacteria and at higher concentrations fungi, too. Chlorhexidine is not sporicidal and not lethal to acid-fast bacteria or viruses. The optimum pH for the effectiveness of Chlorhexidine is 8.

Chlorhexidine is mainly used as an active ingredient in disinfectants, deodorants and antiseptics and as a preservative in cosmetics and pharmaceuticals. The EC positive list of preservatives permitted for use in cosmetics mentions Chlorhexidine and its salts with a maximum authorized concentration of 0·3%. Formulating and using the active ingredient one has to remember its cationic nature which causes reduction of activity in the presence of organic matter, e.g. blood, serum, soaps and other anionic compounds. Another cause of activity loss is based on the very low water-solubility of certain Chlorhexidine salts; borate, citrate, carbonate, bicarbonate, chloride or phosphate salts precipitate in systems containing such anions.

16.3 Ampholytes

There is also a wide variety of microbicidal ampholytes; as they are all very similar in properties and activity only the most important ones are described here.

16.3.1 Dodecyl-di(aminoethyl)glycine

CH_3-$(CH_2)_{11}$-NH-$(CH_2)_2$-NH-$(CH_2)_2$-NH-CH_2-COOH

$C_{18}H_{39}N_3O_2$ M_r 329·51
CAS-No. 6843-97-6

Chemical and physical properties

Appearance	Clear yellow liquid
Solubility	Miscible with water

Toxicity data

LD_{50} oral	3000 mg/kg rat
Non-toxic level for rats	300 mg/kg food (application period 2 years)
ADI value	0·15 mg/kg/day

Not carcinogenic.
1% solutions do not cause skin irritation.

Antimicrobial effectiveness/applications

Dodecyl-di(aminoethyl)glycine is a broad spectrum microbicide. A 1% solution is microbicidal for bacteria, including Pseudomonades and *Mycobacterium tuberculosis*, yeasts and fungi within 1–60 min. The microbicidal ampholyte is not compatible with anionic and non-ionic detergents; its antimicrobial effectiveness is, however, only partly reduced in the presence of proteins.

Dodecyl-di-(aminoethyl)glycine is an active ingredient in disinfectants preferably for the use in the food industry. It is also used as a preservative for certain pharmaceuticals. The optimum pH range for the antimicrobial activity of the ampholyte is 5–9.

16.3.2 Dodecyl-β-alanine — *N*-dodecyl-β-aminopropionic acid

$CH_3\text{-}(CH_2)_{11}\text{-}NH\text{-}CH_2\text{-}CH_2\text{-}COOH$

$C_{15}H_{31}NO_2$ M_r 257·42
CAS-No. 1462-54-0

16.3.3 Dodecyl-β-aminobutyric acid

$$CH_3\text{-}(CH_2)_{11}\text{-}NH\text{-}\underset{\displaystyle CH_3}{\underset{|}{CH}}\text{-}CH_2\text{-}COOH$$

$C_{16}H_{33}NO_2$ M_r 271·45
CAS-No. 10024-28-9

17 ORGANOMETALLIC COMPOUNDS

Among the microbicidal organometallic compounds, the organomercury and organotin compounds are those which should be mentioned first, for they are the ones that have been used for material protection on a large scale throughout the world. Today, however, there is a strong movement throughout in favour of their substitution in view of their toxicity and especially their ecotoxicity. Organomercury compounds, in particular, have already been largely substituted and are no longer of much importance as microbicides for material protection. In the truest sense of the word they may be termed biocides, since they are effective not only against microbes, but also against all forms of life. Although Hippocrates already knew of the toxicity of mercury compounds he could not anticipate the environmental problems that were to arise through their excessive use — problems rooted in the fact that these compounds are biodegradable only to the stage of methyl- or dimethylmercury, both of which are extremely toxic (Lakowitz & Anderson, 1980). They diffuse rapidly across permeability layers such as membranes. Methyl- and dimethylmercury therefore accumulate in living organisms, especially those at the end of the food chain, such as predatory fish and animals dependent on fish. Without this accumulation the lethal doses of mercury compounds would be less problematical. If a quantity of mercury compounds is absorbed by a human, about 10% of it is carried to the brain. Long-term exposure to mercury compounds is particularly dangerous, even accumulation at very low rates causing manifest damage to the brain and nervous system after a number of years.

Mercury compounds, even at extremely low concentrations, are environmentally unsafe for another reason: investigations have shown that at concentrations as low as 1 ppb mercury-based microbicides cause 50% inhibition of the photosynthesis of phytoplankton and that at 50 ppb this photosynthesis stops entirely (Harries *et al.*, 1970). Concentrations of this order are also toxic to fish.

The arguments against the use of organomercury compounds go beyond toxicity and ecotoxicity; the chemico-physical properties of these products are not particularly favourable either. For example, organomercury compounds are not ideal as paint film fungicides because they are too volatile and insufficiently stable to light. Furthermore, by reacting with H_2S in the atmosphere — in industrial regions, for example, but also in agricultural regions — they make paint films turn grey.

Organotin compounds, though less toxic to warm-blooded animals than organomercury compounds, are also very toxic in the environment. Even at concentrations in the ppb range they disturb the biocenosis of surface waters considerably and therefore have to be handled with the greatest care (Polster & Halacka, 1971). They are sensitive to light, but in their main applications, wood preservatives and antifouling paints, this does not matter. Yet their ecotoxicity has led to worldwide demands for a tin ban where antifouling paints are concerned. For the reasons just mentioned, scientists throughout the world began to look for less toxic, yet effective, active ingredients for these paints.

Among the organometallic compounds whose substitution is being demanded for reasons of toxicity is also the arsenic compound 10,10'-oxy(bis-phenoxy) arsine (OBPA), a microbicide which has been used predominantly throughout the world to make plastics resistant to microbes.

Lastly, trihexylene glycol diborate should be mentioned. This organometallic compound of relatively low mammalian toxicity is soluble in organic solvents and thus represents a form in which boric acid (Section 6.2.1) can be transported. When, for example, it is applied to wood, penetration is followed by in-situ hydrolysis to boric acid, the wood thus being protected from fungi and insects.

Because, among organometallic compounds, mercury, tin and arsenic compounds are becoming increasingly less important as microbicides for material protection, only the most important prototypes of these compounds will be described in detail below.

Microbicidal metal chelates are described elsewhere:

Section 11.1.4 Zinc-bis-(2-pyridine-thiol-1-oxide) = Zinc Pyrithione
Section 11.5 Copper 8-hydroxyquinoline = Oxine copper.

17.1 10,10'-Oxybisphenoxyarsine (OBPA)

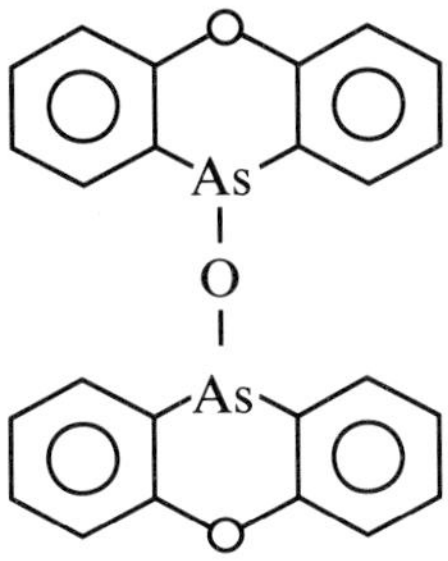

$C_{24}H_{16}As_2O_3$

M_r 502·24
CAS-No. 58-36-6

Chemical and physical properties

Appearance	White, odourless powder
As content	29·84%
Melting point	184–186°C
Vapour pressure	10^{-5} Pa at 25°C
Solubility	5 mg/litre H_2O. Soluble in organic solvents, e.g. propylene glycol, butylbenzyl phthalate, di(2-ethylhexyl) phthalate, diisodecyl phthalate, epoxidized sojaoil
Stability	Heat resistant under processing conditions for plastic material, e.g. PVC

Toxicity data

LD_{50} oral 30–40 mg/kg rat

Corrosive to skin and mucosa.

Rats exposed for 2 weeks to vapours of OBPA heated to 121°C showed no signs of irritation of nose and eyes.

Fish toxicity of a 10% solution of OBPA in propylene glycol for trout:

LC_{50} 0·2 mg/litre (exposure: 96 h)

Antimicrobial effectiveness/applications

OBPA is a biocide, that means it is not only toxic to microbes but also to fish and other living species. Its antimicrobial effectiveness is similar to that of organomercurials covering a broad spectrum of bacteria, yeasts and mould producing fungi (minimum inhibition concentrations approx. 10 mg OBPA/litre). Among the fungal species *Scopulariopsis brevicaule* is relatively resistant to OBPA under use conditions.

OBPA is mainly used for the antimicrobial treatment of plastic material. For that purpose it is offered as a solution in plasticizers, e.g. 2% solutions in epoxidized sojaoil or diisodecyl phthalate or di(2-ethylhexyl) phthalate or polypropylene glycol or butylbenzyl phthalate. Such formulations are easily incorporated into master batches for the production of plastic materials without negatively influencing the properties of the finished products. OBPA is sufficiently heat resistant to stand the processing temperatures. It is recommended to incorporate 1·5% of one of the a.m. 2% OBPA solutions into plastic material for interior application, that is 0·03% OBPA or 0·009% As only. With regard to the small amounts of OBPA which achieve antimicrobial effectiveness in plastic materials, leachability of active ingredient may be a problem, although the solubility of OBPA in water is very low. For exterior application the recommendation therefore is addition of 2·5% of a 2% OBPA solution. Upsher & Roseblade (1984) report about the limited long-term activity of OBPA in plastic material exposed at a jungle site.

17.2 2,2'-[(1,1,3-trimethyl-1,3-propanediyl) bis (oxy)] bis [4,4,6-trimethyl-1,2,3-dioxyborinane] — Trihexylene glycol biborate

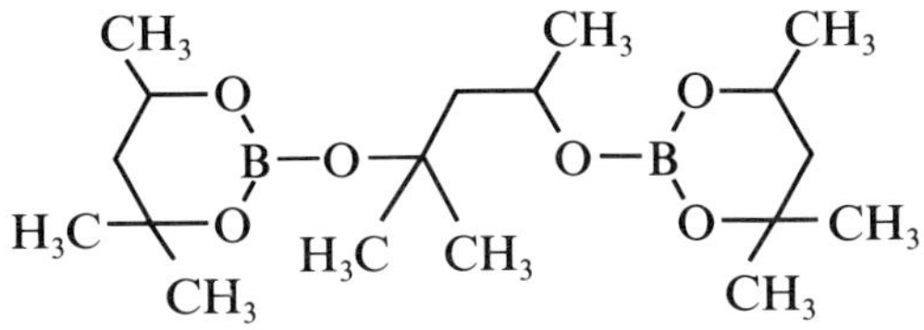

$C_{18}H_{36}O_6B_2$ M_r 370·11
CAS-No. 100-89-0

Chemical and physical properties

Appearance	Clear, colourless to pale yellow easy pourable liquid; odour: similar to hexylene glycol
Boron content	5·6–5·8% (equivalent to 32·0–33·2% boric acid)
Boiling point	314–326°C at 101·3 kPa; 143–149°C at 0·267 kPa
Refractive index	1·4408 at 18·5°C
Flash point	175°C closed cup
Solubility	Soluble in all proportions in white spirit, kerosene, carbon tetrachloride, benzene, toluene, xylene, petroleum ether
Stability	Hydrolyses in contact with water, even when exposed to atmospheric moisture

Toxicity data

LD_{50}	oral	>4000 mg/kg rat
	dermal	>2000 mg/kg rat

Moderate irritant to skin and mucosa; not a skin sensitizer.

Antimicrobial effectiveness/applications

Trihexylene glycol biborate has been developed for the remedial preservative treatment of timber. After penetration *in situ* the hydrolysis to boric acid (Section 6.2.1) occurs providing fungicidal and insecticidal properties. For boric acid on *Pinus sylvestris* sapwood the following toxic limits have been established:

for *Coniophora puteana* 0·43–0·65 kg/m^3
for *Poria xantha* 0·08–0·20 kg/m^3
(Forest Products Research Laboratory, Princes Risborough, 1968).

The biborate is used in solvent based wood preservatives. It is also effective on white rots such as *Polystictus versicolor*.

17.3 Phenylmercury acetate (PMA)

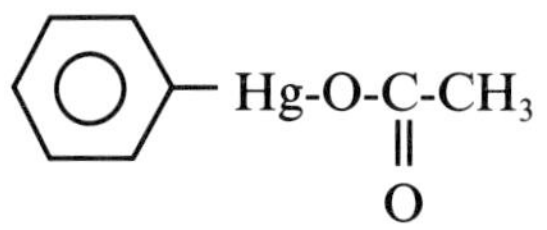

$C_8H_8HgO_2$

M_r 336·74
CAS-No. 353-367-407

Chemical and physical properties

Appearance	Crystalline, colourless powder
Mercury content	59·6%
Melting point	150–152°C
Solubility	In water 1·7 g/litre; soluble in ethanol and acetone
Stability	Due to its cationic character PMA reacts with anionic detergents and sulphides under inactivation

Toxicity data

LD_{50} oral   30 mg/kg rat

Irritant to skin and mucosa; causes skin sensitization; skin contact can result in adsorption of measurable amounts of mercury.

MWC (calcul. as Hg)   0·01 mg/m^3

PMA is highly ecotoxic.

Antimicrobial effectiveness/applications

PMA is extremely effective against bacteria, fungi, yeasts and algae, it acts also sporistactically. The minimum inhibition concentrations for bacteria range between 5 and 15 mg/litre and those for fungi, yeast and algae below 1 mg/litre. In acidic media PMA is not particularly effective; the optimum pH for activity of PMA is approximately 8·8, as at higher pH values dissociation of PMA to the electrophilic active radical Ph-Hg^+ increases. The radical is likely to interfere with a large number of cell processes by binding, for example, to essential thiol groups thus inhibiting glycolysis. But PMA and other organomercury compounds also show a strong affinity for phosphates, histidine side chains of proteins and for purines, pteridines and porphyrines (Corbett *et al.*, 1984). It is therefore not surprising that organomercury compounds, unlike most other microbicides, also inhibit exogenic enzymes, e.g. cellulases.

PMA has been used widely as a microbicide for the in-can/in-tank protection of aqueous functional fluids such as latex paints. The ability of PMA to inhibit enzymes that break down cellulosic thickeners was considered a special advantage in that application. At higher addition rates PMA acts also as a paint film fungicide and algicide. However, the duration of activity of PMA in paint films is limited due to volatility and leachability of the active ingredient. Additionally PMA was used extensively in slimicides, algicides, antifouling paints and wood preservatives, where it also exhibited insecticidal effectiveness.

17.4 Phenylmercury oleate (PMO)

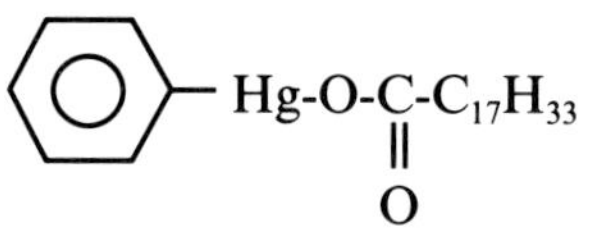

$C_{24}H_{38}HgO_2$ M_r 559·16

Chemical and physical properties

Appearance	White to yellow powder
Mercury content	35·9%
Melting point	186–215°C
Solubility	Practically insoluble in water; soluble in organic solvents; dissolves readily in mineral spirits and aromatic solvents
Stability	Stable between pH 4 and 9. Coloration and inactivation with sulphides

Toxicity data

LD_{50} oral Approx. 60 mg/kg rat

Irritant to skin and mucosa. Highly ecotoxic.

MWC (calcul. as Hg) 0·01 mg/m^3

Antimicrobial effectiveness/applications

PMO is in efficacy similar to PMA. As PMO is virtually insoluble in water and of low volatility only, it has been used mainly as an active ingredient in wood preservatives, antifouling coatings, exterior and interior paints and sealants.

17.5 Sodium ethylmercury thiosalicylate — Thiomersal

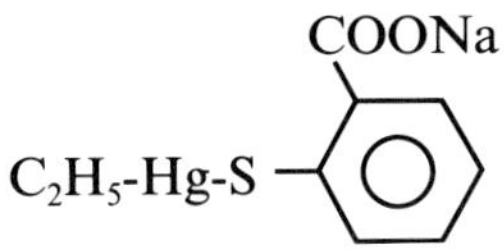

$C_9H_9HgO_2SNa$

M_r 404·82
CAS-No. 54-64-8

Chemical and physical properties

Appearance	Yellowish crystalline powder
Mercury content	49·6%
Melting point	234–237°C (decomposition)
Solubility	Highly soluble in water (1 g in 1 ml H_2O) and ethanol (1 g in 8 ml C_2H_5OH); practically insoluble in ether and benzene
pH (1% solution in H_2O)	6·7
Stability	Sensitive to light, chlorides and iodides

Toxicity data

LD_{50} oral 98 mg/kg rat

Skin sensitizer.
Irritant to skin and mucosa. Highly ecotoxic.

Antimicrobial effectiveness/applications

Thiomersal is highly effective against Gram-positive and Gram-negative bacteria and is mainly used as a preservative in pharmaceutical preparations; addition rates: 100–200 mg/litre. It is not compatible with non-ionic surface active compounds and proteins. In the EC list of preservatives allowed for cosmetics Thiomersal is mentioned with a maximum authorized concentration of 0·007% (of Hg) for eye make-up and eye make-up remover only; on the label there must be printed the warning: Contains Thiomersal.

17.6 Bis(tri-*n*-butyltin)oxide (TBTO) — Hexabutyl-distannoxane

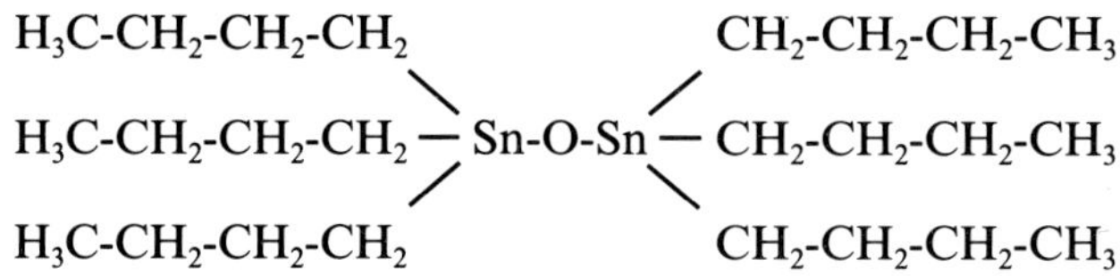

$C_{24}H_{54}OSn_2$ M_r 596·07
CAS-No. 56-35-9

Chemical and physical properties (95% TBTO)

Appearance	Colourless to light yellow liquid of characteristic odour
Sn content	38·6%
Boiling point	220–230°C at 13 hPa; 173°C at 1·3 hPa
Melting point	<-45°C
Density at 20°C	1·17 g/cm^3
Refractive index at 20°C	n_D 1·4865–1·4870
Vapour pressure at 20°C	Approx. 1×10^{-3} Pa
Solubility	20–60 mg/litre H_2O dist.; 3–10 mg/litre artificial sea water; sparingly soluble in ethylene glycol, soluble in most organic solvents
Partition coefficient (n-octanol/H_2O)	Low P_{OW} 3·2–3·8
Stability	Hydrolysis/hydration to $TBTOH_2^+ \cdot OH^-$ in water; the C-Sn bond of TBTO is, however, stable between pH 2·9 and 10·3, but sensitive to UV light (degradation of Bu_3SnOH via Bu_2Sn^{2+} and $BuSn^{3+}$ to Sn^{4+})

Toxicity data (Behret, 1988)

LD_{50} oral	92–194 mg/kg rats or mice
dermal	605–11 700 mg/kg rats or rabbits
intravenous	6·0 mg/kg mice
interperitoneal	5–20 mg/kg rats or mice
LC_{50} on inhalation of aerosols (4 h)	65 mg/m^3 for rats

Corrosive to the skin and mucous membranes, but not a skin sensitizer. From many test results available up to now there is no evidence of mutagenicity, teratogenicity or carcinogenicity.

MWC 0·002 ml/m^3 (0·05 mg/m^3)

Ecotoxicity

LC_{50}	
— (96 h) for fish	6–24 mg/litre
— (48 h) for oysters	1·6–1·9 mg/litre; growth inhibition starts at 0·02–0·06 mg/litre
EC_{50} (48 h) for *Daphnia magna*	1·7 mg/litre
NOEC (21 days) for *Daphnia magna*	0·16 mg/litre

Multiplication of micro-algae in fresh water and sea water is inhibited by concentrations of 0·33–1·03 mg/litre (EC_{50} 72 h).

Biodegradation of TBTO by micro-organisms occurs under debutylation to inorganic tin (Sn^{4+}). However, at a concentration of 5 mg TBTO/ml the activity of activated sludge organisms is inhibited already.

Antimicrobial effectiveness/applications

Although organotin compounds have been known since the middle of the last century, their suitability as microbicides for the protection of materials against biodeterioration was not recognized before the middle of this century. Especially TBTO then quickly gained considerable importance for the microbicidal treatment of a variety of materials due to its high activity and broad spectrum of effectiveness which covers bacteria, yeasts, fungi and algae. Tributyltin compounds such as TBTO belong to the electrophilic active microbicides and work for instance by inhibiting oxidative phosphorylation.

The use of TBTO as an active ingredient in technical preservatives and in disinfectants is no longer of significant importance. Approximately 79% of the TBTO and similar derivatives are used in antifouling coatings and 20% in wood preservatives. However, in these important fields of application the use of TBTO is decreasing because of unfavourable toxicity data. In the most significant field of application for organotins — antifouling coatings (Anderson & Dalley, 1986) — the ecotoxicity of the compounds causes severe problems. TBTO was much in demand as an active ingredient for antifouling paints also because of its activity against different sea animals which attach to ship bottoms.

In wood preservatives an advantage is the strong efficacy of TBTO against wood destroying fungi (Table 123). As the activity of TBTO against staining fungi, e.g. *Aureobasidium pullulans* and *Sclerophoma pityophila*, is not very pronounced, in wood preservatives TBTO is generally combined with other fungicides, e.g. trihalomethylthio compounds (Section 14), benzimidazolyl methylcarbamate (Section 9.4) and iodopropargylcarbamates (Section 9.1), which are highly effective against wood staining fungi. The permanence of TBTO in timber is not unlimited. There may occur losses of TBTO from impregnated timber by evaporation and by

Table 123. Toxic Limits of TBTO, Tested in Pine-wood according to British Standard 838[a]

Test fungi	*Toxic limits (kg/m^3/wood)*
Coniophora cerebella	0·36–0·81
Lenzites trabea	0·02–0·06
Poria monticola	0·02–0·06
Merulius lachrymans	0·02–0·06
Polystictus versicolor	1·18–1·33

[a] Schering AG 1984, Technical Information — Organotin Compounds.

degradation *in situ*, particularly from zones near the surface and from unpainted wood. In the inner parts of impregnated wood TBTO is normally protected against the degrading influence of UV light, but not in any case against fungal detoxification. According to Henshaw *et al.* (1978) especially the white-rot fungus *Coriolus versicolor* is able to degrade TBTO with the help of an efficient oxidase system.

17.7 Tributyltin esters —TBT esters

The esterification of bis(tri-*n*-butyltin) oxide (TBTO) with acids leads to tributyltin esters:

$$(n\text{-but.})_3\text{Sn-O-Sn}(n\text{-but.})_3 + 2\ \text{R-COOH} \longrightarrow \text{R-COO-Sn}(n\text{-but.})_3 + H_2O$$

Among these, carboxylic acid esters have gained importance in the field of wood preservation, because they are of lower volatility and water-solubility, but higher thermal stability than TBTO. From comparison data it can be concluded that the tributyltin (TBT) moiety of the TBT esters determines their antimicrobial effectiveness and also their oral toxicity. This is in line with the fact that the TBT esters tend to dissociate into the hydrated TBT cation and the respective anion. Due to the lower TBT content compared with TBTO greater quantities of TBT carboxylic acid ester therefore have to be used to achieve sufficient antimicrobial effectiveness.

A hydrohalic acid ester of practical importance as a microbicide has been tributyltin fluoride.

17.7.1 Tributyltin naphthenate (TBTN)

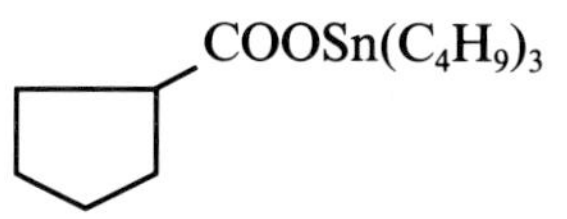

$C_{18}H_{36}O_2Sn$ M_r 403·2
CAS-No. 85409-17-2

Chemical and physical properties

Appearance	Clear, yellow liquid with a characteristic odour
Tin content (technical grade)	Approx. 21%
Boiling point	Approx. 125°C at 0·5 hPa
Density at 20°C	Approx. 1·09 g/cm³
Viscosity at 25°C	Approx. 2900 mPa s
Flash point	78°C
Solubility	In water: 1·3 mg/litre; soluble in organic solvents
Volatility	Weight loss after storage at 65°C for 6 days: 2·9% (TBTO for comparison: 26%)

Toxicity data (Schweinfurth & Günzel, 1987)

LD_{50} oral	225 mg/kg rat
dermal	4600 mg/kg rat
LC_{50} for rats on inhalation of aerosols	152 mg/m³

Irritant to skin and mucosa.

MWC (Calcul. as TBTO) 0·002 ml/m³ (0·05 mg/m³)

Antimicrobial effectiveness/applications

TBTN exhibits fungicidal and algicidal activity and has been used as an active ingredient in antifouling paints and wood preservatives. The fungicidal activity of TBTN is preferably against wood-destroying fungi. Due to the lower tributyltin content compared with TBTO (Section 17.6) approximately double the addition rates have to be applied.

17.7.2 Tributyltin linoleate (TBTL)

$CH_3(CH_2)_4CH = CHCH_2CH = CH(CH_2)_7COOSn(C_4H_9)_3$

$C_{30}H_{58}O_2Sn$ M_r 569·50

Chemical and physical properties

Appearance	Clear, yellow liquid with a characteristic odour
Tin content	Approx. 20·5%
Boiling point	Approx. 140°C at 10·5 hPa
Density at 20°C	1·05–1·06 g/cm^3
Viscosity at 25°C	Approx. 350 mPa s
Flash point	Approx. 105°C
Solubility	In water: 1 mg/litre; soluble in organic solvents
Volatility	Weight loss after storage at 65°C for 6 days: 1·0% (TBTO for comparison: 26%)

Toxicity data (Schweinfurth & Günzel, 1987)

LD_{50} oral 190 mg/kg rat

Irritant to skin and mucosa.

MWC (calcul. as TBTO) 0·002 ml/m^3 (0·05 mg/m^3)

Antimicrobial effectiveness/applications

TBTL is a microbicide of minor importance in the field of material protection; its antimicrobial activity corresponds to that of TBTN (Section 17.7.1).

17.7.3 Tributyltin benzoate (TBTB)

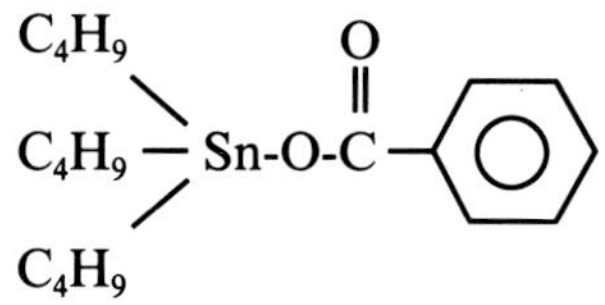

$C_{19}H_{32}O_2Sn$ M_r 411·18

Chemical and physical properties

Appearance	Liquid at temperatures above 25°C
Tin content	28%
Boiling point	Approx. 135°C at 10·3 hPa
Density at 20°C	1·17–1·20 g/cm³
Viscosity at 25°C	Approx. 20 mPa s
Solubility	Sparingly soluble in water, soluble in organic solvents

Toxicity data (Schweinfurth & Günzel, 1987)

LD_{50}	oral	100–200 mg/kg rat;
		108 mg/kg mouse
	dermal	505 mg/kg rat

Irritant to skin and mucosa.

MWC (calcul. as TBTO) 0·002 ml/m³ (0·05 mg/m³)

Antimicrobial effectiveness/applications

TBTB is a microbicide which has not gained much importance; it has been used in disinfectants.

17.7.4 Tributyltin fluoride (TBTF)

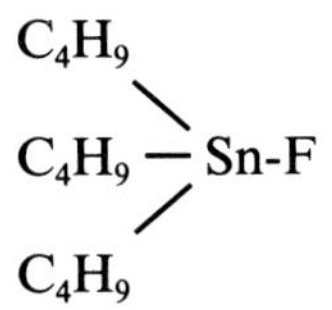

$C_{12}H_{27}FSn$

M_r 279·06
CAS-No. 27615-98-1

Chemical and physical properties

Appearance	White powder
Tin content	Approx. 42%
Melting point	269–271°C
Solubility	Practically insoluble in H_2O; soluble in organic solvents: 4·8% in methanol, 3·4% in isopropanol, 0·9% in xylene, 10% in chloroform

Toxicity data (Schweinfurth & Günzel, 1987)

LD_{50} oral	94 mg/kg rat
dermal	680 mg/kg rabbit

Irritant to skin and mucosa.

MWC (calcul. as TBTO) 0·002 ml/m^3 (0·05 mg/m^3)

Antimicrobial effectiveness/applications

TBTF is effective against bacteria, yeasts, fungi and algae. It has been used mainly as an active ingredient in antifouling paints and wood preservatives.

18 VARIOUS COMPOUNDS

18.1 *N*-Nitroso-*N*-cyclohexyl-hydroxylamine

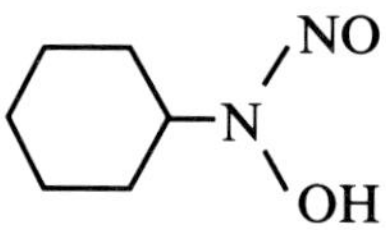

$C_6H_{12}H_2O_2$

M_r 144·18
CAS-No. 4883-72-1

18.1.1 Tris-(*N*-cyclohexyldiazeniumdioxy)-aluminium

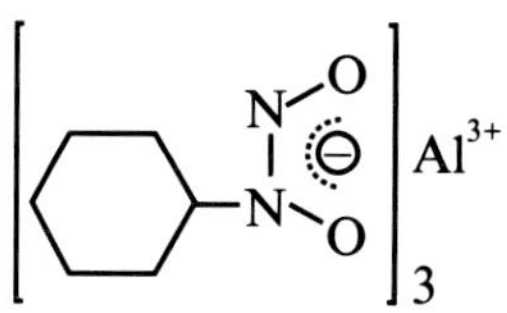

$C_{18}H_{33}N_6O_6Al$

M_r 465·48
CAS-No. 40027-80-3

Chemical and physical properties

Appearance	Yellow, brownish solid (pellets) with a typical smell
Density	0·7 g/ml

Toxicity data

LD_{50} oral	5610 mg/kg rat
dermal	2500 mg/kg rat
on inhalation of aerosol (4 h)	0·39 mg/litre for rats

Non-mutagenic according to the Ames test. Irritant to the skin and mucous membranes.

Tris-(*N*-cyclohexyldiazeniumdioxy)-aluminium is especially effective against wood-destroying fungi.

18.1.2 *N*-Cyclohexyldiazeniumdioxy-potassium hydrate

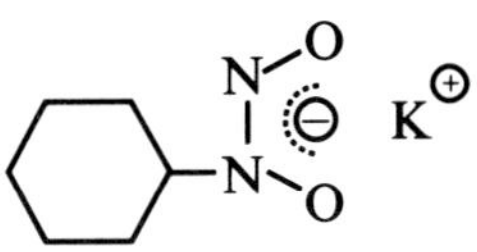

$C_6H_{13}N_2O_3K$

M_r 200·28
CAS-No. 27697-50-3

Chemical and physical properties

Appearance	Colourless solid with a typical smell
Density	0·7 g/ml
Solubility	Highly soluble in water

Toxicity data

LD_{50} oral   440 mg/kg rat

Irritant to the skin and mucous membranes.

N-cyclohexyldiazeniumdioxy-potassium is a powerful fungicide which in particular is active against wood-destroying fungi. It is mainly used for the protection of wooden materials and for those incorporated together with glues and adhesives into such materials.

18.2 2,6-Dimethyl-4-tridecylmorpholine — Tridemorph

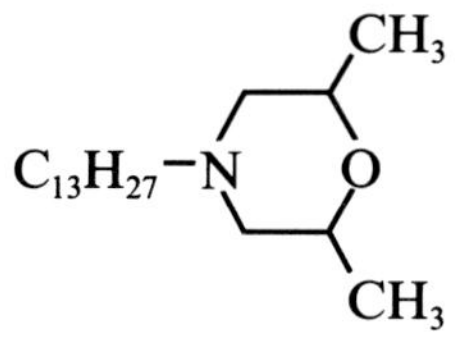

$C_{19}H_{39}NO$ M_r 297·53
CAS-No. 81412-43-3

Chemical and physical properties

Appearance	Yellow liquid with a slight typical odour
Density	Approx. 0·87 g/ml
Crystallization point	Below −20°C (reversible)
Flash point	138°C

Toxicity data

LD_{50} oral approx. 870 mg/kg rat

Irritant to the skin and mucous membranes.

Tridemorph is a broad spectrum microbicide. At application rates of ≤100 mg/litre it has been found to be an effective agent against bacteria, yeasts and fungi, in particular wood colouring and wood-destroying fungi. Accordingly it is used in wood preservatives.

18.3 Methylene bisthiocyanate (MBT)

NCS-CH_2-SCN $C_3H_2N_2S_2$ M_r 130·13
CAS-No. 6317-18-6

Chemical and physical properties

Appearance	Yellow crystalline powder of pungent odour	
Density	2·0 g/ml	
Melting point	105°C	
Solubility at 20°C in g/100 g solvent	acetone	41
	acetonitrile	42
	dimethyl sulphoxide	approx. 80
	dioxane	32
	ethylene glycol	approx. 4
	ethylene glycol-monomethyl ether	approx. 17
	methanol	8
	methyl ethyl ketone	approx. 25
	toluene	approx. 1
	water	0·5
Stability	Stable in acidic systems, if not heated; decomposes in alkaline systems with a pH above 7·5; heat stable to 100°C	

Toxicity data

LD_{50} oral	104·9 mg/kg rat
dermal	4220 mg/kg rabbit
on inhalation	32 mg/m^3

Corrosive to the skin and mucous membranes. Highly toxic to fish.

Antimicrobial effectiveness/applications

MBT is a powerful microbicide of limited stability; it is used to control bacteria, algae, yeasts and fungi in industrial water systems, especially in paper mills.

Table 124. Minimum Inhibition Concentrations (MIC) of MBT in Nutrient Agar

Test organism	*MIC (mg/litre)*
Aspergillus niger	2·5–5
Aerobacter aerogenes	5–10
Clostridium sp.	5–10
Bacillus megaterium	5
Staphylococcus aureus	5
Chlorella sp.	3
'Black algae' (from pools)	12

18.4 Iodomethylthiocyanate

I-CH_2-SCN	C_2H_2INS	M_r 199·02 CAS-No. 73909-00-9

Chemical and physical properties

Appearance	Liquid, red brown concentrate of a faint typical smell
Density	2·2 g/ml

Toxicity data

LD_{50} oral	26 mg/kg rat
dermal	100 mg/kg rat

Extremely irritating to the skin, mucous membranes and the retina of the eye.

Antimicrobial effectiveness/applications

Iodomethylthiocyanate is a powerful microbicide with a broad activity spectrum; it is offered as an experimental compound. Due to its high toxicity it has not gained practical importance as a microbicide for the protection of materials and industrial water systems.

18.5 Chloronaphthalene mixture

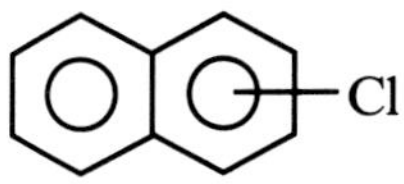

$C_{10}H_7Cl$ M_r 162·62
CAS-No. 25586-43-0

Chemical and physical properties

Appearance	Clear, slightly yellow liquid with a strong characteristic odour
Density at 20°C	1·20 g/ml
Boiling point	259°C
Vapour pressure (hPa)	Approx. 0·05 at 20°C; Approx. 0·07 at 25°C; Approx. 0·4 at 50°C
Crystallization point	Approx. −6°C
Flash point	Approx. 115°C
Solubility	Sparingly soluble in water, very soluble in organic solvents

Toxicity data

LD_{50} oral	2030 mg/kg rat
dermal	>1000 mg/kg rat
on inhalation (1 h)	>20 mg/litre for rats

Moderately irritant to skin and mucosa.

Ecotoxicity

LC_{50} for *Leuciscus idus*	4·15 mg/litre
for activated sludge organisms	>2000 mg/litre

Easily degraded by adapted bacteria.

Antimicrobial effectiveness/applications

Chloronaphthalene is used as a fungicide and insecticide in wood preservatives, mainly for wood in ground contact. Its strong odour inhibits broad application.

18.6 2-Methylthio-4-tert.-butylamino-6-cyclopropylamino-*s*-triazine

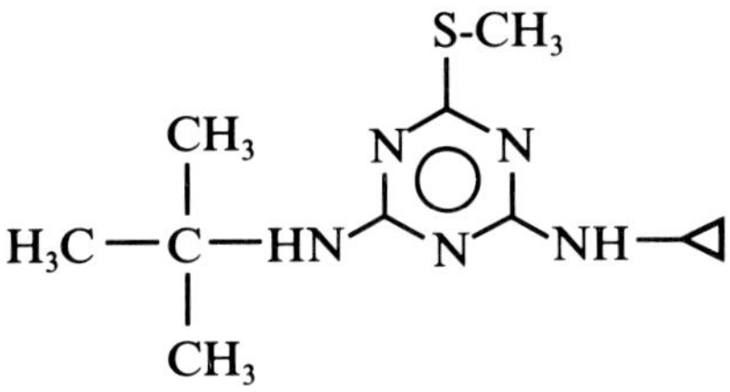

$C_{11}H_{19}N_5S$ M_r 253·36
CAS-No. 28159-98-0

Chemical and physical properties

Appearance	Yellowish powder	
Density	1·1 g/ml	
Melting point	130–133°C	
Solubility (g/litre) at 20°C	water	0·007
	xylene	approx. 50
	®Shellsol-A	80
	solvent naphtha	50
	white spirit	10
	methylisobutylketone	120
	butyl acetate	150
	ethylglycol monomethylether	100
	butylene glycol	150
	propylene glycol	10
	octanol	50
	®Texanol	50
Vapour pressure at 25°C	$8{\cdot}8 \times 10^{-7}$ hPa	

Toxicity data

LD_{50} oral	>2000 mg/kg rat
on inhalation (4 h) for rats	>4100 mg/m^3

Not irritant to the skin and eyes.

Ecotoxicity

LC_{50} for *Daphnia* (24 h)	66 mg/litre
for Zebra fish (96 h)	4 mg/litre
for Rainbow trout (96 h)	0·94 mg/litre
for Bluegill	2·9 mg/litre

Antimicrobial effectiveness/applications

The *s*-triazine derivative is not effective against bacteria, yeasts or fungi. It is an inhibitor of photosynthesis and accordingly effective against algae. The activity spectrum is broad; it covers fresh water algae and sea water algae. The minimum inhibition concentrations for these organisms range between 0·01 and 0·1 mg/litre. The very low water-solubility of the compound in combination with its high anti-algal activity and favourable toxicological data have led to the proposal to use the s-triazine derivative as an active ingredient in antifouling coatings for the substitution of ecotoxic organotin compounds (Section 17). The *s*-triazine derivative shows no effect, however, against fouling by sea animals such as barnacles, serpulids and molluscs. Combination with copper compounds therefore may be required.

18.7 Trans-1.2-bis(*n*-propylsulphonyl) ethene

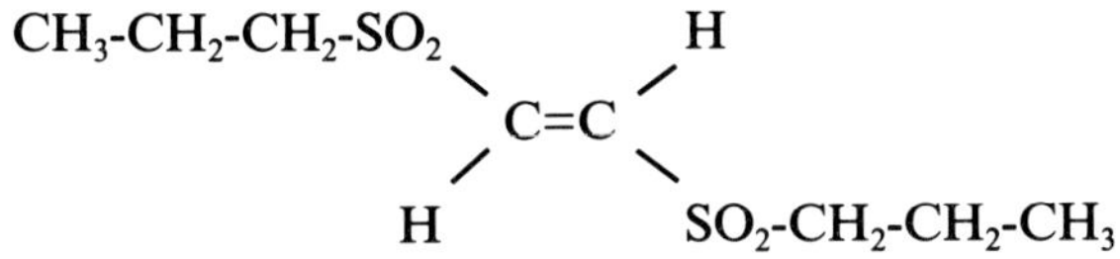

$C_8H_{16}O_4S_2$ M_r 240·35
CAS-No. 93114-09-1

Chemical and physical properties

Appearance	White, crystalline powder
Melting point	153–154°C
Solubility	Virtually insoluble in water; moderately soluble in organic solvents (e.g. 2·5 g/100 g acetone, 6·5 g/100 g dimethyl formamide)
Stability	Not stable in the presence of ammonia or other alkaline compounds

Toxicity data

LD_{50} oral	2070 mg/kg rat
interperitoneal	11·5–16·5 mg/kg rat, mouse or guinea pig
dermal	> 10 g/kg rabbit

Causes eye damage and skin irritation.

Antimicrobial effectiveness/applications

Table 125. Minimum Inhibition Concentrations (MIC) of Trans-1.2-bis(*n*-propylsulphonyl) Ethene in Nutrient Agar

Test organism	*MIC (mg/litre)*
Aspergillus niger	200
Chaetomium globosum	150
Penicillium glaucum	50
Candida albicans	75
Candida krusei	50
Rhodotorula mucilaginosa	50
Saccharomyces cerevisiae	100
Torula rubra	100
Torula utilis	100
Escherichia coli	500
Staphylococcus aureus	500

Because of its broad activity against mould producing fungi it has been recommended to use trans-1.2-bis(*n*-propylsulphonyl) ethene as a paint and adhesive film fungicide (Taylor, 1965). Its applicability is, however, limited to solvent based systems, as the a.i. is not stable in aqueous alkaline media.

18.8 Barium metaborate monohydrate

$Ba(BO_2)_2 \cdot H_2O$ M_r 240·95
CAS-No. 13701-59-2

Chemical and physical properties

Appearance	White powder
Density	3·25–3·35 g/ml
Refractive index	1·55–1·60
Barium content	57%
Boron content	9%
pH (saturated solution)	10–10·5
Fusion temperature	900–1050°C
Solubility	0·3–0·4% in water; insoluble in organic solvents

Toxicity data

LD_{50} oral	850 mg/kg rat
dermal	>2000 mg/kg rat
on inhalation for rats	>21·70 mg/litre

Irritating to eyes; not irritating to skin.

Ecotoxicity

LC_{50} for Rainbow trout and Bluegill sunfish >100 mg/litre (exposure: 96 h)

In the presence of sulphates conversion to insoluble and innocuous barium sulphate.

Antimicrobial effectiveness/applications

Barium metaborate can be regarded as a multifunctional pigment when used in paints. It keeps an alkaline environment in the paint can and the paint film, thus inhibiting mould growth and corrosion. The addition rates for mould resistant coatings are, however, relatively high (5–20%). Not all types of polymer emulsions tolerate such high salt concentrations. The water-solubility of barium metaborate also is not negligible and has to be taken into consideration if the compound is used in exterior coatings as a paint film fungicide.

19 OXIDIZING AGENTS

Oxidizing agents such as hydrogen peroxide, peroxy acids, halogens and halogen-releasing compounds have a microbicidal effect thanks to their strong oxidation power which is also directed, unspecifically, towards organic matter, that is, towards micro-organisms. It is in the nature of this effect not to be selective and to cover Gram-positive and Gram-negative bacteria, yeasts, fungi and algae but also, advantageously, spores and virsues. The aggressiveness of the agents under discussion, not only directed towards microbes, must be taken into account when considering their use as microbicides and does limit their applicability. Another factor to be considered in this connection is the oxidizing agents' intrinsic limited stability — which on the other hand offers the advantage that the agents can be used without ecologically harmful residues being obtained. In consideration of their characteristics, the oxidizing agents are used on quite a large scale as sanitizers in food processing plants, public buildings, hospitals, schools, restaurants and households; another large application field is water treatment: pool and drinking water, process water, sewage and waste water effluents.

19.1 Hydrogen peroxide

H-O-O-H	H_2O_2	*Mr* 34·02 CAS-No. 7722-84-1

Chemical and physical properties (30–32% solution in water)

Appearance	Water clear, colourless fluid
Density at 20°C	1·110 g/cm^3
Solubility	Highly soluble in water
Stability	Decomposes slowly to water and oxygen, if not stabilized, e.g. by sulphuric or phosphoric acid; decomposition is accelerated by traces of (heavy) metals, alkalis and dust

Toxicity data

LC on inhalation	100 ppm for rats and Guinea pigs

Corrosive to mucous membranes.

MWC value	1 ml/m^3 (1·4 mg/m^3)

Antimicrobial effectiveness/applications

Hydrogen peroxide generates hydroxyl radicals (H-O·) which are highly reactive and responsible for the antimicrobial action. The enzymes catalase and peroxidase which are produced by respiring cells to protect the cells from damage by steady-state levels of metabolically generated hydrogen peroxide are overwhelmed by higher hydrogen peroxide concentrations, e.g. 3–6%. Such concentrations are used in disinfectants and sanitizers. They are effective within minutes and therefore not considerably disturbed by inactivation processes occurring simultaneously, e.g consumption of active ingredient by organic matter.

The numbers of living cells in water can be sufficiently reduced by the addition of 1% of a 3% hydrogen peroxide solution. Heavily contaminated products, e.g. water based paints and thickener solutions showing signs of viscosity loss due to enzymatic degradation may be saved by the addition of 0·05–0·1% of a 30% hydrogen peroxide solution with stirring. After 24 h, when microbes and enzymes are inactivated and most of the unpleasant odours are eliminated by oxidation, the viscosity of the material is restored with additional thickener and the in-can/in-tank protection by the incorporation of a suitable preservative.

The antimicrobial activity and decomposition of hydrogen peroxide increases and accelerates with increasing temperature. The pH for optimum efficiency is in the acidic range; in alkaline media hydrogen peroxide decomposes too quickly.

19.2 Peroxyacetic acid — Peracetic acid

$$H_3C\text{-}\overset{\overset{\displaystyle O}{\|}}{C}\text{-}O\text{-}OH \qquad C_2H_4O_3 \qquad M_r\ 76{\cdot}05$$

CAS-No. 79-21-0

Chemical and physical properties (32% solution in diluted acetic acid)

Appearance	Clear, colourless liquid with a pungent odour
Density at 20°C	1·130 g/cm^3
Refractive index	n_D^{20} 1·3876
Flash point	56°C
Solubility	Highly soluble in water, ethanol and ether
Stability	In stabilized solutions there is an equilibrium between H_3C-COOOH, H_2O_2 and acetic acid; decomposes slowly to acetic acid, water and oxygen; its instability is promoted by heavy metal ions; explosive when heated up to 110°C

Toxicity data

Corrosive to skin and mucous membranes.
Non-mutagenic.

Antimicrobial effectiveness/applications

According to Eggensperger (1979) peracetic acid is several orders of magnitude more effective than hydrogen peroxide (Section 19.1). The spectrum of activity is extraordinarily broad; it covers Gram-positive and Gram-negative bacteria, yeasts, fungi, spores and viruses. The following concentrations of peracetic acid (10% a.i.) act microbicidally on bacteria within 2 min, 5 min and 24 h respectively: 20 mg/litre, 10 mg/litre, 1 mg/litre. The sporicidal activity of peracetic acid is characterized by high rapidity. Advantageously peracetic acid is effective, that means breaks down, without leaving residues. In concentrations of 0·1–0·5% it may be used for the cold sterilization of surgical instruments. In detergents for the cold wash of heat sensitive synthetic textiles peracetic acid is used as a sanitizer. As far as ion exchange resins are not affected by peracetic acid the peracid can be used for the disinfection of resins particularly in the food industry.

The optimum pH for the microbicidal action of peracetic acid varies between 2·5 and 4.

19.3 Higher peroxycarboxylic acids

Peroxymonocarboxylic acids having a carbon chain length of 5–8 carbon atoms show in comparison to peroxyacetic acid even, enhanced activity against a broad spectrum of micro-organisms including spores and also they are sufficiently soluble in water to be used in disinfectants. However, aqueous solutions of these peroxyacids have only very limited stability. A solid, substantially water insoluble organic peroxyacid such as 1,12-diperoxy dodecanedioic acid shows effective killing ability with respect to various bacteria, such as *Staphylococcus aureus, Streptococcus faecalis, Proteus mirabilis* and *Escherichia coli*, but fails to kill *Pseudomonas aeruginosa* within an acceptable killing time of less than 30 min at appropriate concentrations of 100 ppm active oxygen content. However, aqueous formulations of 1,12-diperoxydodecanedioic acid containing a synergistically effective amount of a sequestering agent, such as nitriloacetic acid and a surfactant plus a buffer to effect a pH of the composition in the range of from 2 to 7 exhibit a broad spectrum of effectiveness including *Pseudomonas aeruginosa* at reasonable concentrations (Ploumen, 1991).

19.4 Halogens, hypohalogenites and halogen releasing compounds

Halogens and hypohalogenites are strong oxidizing agents and consequently exhibit microbicidal efficacy. Among the halogens, iodine, chlorine and bromine are of practical significance as microbicides. Fluorine is not appropriate for practical applications as it is far too toxic, irritant and corrosive. The most important product is chlorine.

Although the industrial production of chlorine and sodium and calcium hypochlorites already started in 1785, it was not before the first half of the 19th century that the deodorizing and disinfecting properties of chlorine and hypochlorites were detected and led to the use of chloride of lime in hospital wards. Before that time chlorine and hypochlorites were used in textile bleaching. In 1881 Robert Koch demonstrated scientifically the microbicidal effect of hypochlorites by exposing pure cultures of bacteria to chlorine compounds. Thirteen years later Traube proved that hypochlorites can be used successfully for the purification and disinfection of water.

The oxidation potential of halogens is a consequence of their strong affinity to electrons. The introduction of chlorine into water leads to the following equilibrium:

$$Cl_2 + H_2O \longleftrightarrow HOCl + HCl$$

$$HOCl \rightleftarrows H^+ + OCl^-$$

without loss of activity. In the presence of ammonia or other nitrogenous compounds chlorine forms chlorine releasing compounds, so-called chloramines or *N*-chloro compounds according to the following equation:

$$Cl_2 + H\text{-}NR_2 \longleftrightarrow Cl\text{-}NR_2 + HCl$$

From the fact that the microbicidal activity of chlorine and chlorine releasing agents decreases with increasing pH one concludes that the undissociated hypochlorous acid (HOCl) and not the hypochlorite anion (OCl^-) is essential for the effectiveness. With regard to the mechanism of activity of chlorine or hypochlorous acid or *N*-chloramines the experiments of Friberg (1956) using radioactive ^{35}Cl and in 1957 using radioactive ^{32}P are especially informative. In view of the results of these experiments it is concluded that contact oxidation reactions of chlorine/hypochlorous acid at the bacterial cell wall cause destructive permeability changes which lead to the kill of the microbial cell before chlorine or *N*-chloramines accumulate in the microbe cell.

Le Chevallier *et al.* (1988) found that biofilm bacteria grown on the surfaces of granular activated carbon particles, metal coupons, or glass microscope slides were 150 to more than 3000 times more resistant to hypochlorous acid (free chlorine, pH 7·0) than were unattached cells. On the other hand, resistance of biofilm bacteria to monochloramine ($Cl\text{-}HH_2$) disinfection ranged from 2–100-fold more than that of unattached cells. The results suggested that, relative to the inactivation of unattached bacteria, monochloramine was better able to penetrate and kill biofilm bacteria than free chlorine. For free chlorine, the data indicated that transport of the active ingredient into the biofilm was a major rate-limiting factor. Because of this phenomenon, increasing the level of free chlorine did not increase disinfection efficiency. The greater penetrating power of monochloramine apparently compensated for its limited disinfection activity.

An increase in temperature increases the microbicidal activity of chlorine, hypochlorites and *N*-chloramines significantly, but also the consumption of the active ingredients by organic matter such as amino acids, proteins, peptones, body fluids, tissues and vegetable matter when present in a sanitizing solution. By competitive reactions chlorine is withdrawn from microbicidal action unless chlorine releasing compounds, e.g. chloramines, are formed. Sugar and starches apparently do not significantly affect the activity of chlorine nor do, according to Shere (1948), 500 ppm of alkyl aryl sulphonate.

Micro-organisms exhibit different resistances to chlorine and hypochlorites. Vegetative cells may be killed by 0·15–0·25 ppm available chlorine. Gram-positive species are less resistant than Gram-negative species and the spore forming organisms are about 10–1000 times more resistant to chlorine than the vegetative forms. For inactivation of mould spores one needs approximately 150–500 mg hypochlorite/ml. Algae are sensitive to halogens and halogen releasing compounds, too. Growth control can be achieved by 2 mg active chlorine/ml. In swimming pools treated with chlorine problems may occur by the selection of so-called 'black algae' which are resistant to the active chlorine concentrations tolerated in pools.

19.4.1 Chlorine

Cl_2	M_r 70·91
	CAS-No. 7782-50-5

Chemical and physical properties

Appearance	Heavy gas of greenish-yellow coloration with a characteristic irritating and penetrative odour; not inflammable
Boiling point	−34°C
Freezing point	−101°C
Density	Chlorine gas is about 2½ times as heavy as air
Solubility in H_2O	9·5 g/litre at 10°C; 4·5 g/litre at 40°C
	$Cl_2 + H_2O \leftrightarrow HOCl + HCl$
Stability	Chlorine in solution is degraded by UV light; copper, nickel or cobalt catalyse the decomposition; stability is favoured by high alkalinity, low temperature and absence of organic matter

Toxicity data

Highly toxic and corrosive to skin and mucous membranes.

MWC value   0·5 ml/m^3 (1·5 mg/m^3)

Antimicrobial effectiveness/applications

The antimicrobial activity of the chlorine covers all species of micro-organisms including spores and viruses. Among the halogens chlorine exhibits the strongest sporicidal efficiency which is, however, slow. Chlorine is used for water treatment, including sewage and waste water. After treatment with chlorine the detectable concentration of active chlorine in drinking water should be $\geq$1 ppm; in treated swimming pool water 0·3–0·6 ppm active chlorine should be detectable. Using chlorine solutions, e.g. for disinfecting purposes, one has unconditionally to take into account the chlorine consumption by organic matter.

Chlorination of waste water treatment plant effluents and other waters, such as cooling waters and industrial wastes can produce stable chlorine-containing organic compounds which may, if discharged to rivers and lakes, have biotoxic effects at chronic, low level concentrations and may contribute to such phenomena as fish population and species changes that occur below the discharges of chlorinated waste water treatment plant effluents. Jolley (1975) detected stable chlorine-containing organic substances in effluents that had been chlorinated to a 1–2 mg/litre chlorine (Cl_2) residual. The total chlorination yield is reported to be approximately 1% of the chlorine dose; approximately 99% is apparently used in oxidation reactions and converted to chloride.

19.4.2 Salts of hypochlorous acid — Hypochlorites

$NaOCl - Ca(OCl)_2$
KOCl = Eau de Javelle
CaCl(OCl) = Chloride of lime

Sodium hypochlorite — Eau de Labarraque

NaOCl M_r 74·44
CAS-No. 7681-52-9

Chemical and physical properties
(15% aqueous solution corresponding to 5% active chlorine)

Appearance	Yellowish liquid with the characteristic odour of free hypochlorous acid (HOCl)
Density at 20°C	1·097 g/cm^3
Solubility	Soluble in water in any proportion
Stability	Relatively stable in alkaline solution; unstable at lower pH values; sensitive to light; not heat resistant

Toxicity data

Corrosive to skin and mucous membranes.

Antimicrobial effectiveness/applications

Sodium hypochlorite solution is appropriate to control a wide spectrum of micro-organisms. The active ingredient is HOCl which increases if the pH is reduced below 7. Sodium hypochlorite therefore is especially active in neutral or slightly acidic media. It does not leave residuals, is virtually colourless, non-staining, easy to handle and most economical in use. Hypochlorite solutions are proven sanitizers used in households, public buildings, the food industry and hospitals. Water treatment is another application field for sodium hypochlorite. Quantities corresponding to 0·5 ppm active chlorine are sufficient for the treatment of swimming pools.

19.4.3 Chlorine dioxide

ClO_2 M_r 67·46
CAS-No. 10049-04-4

Chemical and physical properties

Appearance	Red to yellow gas with an intense pungent odour or liquid or crystalline
Melting point	−79°C
Boiling point	9·9°C
Solubility	Soluble in water
Stability	Unstable and not storable; decomposition at 30°C

Toxicity data

Corrosive to skin and mucous membranes.

MWC value   0·1 ml/m^3 (0·3 mg/m^3)

Antimicrobial effectiveness/applications

Chlorine dioxide's oxidation potential is 2·5 times higher than that of chlorine; ClO_2 therefore is correspondingly stronger in microbicidal activity. Because of its instability it has always to be prepared shortly before use. ClO_2 can for example be generated from *c.* 7·5% sodium chlorate ($NaOCl_3$) solution after the addition of hydrochloric acid (9·5%). It is occasionally used for the antimicrobial treatment of water and as a slimicide in the paper industry.

19.4.4 Sodium *N*-chloro-toluene-4-sulphonamide — Chloramine T

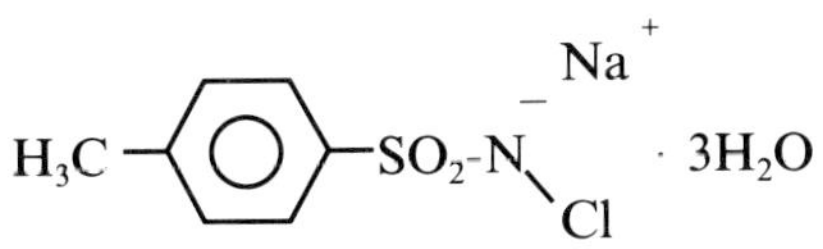

$C_7H_7ClNNaO_2S \cdot 3H_2O$

M_r 281·7
CAS-No. 127-65-1

Chemical and physical properties

Appearance	White to yellowish, crystalline powder with a slight odour of chlorine
Melting point	167–170°C (decomposition)
Solubility	At 25°C 12% in water; sparingly soluble in trichloromethane, virtually insoluble in ether
Stability	Chlorine releasing agent; may decompose violently if heated above 130°C

Toxicity data

Irritant to skin and mucosa, but less irritating than chlorine.

Antimicrobial effectiveness/applications

The microbicidal activity of Chloramine-T is due to the release of chlorine, but the action of Chloramine-T is slower, especially at alkaline pH levels. Chloramine-T therefore is used as a sanitizer at low pH levels and long exposure.

19.4.5 Chlorinated isocyanuric acids

19.4.5.1 Sodium dichloroisocyanurate — Sodium dichloro-s-triazine-2,4,6(1H,3H,5H)-trione — Halane

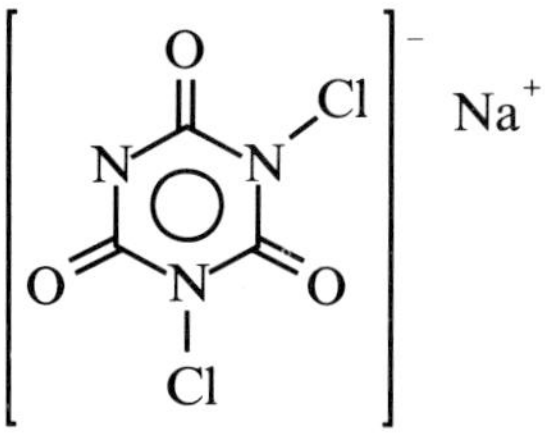

$C_3Cl_2N_3NaO_3$ M_r 219·94
CAS No. 2893-78-9

Chemical and physical properties

Appearance	White, crystalline powder with the odour of chlorine
Solubility	25% in water
pH (1% solution)	5·9
Stability	Hydrolysis in water to HOCl and cyanuric acid

pH optimum for antimicrobial effectiveness: 6–10. The potassium salt of dichloroisocyanuric acid is also available.

19.4.5.2 Trichloroisocyanuric acid — Trichloro-s-triazine-2,4,6(1H,3H,5H)-trione

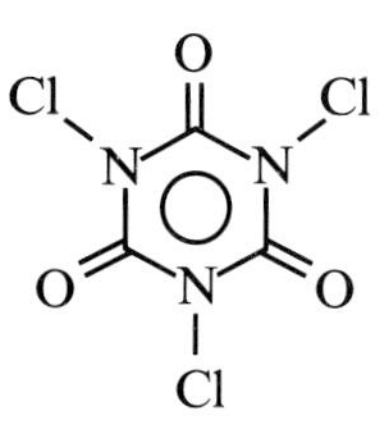

$C_3Cl_3N_3O_3$ M_r 232·41
CAS No. 87-90-1

Chemical and physical properties

Appearance	White, crystalline powder with the odour of chlorine
Melting point	249–251°C
Solubility	1·2% in water at 25°C
pH (1% solution)	2·7–2·9
Stability	Hydrolyses in water to form HOCl and cyanuric acid

pH optimum for antimicrobial effectiveness: 6–9·5.

19.4.6 Other chlorine releasing agents

19.4.6.1 1,3-Dichloro-5,5-dimethyl hydantoine — 1,3-Dichloro-5,5-dimethyl-2,4-imidazolidindione

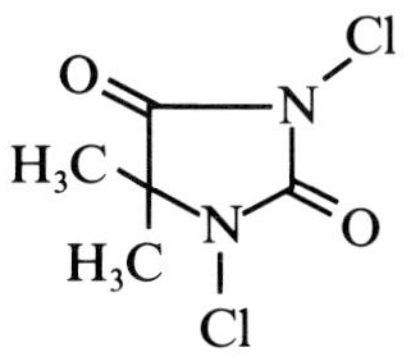

$C_5H_6Cl_2N_2O_2$ M_r 197·02
CAS No. 118-52-5

Chemical and physical properties

Melting point 134–136°C

19.4.6.2 Trichloro melamin — 2,4,6-Triimino-N_2,N_4,N_6-trichloro-*s*-triazine

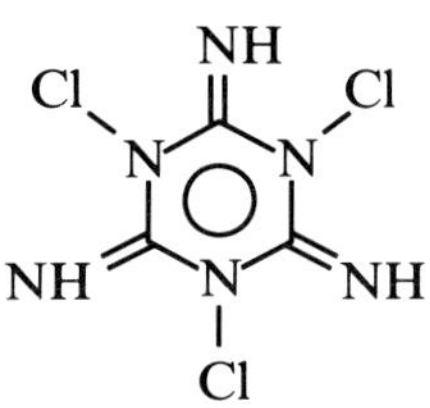

$C_3H_3Cl_3N_6$ M_r 229·46
CAS No. 7673-09-8

Chemical and physical properties

Melting point >300°C

19.4.6.3 Succinchlorimide

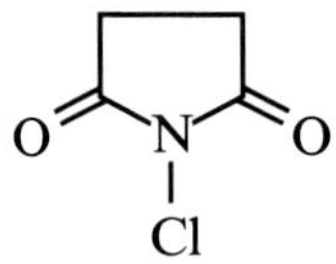

$C_4H_4ClNO_2$

M_r 133·53
CAS No. 128-09-6

Chemical and physical properties

Melting point	150–152°C

19.4.7 Iodo polyvinyl pyrrolidone — Povidone

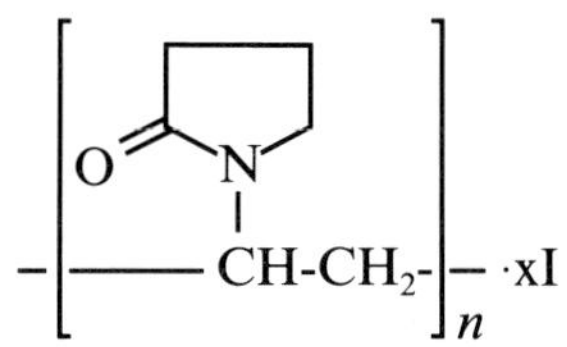

CAS No. 25655-41-8

Chemical and physical properties

Appearance	Brown powder
Effective iodine content	9–12%
Melting point	300°C (decomposition)
Bulk density	430 g/litre
Solubility	Soluble in water and ethanol, virtually insoluble in acetone, ether, trichloromethane
pH (10% aqueous solution)	1·5–2·5
Stability	Iodine vapour is liberated at 25°C; sensitive to moisture

Toxicity data

LD_{50} oral 6 g/kg rat

The 1% solution is slightly (reversibly) irritant to skin and eyes.
Non-mutagenic.

Antimicrobial effectiveness/applications

Povidone is an iodophor, that means an iodine carrier which in solutions and on heating releases iodine. However, the disadvantages of iodine (i.e. unpleasant odour, skin irritation, staining of tissue and corrosions of metal surfaces) are diminished. On dilution in water I_2, HOI, OI^- and I_3 are set free, which are responsible for the antimicrobial action of Povidone. The disinfecting effect extends to different classes of micro-organisms including spores and viruses. The most important application of Povidone is the medical disinfection of skin.

PART IV
Applications

INTRODUCTION

Microbicides for the protection of materials today are the result of controlled development work, carried out in close contact with the user, in which toxicological and ecological considerations were given as much priority as questions relating to practical application technology and economy. The Minamata (Japan) disaster which occurred in the 1950s when industrial mercury discharges reached dangerous levels in local fish, killing or disabling more than 2000 residents, will not happen again.

In the light of the knowledge that the earth's raw material resources are limited, the need for the development of modern preservatives which protect valuable products from loss of quality and deterioration is more urgent than ever. They not only protect a great number of perishable products from depreciation and destruction, but also prevent problems in industrial processes caused by algae and slime formation and microbial induced corrosion. The large number of microbicides for the protection of materials is an inevitable consequence of the great variety of materials which need protection and their varied physical and chemical properties. The active ingredients and their formulations have to be 'tailored' to suit each material; they must also be compatible with a large number of processes in which they are used to prevent problems caused by microbe activity. And last but not least the microbicides must be effective against the vast variety of microbes with widely differing resistances. It will always be an illusion to think that the microbial destruction of materials can be prevented by just a handful of different microbicides. The trend is today in the opposite direction, towards even greater diversification. This is the result of new demands placed on individual fields of application for microbicides with regard to biodegradability and toxicity, for example. In order to meet these demands, the development of 'tailormade' new active ingredients is essential.

The number of fields of application for microbicides is nearly as huge as the number of microbicides. They reach from citrus fruits to fuel and from toothpastes to drilling muds. The selection of a suitable microbicide will depend first on the type of microbes to be controlled and secondly on the environment in which the active ingredient is to be used. There are available bactericides, fungicides, algicides or active ingredients which are especially effective against yeasts or those having a particularly broad spectrum of activity. The composition or the pH of a material to be protected can already provide some information on the type of microbe attack which may be expected.

Products containing sugars, e.g. lignosulphonates, are mainly attacked by yeasts. Mould-producing fungi prefer substrates with a weakly acid or neutral reaction; they like to grow at air–liquid interfaces, on surfaces, and in aerated products. Bacteria mainly decompose products containing protein and with a neutral to weakly alkaline reaction. In so far as acidification is possible, e.g. to pH 4, this in itself limits the growth of bacteria and aids preservation. Microbial slimes build up in process water, and if the water is exposed to light abundant growth of algae also occurs. Microbial growth on surfaces will occur only when sufficient water (relative humidity) is available.

A relative humidity of 80% is conducive to growth, 60–80% may or may not be depending on other factors, but below 60% no growth will occur. For microbial growth on surfaces it is not obligatory that the entire substrate is a nutrient for micro-organisms; it suffices when one component of the substrate, e.g. a plasticizer, can serve as a nutrient or a thin layer of dirt or organic matter which has accumulated on the surface. Although water is essential for microbial growth, microbes can grow even in hydrocarbon fuels which contain only traces of water. This becomes possible if some condensation of water occurs as the temperature decreases; then both bacteria and fungi are capable of growing at the oil–water interface and of producing slimes which may plug filters and lines.

Without regard to details one can distinguish between four general groups of fields of application for microbicides:

1. Industrial aqueous process fluids, e.g. cooling waters, pulp and papermill process waters and suspensions, secondary oil recovery systems, spinning fluids, metal working fluids.
2. In-tank/in-can protection of aqueous functional fluids, e.g. polymer emulsions, water based paints and adhesives, glues, starch slurries, thickener solutions, gelatine, wax emulsions, inks, polishes, pigment and mineral slurries, rubber latexes, concrete additives, drilling muds, aqueous cosmetic and pharmaceutical formulations.
3. Antimicrobial treatment of materials that finally contain little or no water in a free state, e.g. paint and adhesive films, textiles, paper, paperboard, plastics, hoses, cords, rubber products, leather, wood.
4. Disinfection of inanimate surfaces and equipment.

It is self evident that in the first two groups of fields of application one needs microbicides of sufficient water solubility, and as far as two-phase systems are concerned, e.g. emulsions, microbicides with a distribution coefficient or partition coefficient which favours the enrichment of the active ingredient in the water phase. For the oil–water partition coefficient K^{o}_{w} the following equation of Bean *et al.* (1965) is valid:

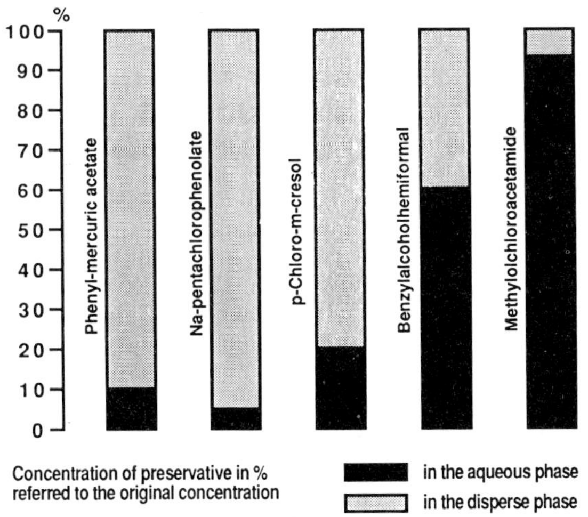

Fig. 37. Distribution of various microbicides between aqueous and organic phase of an aqueous acrylic paint formulation.

$$C_w = C \frac{\Phi + 1}{K^o_w \Phi + 1}$$

C_w = concentration of microbicide in water
C = total microbicide concentration
Φ = oil/water ratio

In Fig. 37 is demonstrated how different microbicides behave differently in an emulsion paint according to their varying partition coefficients.

To fulfil the requirements for the third group of applications it is essential to dispose of microbicides which are virtually insoluble in water and are not volatile, do not cause coloration and are practically odourless. Active ingredients for disinfectants should exhibit microbicidal action within minutes and additionally should have properties which make it possible to prepare liquid, water-dilutable concentrates of the actives.

There are some rules to which one should pay attention in connection with the application of microbicides for the protection of materials.

Perfect starting materials and hygienic practices may not be a substitute for preservatives and disinfectants, but they are essential conditions for their economic use.

The efficacy of a preservative and the concentration level of a microbicide to be added are very much dependent on the germ content of the material to be protected. This is in particular valid for electrophilic active microbicides which in general react irreversibly with nucleophilic components of the microbial cell, that means that they are used up by being effective. But membrane active microbicides which adsorptively coat microbial cell walls are also withdrawn from action at least temporarily if large numbers of microbial cells are present. The exponential growth profile of microbes (p. 2) always has to be taken into consideration.

Damage already inflicted by microbes cannot be undone even by very large additions of microbicides. Since the concentrations of active ingredients normally required for microbe control in general are not sufficient to inactivate exogenic enzymes and thus to prevent enzymatic degradation processes, steps should be taken very early to preserve the material, ensuring that enzyme production by microbes does not take place. Oxidizing agents (III.19) and organomercury compounds (III.17) are examples of active ingredients being able to inhibit enzymes, e.g. cellulosic material degrading cellulases.

The antimicrobial effect of formaldehyde and formaldehyde releasing agents (III.3) may be reduced by ammonium ions (formation of hexamethylene tetramine (III.3.3.1)).

Phenolic compounds (III.5) can irreversibly react with formaldehyde especially at pH value above 9; the reaction products are much lower in activity than the starting products (Fig. 19, p. 56).

Non-ionic surfactants have an adverse effect on the activity of all phenolic microbicides (III.5) including *p*-hydroxybenzoates (III.6) (Fig. 35, p. 376); anionic surfactants have considerably less effect. Cationic active ingredients (III.16.1) are inactivated by anionic components.

Selecting an appropriate microbicide one has always to take into consideration the pH of the medium to be protected. Degradation/hydrolysis of the microbicide may occur at certain pH values. There are microbicides which are effective only within certain pH ranges (optimum pH), e.g. glutaraldehyde (III.2.3), acids (III.6), phenolics (III.5), quaternary ammonium compounds (III.16.1).

Storage of finished or intermediate aqueous functional fluids at varying temperatures will lead to condensation of preservative-free water in the head space of the storage tank unless a volatile microbicide is used. The condensate will then rapidly become contaminated from its environment. Dripping back on to the surface of the stored product a layer of preservative free condensate forms which is highly susceptible to microbial growth. It is therefore advisable to use microbicides which exhibit head space activity or to store the protected aqueous fluids at more or less constant temperatures.

If one is confronted with the task of solving a biodeterioration problem, one can expediently proceed stepwise:

1. Precise identification of the biodeterioration problem.
2. Specification of requirements for an appropriate microbicide.
3. Selection of corresponding microbicides.
4. Pretrials — evaluation.
5. Application of a suitable microbicide under practical conditions.
6. Solution to the biodeterioration problem.

EVALUATION OF PRESERVATIVES

The preservative effects of microbicides in aqueous functional fluids can be ascertained, and reliably effective concentrations established, by a method which, though referred to as classical, has been extensively modified over the years. The basis of the method, which has not itself been changed, comprises three steps:

1. Inoculation of the test medium with micro-organisms.
2. Incubation of the contaminated test medium.
3. Viable cell count or determination of the cell count (cfu = colony forming units) per g or ml.

Each of these steps, however, can be modified in many ways. The test medium can be inoculated with a single, several or numerous micro-organism species. For this one can use pure cultures obtained in the laboratory under optimal conditions of temperature, pH value, humidity, nutrient availability and osmotic pressure, or what are known as acclimated micro-organisms, which are isolated from aqueous functional fluids that have been spoiled by the microbes concerned. Acclimated micro-organisms should be used whenever possible. Non-acclimated micro-organisms have in many cases died after being transferred from their optimal environment to the test medium, even where this is a blank sample and where such a sample is very susceptible to microbial growth. The lethal stress suffered by non-acclimated test organisms in such cases is caused by the sudden change from one medium to a different and contrasting medium (Cooke *et al.*, 1991). If acclimated micro-organisms are not obtainable and a selection of pure cultures is available, one should always include *Pseudomonas aeruginosa* among the organisms tested because it is a troublesome, difficult-to-control organism that is found very often in fluids that have been spoiled by microbes. According to Kempson (1976) it is safe to assume that a treatment which controls *Pseudomonas aeruginosa* reliably will also be effective against other, less resistant species. It is advisable to repeat the inoculation once, or preferably several times, after a definite incubation time. This shows, without a chemical analysis being needed, whether or not the preservative has been inactivated by the test medium — possibly through hydrolysis or because the preservative has reacted with constituents of the test medium. Such reactions can be accelerated with heat, e.g. by maintaining the temperature at 50°C for 1 week. Inoculation times can be varied, just as incubation temperatures can (e.g. 48 h, 27°C). As an alternative to a viable cell count, a streak assessment can be done. For this a given amount of the fluid to be investigated is spread on a sterile nutrient substrate in a petri dish, after which the intensity of the microbe growth arising from the streak is evaluated.

Evaluating antimicrobial treatments of other materials is less easy than checking the preservation of aqueous functional fluids. Help is available, however, from institutes and microbicide suppliers that routinely use a number of standard methods for specific materials, such as plastics, wood, paint films and textiles. The same applies to the testing of disinfectants. The final test, however, is one under practical conditions, which may vary very much according to the place and region and which cannot always be simulated in the laboratory.

ANTIFOULING COATINGS

Antifouling coatings or marine paints comprise a special segment within paint production. Such paints must survive under the harshest environmental conditions on a wide range of immersed man-made structures such as ships, platforms, pontoons, buoys, nets and tanks used in fish farms. It is expected that antifouling coatings on immersed structures remain free from the attachment of bioslime and fouling organisms for years. Marine algae are widely recognized as one of the most important groups of fouling organisms besides sea animals, e.g. barnacles, serpulids. These organisms are able to settle and multiply on submerged coatings. The consequence is destruction of the protective coating and reduction in operativity; the settlement of fouling organisms on ship hulls causes speed reduction, increased fuel consumption, enhanced corrosion, more frequent drydocking operations, etc.

The main active ingredients in antifouling coatings today are still organotin compounds (III.17) in combination with cuprous-I-oxide or cuprous thiocyante. However, as the slow release of organotin compounds out of immersed antifouling coatings is inevitable, there is growing concern about the contamination of the marine environment with such wide spectrum biocides which are highly ecotoxic. In consequence a world-wide 'tin ban' for antifouling coatings is discussed, and the way is open for the development of novel antifouling coatings which are non-toxic. It can be regarded as a first step that in many countries (e.g. USA, Europe, Japan) it is already forbidden to apply tin-containing antifouling coatings on boats and ships of a length below 25 m.

Examples of algicides which may be used in antifouling coatings instead of organotin compounds are:

Diuron (III.8.14)
Carbamates (III.9.1, 9.8.2, 9.9, 9.10)
Heterocyclic N, S compounds (III.13.5)
N-Haloalkylthio compounds (III. 14, e.g. 14.5 and 14.6)
Chlorothalonil (III.15.7)
2-Methylthio-4-tert-butylamino-6-cyclopropylamino-*s*-triazine (III.18.6).

CONCRETE ADDITIVES

With the aid of concrete additives the viscosity and the hardening of concrete is controlled. Most of the concrete additives are based on lignosulphonates which are able to reduce and to control the amount of water needed to make concrete mixes workable. Lignosulphonates are a by-product of wood pulping (with every ton of pulp produced a ton of lignosulphonates is produced). They are by nature highly susceptible to microbial decomposition. Corresponding concrete additives therefore must be treated with preservatives in order to ensure that they function correctly when used.

Examples of microbicides appropriate for the in-can/in-tank protection of concrete additives are:

Aldehydes (III.2)
Formaldehyde releasing compounds (III.3)
Phenolics (III.5.3.1, 5.3.2, 5.4.1, 5.6.1)
Sodium *N*-dimethyldithiocarbamate (III.9.10.1)
Sodium Pyrithione (III.11.1.3)
Heterocyclic N,S compounds (III.13.3, 13.6, 13.7)
1,2-Dibromo-2,4-dicyanobutane (III.15.11).

COOLING WATERS

Cooling is the largest single use of water in industry. Cooling water systems are generally exposed to microbial contamination. Micro-organisms, such as bacteria and fungi, find a lot of nutrients in cooling water which enable them to proliferate. Nutrients are organic compounds that have leaked out into the system, cooling water additives other than microbicides, mineral salts and windborne debris. Microbial fouling in cooling water systems is also caused by algal growth in the parts of the system which are exposed to light. Free floating masses of algae serve as a nutrient for bacteria. Deposits of microbes in cooling water systems inhibit the efficiency of heat transfer and water flow; wood destroying fungi damage the wood of cooling towers. Among the problem causing bacteria one finds slime formers (e.g. Pseudomonades, Aerobacter, Mucoid) and corrosion inducing micro-organisms such as iron depositing (e.g. *Gallionella, Sphaerotilus*), sulphuric acid producers (*Thiobacillus*), nitric acid producers (*Nitrobacter nitrosomonas*), hydrogen sulphide producers (e.g. *Desulfovibrio desulfuricans*). A broad spectrum of mould producing fungi, yeasts and basidiomycetes have been detected, and green algae, blue-green algae and diatoms.

Among the bacteria which are generally associated with slime problems there are species that are highly resistant to changes in their environment; they thrive in water under anaerobic and aerobic conditions, at low and at high salt concentrations, and at temperatures ranging from 0 to 70°C.

A large variety of slimicides and algicides belonging to different substance classes is offered to inhibit microbial growth in cooling water systems. The microbicides are mostly applied as a shock treatment as soon as the viable cell count reaches 10^5–10^6 cfu/ml.

Quaternary ammonium compounds (III.16.1) are the most important microbicides for the treatment of cooling water. They are most effective against bacteria and algae in alkaline pH ranges, but to profit completely from the high efficacy of QACs one needs rather clean systems; in cooling water contaminated with dirt, oil and other debris QACs are deactivated. Another disadvantage of QACs is the fact that the application of these surface active compounds leads to extensive foaming resulting in reduced cooling efficiency.

Another group of active ingredients to be used in slimicides and algicides for cooling waters belongs to the electrophilic active microbicides:

Glutaraldehyde (III.2.3), acrolein (III.2.4)
3,5-Dimethyl-tetrahydro-1,3,5-2H-thiadiazine-2-thione (III.3.3.16)
2,2-Dibromo-3-nitrilopropionamide (III.8.5)
Heterocyclic N,S compounds (III.13.3)
Compounds with activated halogen groups (III.15)
Methylene bisthiocyanate (III.18.3).

Glutaraldehyde and the tetrahydro thiadiazine derivative are especially effective in cooling water systems with pH values above 7. The other active ingredients mentioned here are sensitive to hydrolysis and therefore are not recommended for cooling systems where the pH exceeds 8. An advantage of these compounds is that they are not persistent and in consequence do not cause significant waste water problems.

Oxidizing agents, namely halogens, hypohalogenites and halogen releasing compounds (III.19.4) is another group of microbicides with a broad spectrum of activity due to their oxidation potential. These chemicals are not persistent; they become ineffective as microbicides relatively quickly. Normal use levels must be dramatically increased to maintain activity where cooling waters become contaminated with organic substances (e.g hydrocarbons), ammonia, sulphides and organic matter. In any case the halogen based oxidizing agents are finally transformed to corrosive halogen ions.

Of minor importance as microbicides for the treatment of cooling waters are dithiocarbamates such as sodium dimethyl dithiocarbamate (III.9.10.1) and potassium *N*-methyl dithiocarbamate (III.9.9). They function best in environments of pH 7 and higher.

COSMETICS, PHARMACEUTICALS

Microbial contamination of cosmetics, toiletries and pharmaceuticals can lead to two types of problems:

1. Health hazards.
2. Product spoilage.

Although the chance that contaminated cosmetics or toiletries cause harm is probably low, the risk cannot be excluded. Health hazards may arise from cosmetic and pharmaceutical products which are inadequately preserved and therefore will support the growth of pathogenic organisms.

Cosmetics in particular are multi-component systems containing, besides water, different nutrients for micro-organisms, such as carbohydrates, nitrogenous compounds, biodegradable detergents and emulsifiers and mineral salts. There is always a risk of microbial contamination during the production of cosmetics and pharmaceuticals; practically unavoidable is the contamination of products by the user. The manufacture of sterile products is neither a required nor practical objective in the cosmetic and toiletry industry, but there are limits to the extent and type of microbial contamination that have to be observed. CTPA for instance suggests the following limits:

Total viable cell count (bacteria, yeasts and fungi) less than 1×10^3 cfu per g or per ml. Products intended for direct use in the eye area and those recommended for use on babies: less than 1×10^2 cfu per g or per ml. Harmful microorganisms should be excluded; although the harmful species are not specified, one can assume that *Pseudomonas aeruginosa, Staphylococcus aureus* and *Candida albicans* belong to the harmful species. Those cosmetics, toiletries and pharmaceuticals which are susceptible to microbial spoilage therefore require the incorporation of an appropriate preservative for in-can protection during storage and use. The range of end products which have been developed and marketed by the cosmetic and pharmaceutical industries is vast and the composition of the products often very complex. This is reflected in the diversity of microbicides which fulfil the corresponding requirements.

Not every highly effective preservative is compatible with the formulation to be protected. Formaldehyde and formaldehyde releasing compounds, for example, may react with proteins and cause coagulation. It has to be noted that acids exhibiting antimicrobial effectiveness are active only in their undissociated state (III.6); that means they can successfully be applied only in acidic formulations. Many cosmetics and toiletries contain detergents in concentrations above the critical micelle concentration; phenolic microbicides including *p*-hydroxy-benzoic acid esters are then deactivated (Fig. 35, p. 376). Non-ionic surfactants are especially detrimental for these microbicides. Anionic surfactants react with cationic microbicides with inactivation and precipitation, and by analogy cationic surfactants reduce the antimicrobial activity of phenolic compounds. Many cosmetics and pharmaceuticals can be regarded as two-phase systems. Appropriate preservatives for such systems should have an oil–water partition coefficient in favour of the water phase.

Examples of preservatives for cosmetics, toiletries and pharmaceuticals corresponding to the EC list of preservatives which cosmetic products may contain are:

Alcohols (III.1)
Aldehydes (III.2.1 and 2.3)
Formaldehyde releasing compounds (III.3)
Phenolics (III.5)
Acids (III.6)
Dibenzamidines (III.10)
Pyridine derivatives (III.11.1.2, 11.1.3, 11.1.4)
Heterocyclic N,S compounds (III.13.3)
1,2-Dibromo-2,4-dicyanobutane (III.15.11)
Surface active agents (III.16.1, 16.2)
Thiomersal (III.17.5).

DETERGENT SOLUTIONS

In detergent solutions containing less than 30–35% surface active agent bacteria, yeast and fungi may proliferate. In-can/in-tank protection with the aid of appropriate microbicides therefore is necessary to maintain the quality and hygienic state of the solutions. On selection of suitable preservatives one has to note the pH of the detergent solution, as a lot of preservatives exhibit optimum effectiveness only in a certain pH range; others are deactivated through inclusion in micelles above the critical micelle concentrations (see Fig. 35, p. 376). It has also to be borne in mind that detergent solutions are very often used for the manufacture of cosmetics.

Examples of microbicides for the in-can/in-tank protection of detergent solutions are:

Aldehydes (III.2)
Formaldehyde releasing compounds (III.3)
Acids (III.6)
Heterocyclic N,S compounds (III.13.3)
1,2-Dibromo-2,4-dicyanobutane (III.15.11).

DISINFECTANTS, SANITIZERS

The task of disinfectants is to kill or irreversibly inactivate infectious or other undesirable bacteria, pathogenic fungi, or viruses on surfaces or inanimate objects. Sanitizers are applied to reduce the number of living bacteria, fungi or viable viruses on inanimate surfaces, in water or in the air. Disinfectants and sanitizers may additionally have cleaning properties, but contrary to the large number of normal industrial and institutional cleaning products they contain microbicides.

The most important microbicides for the use as active ingredients in disinfectants and sanitizers are found in three substance classes:

Phenolics (III.5)
Surface active agents (III.16)
Oxidizing agents (III.19)
Aldehydes (III.2, III.3) and alcohols (III.1) are of minor importance, but distinguished by sporicidal activity.

As there are among the phenolic microbicides compounds of low toxicity, good skin compatibility, and excellent biodegradability accompanied by a broad spectrum of effectiveness, stability, and resistance to organic matter, they continue to remain the most important active ingredients for heavy duty disinfectants to be used above all in hospitals and nursing homes. They are also used in household disinfectants and for application in animal stables. In the food industry the use of phenolic disinfectants and sanitizers is very restricted, because there is a big risk that traces of phenolic compounds negatively influence the taste and quality of food. The favoured active ingredients for disinfectants and sanitizers to be used in the food industry and also in swimming pools are surface active microbicides and oxidizing agents. The latter are very unstable, highly sensitive to organic matter and quickly converted to inactive compounds. Surface active microbicides are much more stable, but also disturbed in their activity by organic soil. Nevertheless oxidizing agents and surface active microbicides are the most common active ingredients for hard-surface cleaners used for cleaning schools, colleges, lodgings and commercial buildings.

FOOD

The application of food additives most probably began when man became sessile approximately 10 000 B.C. An overview about the history of the use of food additives is given by Lück (1988). Today food manufacture and distribution without the use of food preservatives and antioxidants is unthinkable. Processed foods now make up about three-quarters of the average diet for most people. Fortunately yesterday's somewhat unreasoning blanket rejection of food additives is being replaced by the understanding that the very much greater hazard can arise from unprotected food if it becomes microbiologically contaminated (poisoned). However, it is worth mentioning that preservatives and antioxidants together comprise only about 1% of all food additives. And the use of food preservatives today is strictly regulated. The preservatives which are permitted are precisely defined and the foods in which they can be employed and the maximum authorized concentrations, too. The development and registration of new food preservatives is an extremely difficult and expensive procedure. The need for and safety of, new preservatives have to be clearly demonstrated; the obstacles are high and inhibit innovation.

Besides synthetic food preservatives there are also natural ones which were in an advantageous position in so far as they did not require notification as additives. However, it is expected that a system for total notification of additives, including natural ones will be installed.

Examples of natural food preservatives

Polylysine, which is a homopolymer of L-lysine, is one of the essential amino acids. It is used for daily dishes. As it is relatively heat resistant food can be processed by heating after the addition of polylysine.

Protamine (basic protein extracted from salmon roe) exhibits higher antimicrobial activity than sorbic acid (III.6.1) and is used for boiled fish paste, daily dishes, sauce and noodles.

Hinokitiol is an antibacterial ingredient isolated from plants such as cypress and *Hiba arborvita*. It is recommended to knead hinokitiol into non-woven fabric, polyethylene and others which are used in the form of a film and paper for the preservation of the freshness of strawberries, peaches and other fruits; prolonged storage periods are achieved by the use of hinokitiol-containing coverings.

Some of the synthetic food preservatives also occur naturally, e.g. benzoic acid (III.6.1.8), propionic acid (III.6.1.3), sorbic acid (III.6.1.5).

Permitted synthetic food preservatives

Benzoic acid and its salts (sodium, potassium, calcium) (III.6.1.8)
Propionic acid and its salts (sodium, potassium, calcium) (III.6.1.3)
Sorbic acid and its salts (sodium, potassium, calcium) (III.6.1.5)
Sulphur dioxide, sulphites and hydrogen sulphites (sodium, potassium, calcium) (III.6.2.2)
4-Hydroxybenzoates (methyl, ethyl and propyl) (III.6.1.10)
Dimethyl dicarbonate (III.7.7)
2-Phenyl-phenol and is sodium salt (III.5.4.1)
Thiabendazole (III.13.8)
Hexamethylene tetramine (III.3.3.1).

FUEL

The microbial contamination of fuel — hydrocarbon fuel such as aviation fuel, gasoline, diesel oil — is a well-known phenomenon as are the problems associated with such an infection. The basis for the growth of bacteria and fungi in fuel is water which may enter either through seepage or through condensation. The microbes, e.g. Pseudomonades, sulphate reducing bacteria, *Aspergillus fumigatus, Cladasporium resinae*, proliferate in the water phase of the fuel/water mixture by feeding on the nutrients in the hydrocarbon phase. The aerobes which require oxygen for survival are mainly responsible for the formation of sludge, thick slimy mats plugging fuel filters and blocking fuel lines. The anaerobes among the invading microbe species which preferably multiply where oxygen is absent produce and secrete corrosive acids. Corrosion leads to the accumulation of debris and dirt in fuel tanks and lines and can cause leakage. As it is most difficult, if not impossible, to prevent the entry of water into fuel systems and the contamination with microbes, the solution to the problem is found only by the use of suitable microbicides. Such microbicides have to fulfil the following demands:

— Activity at relatively low doses which do not essentially dilute the fuel and do not interfere with the combustion of the fuel;
— Spectrum of effectiveness covering the various microbe species occurring in fuel;
— Solubility in fuel and water and a partition coefficient in favour of the water phase;
— Non-corrosive in use concentrations.

Examples of suitable microbicides are:

Formaldehyde releasing compounds (III.3)
Pyridine derivatives (III.11.1, 11.1.4)
Heterocyclic N,S compounds (III.13.3).

GLUES, ADHESIVES, THICKENERS

Aqueous glues, adhesives and thickeners are more or less sensitive to microbial deterioration depending on their composition and therefore need adequate protection. Although glues and adhesives in powder or solid form do not deteriorate as long as they are dry, microbicides are often incorporated so that aqueous use-preparations are protected. Loss of viscosity caused by the activity of micro-organisms or their enzymes respectively causes severe irreversible damage to glues, adhesives and thickener solutions making the preparations useless. Another problem is that glue and adhesive films under humid conditions are contaminated by mould producing fungi with loss of their function. Hence there exists also a demand for non-volatile fungicides for the protection of glue and adhesive films against mould growth.

In selecting suitable microbicides one has to note the type or composition of the glue, adhesive or thickener, the manufacturing process and the end-use. Adhesives coming into direct contact with foodstuffs require preservatives of low toxicity which are especially odourless and tasteless. Microbicides for incorporation into dry, solid glues and adhesives or thickeners should stand the drying process without degradation or volatilization.

Proteinaceous glues are in particular sensitive to microbial degradation, e.g. skin, leather, boneglues, casein and albumin solutions. Such preparations are not compatible with aldehydes, e.g. formaldehyde releasing compounds; coagulation may occur.

Examples of appropriate microbicides for the in-tank/in-can protection of glues, adhesives and thickeners are:

Aldehydes (III.2)
Formaldehyde releasing compounds (III.3)
Dimethoxane (III.4)
Phenols (III.5)
Acids (III.6.1.8, 6.1.10)
Amides (III.8)
Carbamates (III.9)
Pyrithiones (III.11.1)
Heterocyclic N,S compounds (III.13)
Compounds with activated halogen groups (III.15).

Examples of microbicides for the fungicidal treatment of glues and adhesives are:

Dichlorophen (III.5.7.3)
Carbamates (III.9)
Pyridine derivatives (III.11.1, 11.3, 11.5)
Heterocyclic N,S compounds (III.13)
N-Haloalkylthio compounds (III.14)
Compounds with activated halogen groups (III.15).

LEATHER

There is a lot of water, humidity and organic material involved in the process of leather manufacture, all favouring the growth of micro-organisms. Microbicides are therefore indispensable aids for the leather industry in the fight against damage and losses caused by the effects of bacteria (decay) and mould fungi. Due to different pH values which are predominant during the different steps of leather production various populations of microbe species are responsible for the trouble. This has to be taken into consideration in the selection of appropriate microbicides.

Tannage takes place in an acidic medium. The tanned hides (wet blues) are therefore especially sensitive to the growth of mould fungi such as:

Aspergillus niger
Mucor mucuedo
Paecilomyces variotii
Penicillium spp.
Rhizopus nigricans
Trichoderma viride

Raw hides and soaking baths offer conditions for abundant bacterial growth. Often found are:

Bacillus subtilis
Escherichia coli
Proteus vulgaris
Pseudomonas aeruginosa

In consequence one needs bactericides for the protection of salted and dried raw hides during storage and soaking and fungicides for the protection of wet blues, pickled pelts and vegetable-tanned leather. There are also some microbicides available with suitable properties and active against bacteria and fungi; these microbicides may be used during all steps of the manufacture of leather.

Biodegradability is another essential requirement today for microbicides to be applied in the leather industry.

Examples of suitable microbicides are:

Phenolics (III.5, e.g. 5.3.1, 5.4.1)
Amides (III.8, e.g. 8.1, 8.10)
Carbamates (III.9)
Heterocyclic N,S compounds (III.13, e.g. 13.4, 13.10)
Compounds with activated halogen groups (III.15.2)
Surface active agents (III.16.2).

METAL WORKING FLUIDS (MWF)

These mean water based MWF, not neat oils (lubricants) in which microbes in general cannot proliferate because of the lack of water. Water is a universal and cheap solvent or diluent for MWF concentrates, it offers excellent heat transfer ability (specific heat 1, oils 0·4) and will not burn. Water based MWF can be categorized as follows:

(a) Soluble oils which after dilution with water form milky-white dispersions of oil in water.
(b) Semi-synthetic fluids; the concentrates contain <20% oil and the dilutions are translucent dispersions of oil in water.
(c) Synthetic MWF. These fluids are transparent, true chemical solutions, containing no oil.

The use dilutions contain between 2·5 and 5% MWF concentrate.

Metal working fluids are expected to fulfil a large number of requirements — cooling and lubricating, corrosion protection, electrolyte resistance, skin compatibility, biodegradability, economy through long-lasting effectiveness and, last but not least, microbe resistance. The last named but very important requirement can be fulfilled with the help of microbicides which meet the requirements of today's environmentally conscious society. Microbicides for MWF should combine high efficacy with low toxicity, especially good skin compatibility and biodegradability.

The microbe resistance of metal working fluids is important because it governs the retention of cooling and lubricating performance, anti-corrosion properties (pH), skin compatibility and long-term effectiveness without interruption in production.

The ready-for-use dilutions of MWF concentrates are aqueous systems whose pH permits the survival and multiplication of bacteria, yeasts and fungi (including pathogenic and slime-forming species). They contain organic substances on which micro-organisms can feed, substances originating from the composition or from contamination during use. It is practically impossible to prevent the fluids from contamination and from being infected with micro-organisms. Sources of infection include dirty systems and work pieces, waste and sometimes the dilution water or even people working with the fluids.

The contamination/infection of non-preserved metal working fluids results in microbial decomposition. The consequences of this are an unpleasant odour, clogged filters, the generation of acid metabolic products, a drop in pH, corrosion and creaming. The MWF finally becomes unusable necessitating unscheduled cleaning operations involving shut-down and loss of production.

It is also worth noting that pathogenic micro-organisms in non-protected metal working fluids can multiply to such an extent that the people working with them can contract dangerous skin infections.

A preservative for ensuring the microbe resistance of metal working fluids should not impair the other properties of the solution — lubricant performance, corrosion protection, electrolyte stability, biodegradability and skin compatibility must remain unaffected. The preservative must also be compatible with the emulsifiers without any reduction in its effectiveness. Furthermore, the microbicide should have a broad spectrum of activity and a favourable partition coefficient, that means in favour of the water phase.

The so-called 'bioresistant' MWF need protection, too; these are synthetic fluids based on special amine borates and boric acid esters making the fluids relatively resistant to bacterial growth. This, however, is not equivalent to 'bioresistant', as the suppression of bacterial activity favours the proliferation of yeast and moulds in such fluids to an uncontrollable point, causing clogging of filters and pipes. In addition, when 'bioresistant' MWF are diluted beyond a certain concentration (e.g. 1:50), they lose their bacterial resistance. Therefore at least fungicides have to be used for the in-tank protection of so-called bioresistant metal working fluids. The development of a completely bioresistant MWF which will require no protection with microbicides cannot be an aim, as the waste treatment after disposal of such a fluid will cause problems and high expenses. The disposal costs of used up MWF are in any case exorbitant, thus making fluid protection with biodegradable microbicides a priority. Uneconomically high additions of preservative will be unnecessary, if the dilution of the metal working fluid is carried out in clean and hygienic conditions. Broadly speaking, the concentrates are microbe-free. However, a freshly diluted solution will often contain several thousand microbes per millilitre. After just a few hours' use, a non-preserved metal working fluid will have a microbe content of several million per millilitre. Microbial decomposition will be accelerated by shut-downs and by oil floating on the surface of the metal working fluid. The oil may either have separated from the fluid itself or come from other sources, e.g. washed off a workpiece. These are favourable conditions for the multiplication of anaerobic micro-organisms such as sulphate-reducing bacteria which cause unpleasant 'Monday morning' smells. It is advisable that shut-downs are kept as short as possible and that the oil floating on the surface of the fluid is skimmed off every so often. Problems may also be caused by the presence of sulphurous metal shavings, using water heavily contaminated with microbes to prepare the dilutions, the failure to clean plant thoroughly before filling and many other types of contamination. Plant can be cleaned using a system cleaner with a microbicide added to give it disinfectant properties. It may be sufficient to rinse the cleaned plant with a microbicide solution or to decontaminate the plant before cleaning by adding an effective amount of microbicide to the unusable metal working fluid. This is then circulated through the plant for several hours before being drained out.

Microbicides for the protection of MWF can be added to the concentrate or to the dilution at the tankside. On adding microbicide to the MWF concentrate problems may arise by enhanced incompatibility in concentrates, by the inability to predict all use-dilutions; the latter may cause overdosing and corresponding hazards or underdosing with the risk of selecting resistant microbe species. From a technical standpoint therefore tankside addition is the most rational procedure. It allows the application of the most suitable microbicide or a mixture of microbicides at correct concentrations.

When selecting appropriate microbicides one should pay attention to the fact that Pseudomonades are generally found in non-protected use-dilutions. Among the microbicides for the protection of MWF formaldehyde releasing compounds represent the majority, although they are less effective against fungi than bacteria.

In addition, on the extensive use of formaldehyde releasing compounds alone problems may occur with the selection of formaldehyde resistant bacteria (Paulus, 1976). It therefore makes sense to use different types of active ingredients to increase the spectrum of activity or even better, combinations which are characterized by synergistic effects. Fungicides are especially suitable for the so-called bioresistant metal working fluids.

Examples of appropriate microbicides for the protection of metal working fluids are:

Aldehydes (III.2)
Formaldehyde releasing compounds (III.3)
Phenolics (III.5)
Carbamates (III.9)
Pyrithiones (III.11.1)
Heterocyclic N,S compounds (III.13)
1,2-Dibromo-2,4-dicyanobutane (III.15.11).

PLASTIC MATERIAL (SYNTHETIC POLYMERS)

Plastics in general are considered to be durable or persistent materials. Indeed many commercial synthetic polymers are inert and resistant to biodegradation; others, e.g. polyurethanes, are easily decomposed by micro-organisms. As a rule, this is the result of the polymer chains being split by enzymes secreted by microbes. The fragments serve as a source of carbon, nitrogen and hydrogen for the micro-organisms.

Micro-organisms can also cause damage to plastics by decomposing not the polymer itself, but other components, such as the plasticizer. The mechanical properties of the plastic are thus changed in such a way as to make it unusable. In addition, many fungi capable of growing on plastics secrete coloured metabolic products (e.g. *Streptoverticillium reticulum*) which dissolve in the plasticizer and discolour the plastic. This 'decorative' damage can also impair the utility of certain plastics. Bacteria can also adhere to the surface of synthetic polymers which are exposed to high humidity or aqueous systems. Many freshwater and marine bacteria are able to produce bioslime or so-called biofilms then without using the plastic as a primary food source. Not only is bioslime a severe problem in plastic pipelines or tanks for drinking water, but it also causes such operational problems as decreasing flow rates in cooling systems and other industrial equipment.

The intensity of microbial growth on plastic material depends on the type of synthetic polymer, but also on the type of additives, especially plasticizers used to improve the processing characteristics of the polymer. There are plasticizers such as fatty acid esters and long-chain dicarboxylic acid esters which are most susceptible to fungal growth while phthalate and phosphate esters with alkylol substituents are generally resistant. However, as plastic materials in operation are always subject to contamination with traces of nutrients for micro-organisms, microbial growth will occur even on synthetic polymers resistant to biodegradation and containing resistant plasticizers. To prevent the growth of destructive micro-organisms, suitable microbicides are incorporated in finished plastic material. Only thermally stable, non-volatile, non-discolouring and non-leachable microbicides can be used for that.

Examples of microbicides which fulfil these requirements are:

N-(2-methylnaphthyl)maleinimide (III.8.12)
Benzimidazolylmethylcarbamate (III.9.4)
N-octyl-4-isothiazolin-3-one (III.13.4)
N-Haloalkylthio compounds (III.14.1, 14.2)
10,10'-Oxybisphenoxyarsine (III.17.1).

A review of the biodeterioration and biodegradation of naturally occurring and synthetic plastic polymers has been presented by Seal (1988).

OIL RECOVERY

Secondary oil recovery by the injection of vast quantities of water, mainly sea water, usually follows primary recovery, when the natural reservoir energy subsides and is no longer able to drive the crude oil to the surface. The injection water forces an oil/water mixture out of subterranean rock formations through the wells of the primary recovery. The oil/water mixture is separated and the separated water is introduced to the process again. The presence of micro-organisms in the water can lead to serious problems such as corrosion of metallic equipment and lines, generation of toxic hydrogen sulphide gas and plugging of rock pores in the formation. A main problem among the microbe species involved is *Desulfovibrio desulfuricans*, an anaerobic organism which reduces sulphates to hydrogen sulphide; this reacts with iron to produce black sediments of iron sulphide. Additionally iron bacteria remove iron from the water and surround themselves with a sheath of slimy iron hydroxide, clogging filters, rock formations and sand beds and providing anaerobic conditions favouring the growth of sulphate reducing bacteria. Clogging and corrosion are also caused by slime and biofilm forming bacteria, e.g. Pseudomonades and *Aerobacter aerogenes*. They live predominantly as sessile organisms in biofilm matrices (see p. 20), retaining their corrosive products near the metal surface.

Drilling fluids or muds are other aqueous functional fluids used in oil recovery. Muds are used to cool the drill bit and to keep underground formation pressure stable. The corresponding formulations are based on thickeners such as guar, starch and modified cellulose. These compounds are easily degraded by a great variety of microbe species. The consequence is that the mud's viscosity decreases and the formulation finally loses its functionality. As the aqueous functional fluids applied in oil recovery operations are used in vast quantities, one looks for microbicides to overcome the microbial problems which are effective at very low concentrations, inexpensive and environmentally safe.

Examples of appropriate microbicides are:

Aldehydes (III.2.1, 2.3, 2.4)
Formaldehyde releasing compounds (III.3.3)
Amides (III.8.5)
Heterocyclic N,S compounds (III.13.3)
Surface active agents (III.16.1, 16.2)
Oxidizing agents (III.19.4.3).

POLYMER EMULSIONS

Polymer emulsions, such as styrene-butadiene-rubber (SBR latex), acrylics, polyvinylalcohol, polyvinylacetate and vinyl-acrylics are used in the manufacture of a wide variety of products. Approximately 70% of the polymer emulsions are applied in four major end uses:

1. paint and coatings,
2. adhesives,
3. paper,
4. carpet backing.

Polymer emulsions are aqueous fluids which present a medium for the multiplication of micro-organisms (bacteria, fungi, yeasts). As the emulsions are used for the production of high quality products, including those coming in contact with food, they have to be in a clean condition when used, that means in-tank protection to prevent microbes from proliferating is necessary.

When selecting an appropriate microbicide for in-tank protection one has to note the type of polymer emulsion, its pH and its final use. The temperature of the polymer emulsions at the time of incorporation of the preservative is another point which has to be taken into consideration; the heat resistance of the microbicide may be a requisite. It is self-evident that suitable preservatives for polymer emulsions are characterized by a favourable partition coefficient which guarantees that the active ingredient does not increase in the organic phase.

Examples of appropriate microbicides for the in-tank protection of polymer emulsions are:

Aldehydes (III.2)
Formaldehyde releasing compounds (III.3)
Dimethoxane (III.4)
Amides (III.8)
Heterocyclic N,S compounds (III.13)
1,2-Dibromo-2,4-dicyanobutane (III.15.11).

Microbicides of more limited suitability are:

Phenolics (III.5)
Carbamates (III.9)
Pyrithiones (III.11.1).

PULP AND PAPER

The working conditions in paper-making favour the growth of many species of micro-organisms in pulp, coating mixes, sizes and other materials used in the industry. Mildew stains on pulp, slime formation in the pulp circulation system and the decomposition of coating mixes are examples of this.

These problems can be avoided by using slimicides and other microbicides. Microbicides have become an indispensable part of paper production, particularly because modern processes and the water supply conditions in most plants greatly promote the growth of micro-organisms. In many factories, trouble-free working with largely closed water circuits and continuous production of coated papers and boards have been made possible only by the systematic use of slimicides and preservatives. Since the different types of micro-organism have varying degrees of resistance to microbicides, there is no single preparation which can solve all the preservation problems occurring in the paper industry.

The properties demanded of microbicides, in addition to high activity and cost-effectiveness, also vary according to their specific field of application. In their use concentrations, slimicides must not disturb the paper manufacturing processes, not even if they are added in massive doses. That is to say these products must be compatible with the many auxiliaries used in paper-making. Furthermore, slimicides must have low toxicity, in particular low ecotoxicity and no appreciable amounts must be present in the finished product, so that the paper can also be used in food packaging. Preservatives for coating mixes, filler suspensions and sizes and active ingredients for the antimicrobial finishing of paper and board have to meet stringent requirements with regard to the absence of colour and odour, compatibility and physiological harmlessness in the use concentrations.

Appropriate active ingredients for non-persistent slimicides one finds above all among the compounds with activated halogen groups (III.7 and III.15). Other active ingredients belong to the carbamates (III.9 and III.3.4.10.1) and the heterocyclic N,S compounds (III.13). It is also proposed to use glutaraldehyde (III.2.3) or 3,5-dimethyl-tetrahydro-2-thiono-thiadiazine (III.3.3.16).

Oxidizing agents (III.19) are recommended, too, for slime prevention as far as the oxidizing properties of these agents can be tolerated. Using chlorine, hypochlorites or chlorine dioxide problems may arise by the enrichment of corrosion causing chloride ions in the system.

Moist pulp, papers with an antimicrobial finish, hygienic papers (e.g. paper towels) and mould resistant packaging material (e.g. soap wrappers) are produced with the addition of non-volatile microbicides which are virtually insoluble in water, so that good retention is guaranteed. Water soluble microbicides can be applied together with size solutions and coating mixes. Proposals are 2-phenylphenol and Dichlorophen (III.5.4.1 and III.5.7.3) and carbamates (III.9).

For the in-tank protection of coating mixes, filler suspensions and size solutions preservatives which are used for aqueous coatings or glues and adhesives are suggested.

SUGAR PROCESSING

Severe problems are also caused through micro-organisms in sugar mills where they find a rich variety of preferred nutrients. Micro-organisms present in sugar mills are responsible for a loss of sucrose from the sugar juices. Bacteria also cause the formation of dextran slime and soluble dextran, which lead to increased viscosities and significant processing problems.

The strategy to overcome the microbial problems starts with an improvement of the hygienic conditions in sugar mills, by using disinfectants on surfaces and equipment. Microbicides which degrade and/or evaporate during the process can be used in sugar juices to improve sugar yields without being present in the sugar.

A special problem, difficult to overcome, is the presence of lactic acid bacteria in sugar juices which are fermented with yeasts for the production of alcohol. The yield of alcohol is then reduced by the competitive activity of the lactic acid bacteria producing lactic acid instead of alcohol.

Examples of microbicides used in sugar mills are:

Formaldehyde (III.2.1)
Sodium salts of dithiocarbamates (III.3.4.10.1, 9.9, 9.10.1)
Surface active agents (III.16.1, 16.2).

SURFACE COATINGS

With regard to the protection of material in the surface coatings industry against infestation, destruction and decomposition by microbes, there are two main areas:

(A) In-tank/in-can protection of aqueous coatings against microbial spoilage.
(B) Fungicidal/algicidal treatment of coatings and impregnating agents.

(A) In-tank/in-can protection

Aqueous coatings always contain organic components which can be degraded by microbes. The results of decomposition by microbes are well known: reduction of pH, loss of viscosity, evolution of gas, coagulation and foul smells, until finally the paint becomes unfit for use and complaints result. This damage is irreversible. Therefore steps to control microbes must be taken at the earliest possible opportunity. Bacteria are the major group of causative organisms, but other causes of problems are yeasts and moulds. For the prevention of microbial spoilage of aqueous coatings one therefore needs microbicides with a broad spectrum of effectiveness. They should also dissolve well in water and be characterized by a favourable partition coefficient (see Fig. 37, p. 447). Liquefaction of cellulosic thickeners and aqueous paints can be caused by enzymes which are produced by bacteria and fungi (Springle, 1989). Such enzymes, e.g. cellulases, are capable of exhibiting their bio-catalytical activity at concentrations as low as 10^{-5} units per ml. Since enzymes cannot be inactivated by subsequently adding preservatives in the usual doses, it is recommended that preventative measures be taken at an early stage in production, so that filler and pigment slurries, thickener solutions, polymer emulsions and other materials stored in tanks are all preserved.

For maximum effectiveness, all in-can preservatives must be dissolved and evenly distributed in the paint formulation to be protected. In order to achieve this, it may be necessary to initially dissolve or dilute the active ingredients preferably in water or an aqueous solvent mixture before incorporating them into the paint. There is also a risk that the incorporation of a microbicide in a concentrated form causes incompatibilities, e.g. coagulation.

Examples for appropriate in-tank/in-can preservatives are:

Formaldehyde releasing compounds (III.3)
Carbonic acid esters (III.7.2, 7.3)
Amides (III.8.1, 8.2, 8.3)
Carbamates (III.9.3, 9.8.1)
Heterocyclic N,S compounds (III.13.3, 13.6, 13.7)
1,2-Dibromo-2,4-dicyano butane (III.15.11)
Phenylmercury acetate (III.17.3).

(B) Fungicidal/algicidal treatment of coatings

Mould-resistant coatings are of particular significance in 'damp rooms' such as kitchens, bathrooms, cellars, dairies, breweries and drinks bottling plants as well as in many areas of the textile industry. This significance has grown with the development of fungicides which remain effective for long periods even in the most adverse conditions. Mould-infested paintwork is not only unattractive, it is unhygienic, an important consideration in the food processing industry. Under conditions of high humidity, e.g. in tropical climates, exterior coatings are also susceptible to mould growth and growth of algae additionally (Whitely, 1966; Wee, 1988). Dampness is the supposition for microbial growth on coatings, but it is also promoted by the coating itself. The polymer, plasticizer and other auxiliaries in the coating are nutrients for microbes (primary infestation) and the finished paint film gathers a layer of dirt on which microbes can also thrive (secondary infestation). Primary infestation can be prevented fairly easily by incorporating microbicides in the coating. Microbes come into direct contact with the paint film, and hence the microbicide, and are killed or prevented from growing. Secondary infestation occurs mainly in food processing and textile plants. To prevent secondary infestation, the activity of the fungicide in the coating must be able to penetrate through to the surface of the layer of dirt on the paint film. This effect cannot, however, be based on the use of a highly volatile active ingredient as the paint film would lose its antimicrobial effect after only a few months.

A very thick layer of dirt can impair or even destroy the activity of the best fungicidal coating. Washing the dirt off the surface of the coating restores its effectiveness. For this reason mould-repellent coatings should always be sufficiently resistant to washing.

The demands made by a coatings specialist on microbicides are tough: they must have a low toxicity, good light stability and good compatibility with the other components in a formulation; they must also be non-volatile, colour-fast, odourless, insoluble in water (and thus non-leaching) and economical. As microbicides which satisfy all these requirements are not available, one has to find compromises.

Examples of paint film fungicides/algicides are:

Amides (III.8.4, 8.10, 8.13, 8.14)
Carbamates (III.9.1, 9.2, 9.4, 9.8.2, 9.9, 9.10)
Pyridine derivatives (III.11.3)
Azoles (III.12.1)
Heterocyclic N,S compounds (III.13.4, 13.7, 13.8)
N-haloalkylthio compounds (III.14.1, 14.2, 14.3, 14.5, 14.6, 14.8)
Various compounds (III.19 ...).

TEXTILES

Microbicides are used in the textile industry to effectively prevent mould stains and rotting in textiles and also to prevent textile auxiliaries from decomposition and mould formation.

Cellulosic fibres, such as cotton, are the most susceptible to microbial attack and most uses of microbicides are directed at protecting cotton fabrics. Problem causers are cellulose-decomposing fungi and bacteria. Synthetic fibres in general do not as readily biodeteriorate as cellulosic fibres, but staining by microbes may occur without loss of strength. Protection of fabrics with microbicides is also needed to keep textile materials in a hygienic condition.

With continuous use of fabrics under damp conditions, for example in contact with soil or under direct exposure to water, high demands are made on the wetting-out stability of microbicides used for the microbicidal treatment of the fabrics. Such demands can be met by the application of active ingredients, that are difficult to dissolve in water combined with a waterproof impregnation or subsequent rubberizing treatment. One can also profit from the affinity of quaternary ammonium compounds to textile material, especially to woven fabrics of natural fibres. Phenolic microbicides may be applied in the form of their water-soluble sodium or potassium salts; after impregnation of the textile material with the corresponding alkaline solutions the textile material is transferred into a second acid bath, containing acetic acid for example, fixing the phenolic microbicides on the fabric in their water-insoluble form. Phenolic microbicides react with quaternary ammonium compounds to form water-insoluble quaternary ammonium phenolates. Emulsions or solutions of such phenolates in organic solvents may be used for the protective impregnation of textile material.

Textile finishers apply microbicides mainly for the protection of commercial carpeting and military textiles. Smaller quantities of microbicides are used for health-care textiles and domestic fabrics. Microbicides are also used for the protection of yarns, cordage and rope, e.g. prepared from sisal.

Textile auxiliaries, such as spinning lubricants, sizes, finishes, yarn humidifying agents and print thickeners often provide an excellent nutrient medium for mould fungi and putrefactive bacteria; they can be regarded as aqueous functional fluids which need in-tank or in-can protection with suitable microbicides.

Examples of microbicides to be used in the textile industry are:

Phenolics (III.5.4.1, 5.7.1)
Copper naphthenate (III.6.1.11) — due to its objectionable odour fabrics treated with copper naphthenate are mainly used for outdoor applications.
Fatty acid esters of Dichlorophen (III.7.9)
Salicylanilide (III.8.6)
Oxine copper (III.11.5)
Dyrene (III.15.8)
Quaternary ammonium compounds (III.16.1) including 3-(trimethoxysilyl)-propyldimethyl-octadecyl ammonium chloride (III.16).

VARIOUS AQUEOUS INDUSTRIAL FLUIDS REQUIRING IN-CAN/ IN-TANK PROTECTION BY PRESERVATIVES

Air washer — bitumen emulsions
Ceramic glazes — enzyme solutions
Electrodeposition paints, bath, rinses
Fabric softeners — fountain solutions (printing)
Humidifiers — inks
Mineral slurries — natural rubber latex
Pigment slurries — photographic processing fluids
Polishes — textile processing fluids
Water bed filling — wax emulsions.

WOOD

Only organic fungicides which may be used in wood preservatives are discussed here.

The range of fungi causing economic damage to wood is wide and varied. Wood destroying fungi (basidiomycetes) are able to break down the polymeric matrix of wood; such decay of wood is known as brown rot or white rot. Soft rot microfungi decay wood by softening wood exposed in contact with soil or water superficially.

Blue stain and sapstain fungi colonize and penetrate into wood exposed to humidity without decaying the wood, but causing a blue, black or brown coloration in spots or streaks.

Mould fungi spread on the surface of wetted timber where they feed on simple sugars, and can cause different colorations, e.g. black, brown, green, pink or yellow.

Fungicides for wood preservatives are selected according to the function they have to fulfil. For the temporary protection of freshly felled and newly sawn timber one needs fungicides which are effective against sapstain fungi and mould-producing fungi; long lasting efficacy (persistence) over many years is not an essential requirement for such fungicides. However, to prevent wood in service from disfigurement by blue stain fungi one looks for suitable fungicides the activity of which will last a long time. Fungicides for the protection of wood against wood destroying fungi have to be long-lasting in use, that means stable in the wood, non-volatile and leach resistant.

Examples of fungicides appropriate for wood preservatives are:

2,3,3-Triiodallyl alcohol (III.1.10)
(3-Iodopropargyl)-(4-chlorophenyl) formal (III.3.1.5)
Phenolics (III.5)
Naphthenic acid salts (copper and zinc salts) (III.6.2.1)
(2-Bromo-1,2-diiodoacryl-)ethylcarbonate (III.7.6)
Carbamates (III.9, e.g. 9.1 and 9.4)
Oxine copper (III.11.5)
Azoles (III.12)
Heterocyclic N,S compounds (III.13.4, 13.5, 13.10)
N-Haloalkylthio compounds (III.14)
Compounds with activated halogen groups (III.15.2 and 15.7)
Surface active agents (III.16.1 and 16.2)
Organometallic compounds (III.17)
Various compounds (III.18).

The pressure for substitution on organotin compounds and polyhalogenated phenols, especially on pentachlorophenol, its sodium salt and lauryl ester has increased the use of the other organic microbicides mentioned above. Most of these are newer microbicides having a specific and sometimes narrow spectrum of effectiveness. In modern wood preservatives one therefore finds today very often combinations of fungicides which are effective against all target organisms.

References and Bibliography

Adroer, N., Casas, C., de Mas, C. & Solá, C. (1990). Mechanism of formaldehyde biodegradation by *Pseudomonas putida. Appl. Microbiol Biotechnol.*, **33**, 217–20.

Albert, A. (1968). *Selective Toxicity: The Physicochemical Basis of Therapy*, 5th edn. Chapman and Hall, London.

Albert, A., Rubber, S., Goldacre, R. & Balfour, B. (1947). The influence of chemical constitution on antibacterial activity. Part III. A study of 8-hydroxyquinoline (oxine) and related compounds. *Brit. J. Exp. Pathol.*, **28**, 69–87.

Allwood, M. C. & Meyers, E. R. (1981). Formaldehyde releasing agents. Society of Applied Bacteriology Technical Series 16. Academic Press, London, pp. 69–76.

Anderson, C. B. & Dalley, R. (1986). Use of organotins in antifouling paints. Ocean's 86 Conference Record, Vol. 4, Organotin Symposium, Washington, DC, pp. 1108–13.

Austin, W. G. L. (1964). Control of aquatic weeds. *Outlook on Agriculture*, **4**(1), 35–43.

Baasner, B., Heywang, G., Kühle, E., Paulus, W. & Schmitt, H. G. (1988). Fluoro containing dimethylolnitromethanes. EP-A 257251.

Bansemir, K., Disch, K. H. & Hackmann, K. (1987). Disinfectants. EP 226081 B 1.

Barnes, C. P. & Eagon, R. G. (1986). The mechanism of action of hexahydro-1,3,5-triethyl-*s*-triazine. *J. Ind. Microbiol.*, **1**, 105–12.

Barrueco, C. & de la Pena, E. (1988). Mutagenic evaluation of the pesticides Captan, Folpet, Dichlofluanid and related compounds with the mutants TA 102 and TA 104 of *Salmonella typhimurium. Mutagenesis*, **3**, 467–80.

Bean, H. S., Herman-Akah, S. M. & Thomas, S. (1965). The activity of antibacterials in two-phase systems. *J. Soc. Cosmet. Chem.*, **16**, 15–27.

Becker, F. C. & Gurnee, S. P. (1979). *N*-(2-methyl-1-naphthyl)-maleimide. USP 4, 141, 905.

Behret, H. (1988). Tributylzinnoxid. BUA-Stoffbericht 36. VCH Verlagsgesellschaft, Weinheim, Germany, 89 pp.

Berke, P. A. & Rosen, W. E. (1982*a*). Germall II — A new broad spectrum cosmetic preservative. *Cosmet. Toiletr.*, **97**(6), 49–53.

Berke, P. A. & Rosen, W. E. (1982*b*). Preservative compositions. USP 4, 337, 269.

Block, S. S. (1983). *Disinfection, Sterilization and Preservation.* Lea & Febiger, Philadelphia, PA, p. 820.

Bloomfield, S. F. & Uso, E. E. (1985). The antibacterial properties of sodium hypochlorite and sodium dichloroisocyanurate as hospital disinfectants. *J. Hospit. Infect.*, **6**, 20–30.

Botham, P. A., Hilton, J., Evans, C. D., Less, D. & Hall, T. J. (1990). Assessment of the relative skin sensitizing potency of three biocides using the murine local lymph node assay. *Contact Dermatit.*, **22**, 1–6.

Brake, B. L. (1974). Mildewcide for paint. USP 3, 817, 761. Anol 3, 817, 762.

Bruckner, N. I. Gordon, M. D. & Howell, R. G. (1988). Odorless aromatic dialdehyde disinfecting and sterilizing composition. Europ. Pat. Applic. 0292301.

Büchel, K. H., Holmwood, G., Paulus, W. & Schmitt, H. G. (1988). Use of 1-aryl-3-hydroxy-3-alkyl-4-(1,2,4-triazol-1-yl)-butane derivatives as microbicides for the protection of materials. DE-053621494.

Burk, G. A. & Reineke, C. E. (1979). Stabilized aqueous antimicrobial composition. USP 4, 163, 797.

Chandler, C. S. & Segel, I. H. (1978). Mechanism of the antimicrobial action of pyrithione: effects on membrane transport, ATP levels and protein synthesis. *Antimicrob. Agents Chemother.*, **14**(1), 60–8.

Chiba, M. & Singh, R. P. (1986). High-performance liquid chromatographic method for simultaneous determination of Benomyl and Carbendazim in aqueous media. *J. Agric. Food Chem.*, **34**, 108–12.

Chi-Tung Hsu, A. (1991). Novel benzoxazolone compounds and the use thereof as microbicides. EP 427, 484.

Clark, N. G., Crowshaw, B., Leggetter, B. E. & Spooner, D. F. (1974). Synthesis and antimicrobial activity of aliphatic nitro compounds. *J. Medic. Chem.*, **17**, 977–81.

Clarkson, D. & Clifford, R. P. (1991). Biocide. EP 0277402.

Clemons, G. P. & Sisler, H. B. (1971). Localization of the site of action of a fungitoxic benomyl derivative. *Pestic. Biochem. Biophys.*, **1**, 32–43.

Cook, F. K. & Pierson, M. D. (1983). Inhibition of bacterial spores by antimicrobials. *Food Technol.*, **37**(11), 114–26.

Cooke, P. K., Gandhi, U. R., Lashen, E. S. & Leasure, E. L. (1991). Preservative evaluation: Designing an improved test system. *J. Coat. Technol.*, **63**(796), 33–8.

Cooney, J. J. & Felix, J. A. (1972). Inhibition of *Cladosporium resinae* in hydrocarbon–water systems by pyridinethiones. *Int. Biodet. Bull.*, **8**(2), 59–63.

Corbett, S. R., Wright, K. & Baillie, A. C. (1984). *The Biochemical Mode of Action of Pesticides*. Academic Press, London.

Cramer, G. M. & Ford, R. A. (1978). Estimation of toxic hazard — A decision tree approach. *Food Cosmet. Toxicol.*, **16**, 255–76.

Croshaw, B., Groves, M. S. & Lessel, B. (1964). Some properties of Bronopol, a new antimicrobial agent active against *Pseudomonas aeruginosa*. *J. Pharm. Pharmac.*, **16**, Suppl., 127 T.

Cutler, R. A., Cimijotti, E. B., Okolowich, T. J. & Wetterau, W. F. (1966). Alkyldimethylbenzylammonium chlorides — a comparative study of the odd and even chain homologues. *CSMA Proceedings of the 53rd Annual Meeting*, pp. 102–13.

Dalton, B. L. (1988). Novel non-metallic fungicide for aqueous coatings. *J. Coat. Technol.*, **60**(761), 43–53.

Davis, A., Bentley, M. & Field, B. S. (1968). Comparison of the action of Vantocil, Cetrimide and Chlorhexidine on *Escherichia coli* and the protoplasts of Gram-positive bacteria. *J. Appl. Bacteriol.*, **31**, 448–61.

Davis, J. P. & Doherty, F. G. (1985). Method for controlling mollusks. USP 4, 561, 983.

Deiner, H. (1983). Verfahren zur Herstellung eines Gemisches von Estern des 2,2'-dihydroxy-5,5'-dichlordiphenylmethans und deren Verwendung. DE-Pat. 3031933.
Domagk, G. (1935). Eine neue Klasse von Desinfektionsmitteln., *Deutsche Medizinische Wochenschrift*, **61**, 829–32.
Dryce, D. M., Croshaw, D., Hall, J. E., Holland, V. R. & Lessel, B. (1978). The activity and safety of the antimicrobial agent Bronopol (2-Bromo-2-nitropropane-1,3-diol). *J. Soc. Cosmet. Chem.*, **29**, 3–24.
Eagon, R. G. & Barnes, C. P. (1986). Mechanism of microbial resistance to hexahydro-1,3,5-triethyl-*s*-triazine. *J. Ind. Microbiol.*, **1**, 113–18.
Eggensperger, H. (1979). Desinfektionsmittel auf der Basis persäureabspaltender Verbindungen. *Zentralbl. Bakteriol. Hyg., I. Abt. Orig. B.*, **168**, 517–24.
Elsmore, R. & Guthrie, W. G. (1991). The development of a new active ingredient for the control of biodeterioration. In *Biodeterioration and Biodegradation 8*, ed. H. W. Rossmore. Elsevier Applied Science Publishers, London, pp. 507–9.
Exner, J. H., Burk, G. A. & Kyriacon, D. (1973). Rates and products of decomposition of 2,2-dibromo-3-nitrilopropionamide. *J. Agric. Food Chem.*, **21**(5), 838–42.
Falter, W., Lichtenbusch, B. E. & Schmitz, K. H. (1991). Verfahren zur Desinfektion und Reinigung in der Lebensmittel- und Gebäudeindustrie. DE 05 4000982 A 1.
Fiedler, H. P. (1989). *Lexikon der Hilfstoffe für Pharmazie, Kosmetik und angrenzende Gebiete*. Ed Cantor, Aulendorf, Germany.
Franklin, T. J. & Snow, G. A. (1989). *Biochemistry of Antimicrobial Action*, 4th edn. Chapman & Hall, London.
Franz, K. H. (1963). Clinical trials with thiabendazole against human strongyloidosis. *Am. J. Trop. Med. Hygiene*, **12**, 211–14.
Friberg, L. (1956). Quantitative studies on the reaction of chlorine with bacteria in water disinfection. *Acta Pathol. Microbiol. Scand.*, **38**, 135–44.
Friberg, L. (1957). Further quantitative studies on the reaction of chlorine with bacteria in water disinfection. *Acta Pathol. Microbiol. Scand.*, **40**, 67–80.
Gadher, P., Mercer, E. I., Baldwin, B. C. & Wiggins, T. E. (1983). A comparison of the potency of some fungicides as inhibitors of sterol 14-demethylation. *Pestic. Biochem. Physiol.*, **19**, 1–10.
Gocke, E., King, M.-T., Eckhard, K. & Wild, D. (1981). Mutagenicity of cosmetics ingredients licensed by the European Communities. *Mutat. Res.*, **90**, 91–109.
Gold, L. S., Sawyer, C. B., Magaw, R., Backman, G. M., de Viciana, M., Levinson, R., Hoopes, N. K., Havender, W. R., Bernstein, L., Peto, R., Pike, M. C. & Ames, B. N. (1984). A carcinogenic potency database of the standardized results of animal bioassays. *Environ. Health Perspect.*, **58**, 9–319.
Gorman, S. P., Scott, E. M. & Russel, A. D. (1980). Antimicrobial activity, uses and mechanism of action of glutaraldehyde. *J. Appl. Bacteriol.*, **48**, 161–90.
Grier, H., Witzel, B. E., Jakubowsky, J. A. & Dulaney, E. L. (1980). A broad spectrum hexahydro-*s*-triazine-inhibitor of microbial deterioration in cutting fluids. *Dev. Indust. Microbiol.*, **21**, 411–18.
Gründlinger, R. & Exner, O. (1990). Tebuconazole — A new triazole fungicide for wood preservation. IRG/Document No. IRG/WP/3629.
Hall, E. & Eagon, R. G. (1985). Evidence for plasmid-mediated resistance of *Pseudomonas putida* to hexahydro-1,3,5-triethyl-*s*-triazine. *Curr. Microbiol.*, **12**, 17–22.

Harries, R. C., White., D. B. & Macfarlane, R. B. (1970). Mercury compounds reduce photosynthesis by plankton. *Science*, **170**, 736–7.

Heil, J., Reifferscheid, G., Hellmich, D., Hergenröder, M. & Zahn, R. K. (1991). Genotoxicity of the fungicide Dichlofluanid in seven assays. *Environ. Molec. Mutagen.*, **17**, 20–6.

Henshaw, B. G., Laidlaw, R. A., Orsler, R. J., Carey, J. K. & Savory, J. G. (1978). The permanence of tributyltinoxide in timber. Record of the 1978 annual convention of the British Wood Preserving Association, pp. 19–29.

Hollis, C. G. & Lutey, R. W. (1988). Method for the control of molluscs. USP 4, 789, 489.

Hollis, C. G., Oppony, D. & Bayudu, S. R. (1991). Ionenes. A new class of biocides. In *Biodeterioration and Biodegradation 8*, ed. H. W. Rossmore. Elsevier Applied Science Publishers, London, pp. 503–4.

Hugo, W. B. (1965). Some aspects of the action of cationic surface-active agents on microbial cells with special reference to their action on enzymes. SCI Monograph No. 19. *Surface-Active Agents in Microbiology*, Society of Chemical Industry, London, pp. 67–82.

Hugo, W. B. & Longworth, A. R. (1964*a*). Some aspects of the mode of action of Chlorhexidine. *J. Pharm. Pharmacol.*, **16**, 655–62.

Hugo, W. B. & Longworth, A. R. (1964*b*). Effect of Chlorhexidine on 'protoplasts' and 'spheroplasts' of *E. coli*, protoplasts of *B. megaterium* and the Gram staining reaction of *Staph. aureus. J. Pharm. Pharmacol.*, **16**, 751–8.

Hugo, W. B. & Longworth, A. R. (1966). The effect of Chlorhexidine on the electrophoretic mobility, cytoplasmic constituents, dehydrogenase activity and cell walls of *E. coli* and *Staph. aureus. J. Pharm. Pharmacol.*, **18**, 569–78.

Isquith, A. A., Abbot, E. A. & Walters, P. A. (1972). Surface-bonded antimicrobial activity of an organosilicon quaternary ammonium chloride. *Appl. Microbiol* **24**(6), 859–63.

Jacobs, W. A., Heidelberger, M. & Bull, C. G. (1916). The bactericidal properties of the quaternary salts of hexamethylenetetramine. III. The relation between constitution and bactericidal action in the quaternary salts obtained from halogenacetyl compounds. *J. Exp. Med.*, **23**, 577–99.

James, A. N. (1965). Surface activity and the microbial cell. SCI Monograph No. 19. *Surface-Active Agents in Microbiology*, Society of Chemical Industry, London, pp. 3–23.

Jolley, R. L. (1975). Chlorine containing organic constituents in chlorinated effluents. *J. WPCF*, **47**(3), 601–18.

Kato, T. (1986). Sterol — biosynthesis in fungi, a target for broad spectrum fungicides. In *Sterol Biosynthesis — Inhibitors and Anti-feeding Compounds*. Springer-Verlag, Berlin, Germany, pp. 1–24.

Kato, K. & Fukumura, T. (1962) Degradation of ϵ-caprolactam by bacteria. *Chem. Ind.*, **25**, 1146.

Kato, N., Yamagami, T., Kitayama, K., Shimao, M. & Sakazawa, O. (1984). Dismutation and cross dismutation of aldehydes and alcohols: aldehyde oxidoreduction by resting cells of *Pseudomonas putida* F 61-a. *J. Biotechnol.*, **1**, 273–95.

Kempson, A. K. (1976). The microbiological deterioration of paints. *Australian OCCA Proceedings and News*, 5–14.

Klarmann, E. G., Shternov, V. A. & Gates, L. W. (1933). The alkyl derivatives of halogen phenol derivatives and their bactericidal action. I. Chlorophenols. *J. Am. Chem. Soc.*, **55**, 2576–89.

Kühle, E., Paulus, W., Klauke, E. & Genth, H. (1981). Use of *N,N*-dimethyl-*N*'-dichlorofluoromethylthio-*N*'-tolyl-sulphamide for combating fungi which damage wood. EP 10635.

Kühle, E., Paulus, W., Genth, H., Brandes, W. & Reinecke, P. (1985). N-sulfenylierte Harnstoffe, ein Verfahren zu ihrer Herstellung, diese enthaltende mikrobizide Mittel und ihre Verwendung. EP 0087704.

Kull, F. C., Eisman, P. C., Silverstrowicz, H. D. & Mayer, R. L. (1961). Mixtures of quaternary ammonium compounds and long-chain fatty acids as antifungal agents. *Applied Microbiol.*, **9**, 538–41.

Lakowicz, J. R. & Anderson, C. I. (1980). Permeability of lipid bilayers to methylmercury chloride. *Chem. Biol. Interact.*, **30**, 309–23.

Le Chevallier, M. W., Cawthon, C. D. & Lee, R. G. (1988). Inactivation of biofilm bacteria. *Appl. Environ. Microbiol.*, **54**, 2492–9.

Lederer, S. L., Jakubowsky, J. A. & Birnbaum, H. A. (1982). An effective preservative for adhesives with reduced health hazards. *Adhesive Age*, **25**(4), 28–32.

Lee, B. H., Tsunoda, K. & Takahashi, M. (1990). Laboratory evaluation of triiodoallyl alcohol as a wood preservative. *Material und Organismen*, **25**(2), 145–59.

Leinen, H. T., Lehmann, R. & Klüppel, H. J. (1988). Salicylsäureamide, Verfahren zu ihrer Herstellung und ihre Verwendung. EP 262587.

Lewis, S. N., Miller, G. A., Hausman, M. & Szamborski, E. C. (1971). Isothiazoles I: 4-isothiazolin-3-ones. A general synthesis from 3,3'-Dithiodipropionamides. *J. Heterocycl. Chem.*, **8**, 571–80.

Liu, D. (1989). Biodegradation of pentachlorophenol and its commercial formulation. *Toxicity Assessment*, **4**, 115–27.

Loew, O. (1886). Ueber Formaldehyd und dessen Condensation. *J. prakt. Chemie*, **33**, 321–51.

Lück, E. (1980). *Antimicrobial Food Additives*. Springer Verlag, Berlin.

Lück, E. (1988). Geschichte der Verwending von Lebensmittelzusatzstoffen. *Deutsche Lebensmittel-Rundschau*, **84**(9), 277–81.

Ludwig, R. A. & Thorn, G. D. (1960). Chemistry and mode of action of dithiocarbamate fungicides. *Ado. Pest Cont. Res.*, **3**, 219–52.

Luloff, J. S. & Eilender, A. L. (1975). Antibacterial composition and method employing a certain hexamethylenetetramine adduct. USP 3, 928, 607.

Maddox, B. (1988). Development of a new antimicrobial preservative for toiletries. *Specialty Chemicals Dec.*, 472–4.

McCann, J. & Ames, B. N. (1975). Detection of carcinogens as mutagens in the Salmonella microsometest: assay of 300 chemicals. *Proc. Nat. Acad. Sci. USA*, **72**, 5135–9.

Miller, G. A. & Lovegrove, T. (1980) 3(2H)Isothiazolone, a new class of antifouling toxicant. *J. Coat. Technol.*, **52**(661), 69–72.

Mizutani, T., Ito, K., Nomura, H. & Nakanishi, K. (1990). Nephrotoxicity of thiabendazole in mice depleted of glutathione by treatment with DL-buthionine sulphoximine. *Food Chem. Toxic.*, **28**(3), 169–77.

Moriya, M., Ohtha, T., Watanabe, K., Miyazawa, T., Kato, K. & Shirasu, Y. (1983). Further mutagenicity studies on pesticides in bacterial reversion assay systems. *Mutat. Res.*, **116**, 185–216.

Müller, E. (1951). In *Neuere Anschauungen der Organischen Chemie.* Springer Verlag, Berlin, Germany, pp. 18–19.

Myers, S. A., Allwood, M. C., Gidley, M. S. & Sanders, S. K. M. (1980). The relationship between structure and activity of Taurolin. *J. Appl. Bacteriol.*, **48**, 89–96.

Neely, W. (1963*a*). Action of formaldehyde on micro-organisms. I. Correlation of activity with formaldehyde metabolism. *J. Bacteriol.*, **85**, 1028–31.

Neely, W. (1963*b*). Action of formaldehyde on micro-organisms. II. Formation of 1,3-thiazane-carboxylic acid in *Aerobacter aerogenes* treated with formaldehyde. *J. Bacteriol.*, **85**, 1420–22.

Newton, B. A. (1960). The mechanism of the bactericidal action of surface active compounds. A summary. *J. Appl. Bacteriol.*, **23**, 345–9.

Nolen, G. A. & Dieckman, T. A. (1979). Reproduction and teratology studies of zinc pyrithione administered orally or topically to rats and rabbits. *Food. Cosmet. Toxicol.*, **17**, 639–49.

Paulus, W. (1976). Problems encountered with formaldehyde releasing compounds used as preservatives in aqueous systems, especially lubricoolants. Possible solutions to the problems. In *Proceedings of the 3rd Int. Biodegrad. Symp.*, eds. J. M. Sharpley & A. M. Kaplan, Applied Science Publishers, London, pp. 1075–82.

Paulus, W. (1980). Formaldehyde releasing compounds and their utility as microbicides. In *Biodeterioration*, eds T. A. Oxley, G. Becker, & D. Allsopp. Pitman Publishing Ltd, and The Biodeterioration Society, London, pp. 307–14.

Paulus, W. (1988). Development in microbicides for the protection of materials. In *Biodeterioration 7*, eds. D. R. Houghton, R. N. Smith & H. O. W. Eggins. Elsevier Applied Science Publishers, London, pp. 1–19.

Paulus, W. (1991). Microbicides for the protection of materials: yesterday, today and tomorrow. In *Biodeterioration and Biodegradation 8*, ed. H. W. Rossmoore. Elsevier Applied Science Publishers, London, pp. 35–52.

Paulus, W. & Genth, H. (1983). Microbicidal phenolic compounds — A critical examination. In *Biodeterioration 5*, eds. T. A. Oxley & S. Barry. John Wiley, Chichester, pp. 701–12.

Paulus, W. & Genth, H. (1986). Microbicidal agent. DP 2607033.

Paulus, W. & Kühle, E. (1986). Tailoring of microbicides for material protection — The trihalomethylthio group in various microbicides. In *Biodeterioration 6*, eds. S. Barry & D. R. Houghton. CAB International, Slough, pp. 79–88.

Paulus, W. & Pauli, O. (1970*a*). Synergistic microbicidal composition for the selective control of microbes. GB-Pat. 1264257.

Paulus, W. & Pauli, O. (1970*b*). Antimikrobielle Ausrüstung von Textilmaterial mit Hilfe von Reaktivwirkstoffen. *Textilveredlung*, **5**, 247–55.

Paulus, W. & Pauli, O. (1971). Permanente antimikrobielle Ausrüstung von anionisiertem Textilmaterial. *Textilveredlung*, **6**(4), 217–24.

Paulus, W., Genth, H. & Pauli, O. (1967). Disinfectants and preservatives. USP 3, 328, 240.

Paulus, W., Genth, H. & Pauli, O. (1975). New *N*-phenyl-carbamates. GB-Pat. 1465226.

Ploumen, J. J. H. (1991). Aqueous disinfectant compositions, and concentrates for preparing the same. EP 0461700 A 1.

Polster, M. & Halacka, K. (1971). Beitrag zur hygienischtoxikologischen Problematik einiger antimikrobiell gebrauchter Organozinnverbindungen. *Ernährungsforschung XVI*, **4**, 527–35.

Price, P. B. (1950). Reevaluation of ethyl alcohol as a germicide. *Arch. Surg.*, **60**, 492–502.
Rayudu, S. R. (1988). Ester of carbamic acid useful as a microbicide and preservative. EP Applic. 0365121.
Renbaum, A. (1973). Biological activity of ionene polymers. *Applied Polymer Symposium*, **22**, 299–317.
Reiss, J. (1976). Prevention of the formation of mycotoxins in whole wheat bread by citric acid and lactic acid (mycotoxines in food stuffs, IX). *Experientia*, **32**, 168–9.
Richards, R. M. E. & McBridge, R. J. (1973). Enhancement of benzalkonium chloride and chlorhexidine acetate activity against *Pseudomonas aeruginosa* by aromatic alcohols. *J. Pharm. Sci.*, **62**, 2035–7.
Riha, V. F., Sondossi, M. & Rossmoore, H. W. (1990). The potentiation of industrial biocide activity with Cu^{2+}. II. Synergistic effects with 5-chloro-2-methyl-4-isothiazolin-3-one. *Int. Biodet.*, **26**, 303–13.
Robinson, J. H., Stoertz, H. C. & Graessel, O. E. (1965). Studies on the toxicologic and pharmacologic properties of thiabendazole. *Toxicol. Appl. Pharmacol.*, **7**, 53–63.
Rossmoore, H. W. & Holtzmann, G. H. M. (1974). Growth of fungi in cutting fluids. *Dev. Ind. Microbiol.*, **15**, 273–80.
Rossmoore, H. W., Sieckhaus, F. J., Rossmoore, L. A. & De Fonzo, D. (1978). The utility of biocide combinations in controlling mixed microbial populations in metalworking fluids. *Lubr. Engng.*, **35**(10), 559–63.
Russel, A. D. (1971). The destruction of bacterial spores. In *Inhibition and Destruction of the Microbial Cell*, ed. W. B. Hugo. Academic Press, London, pp. 451–612.
Russel, A. D. (1983). Mechanisms of action of chemical sporicidal and sporistatic agents. *Int. J. Pharmaceut.*, **16**, 127–40.
Russel, A. D. & Chopra, I. (1990). *Understanding Antibacterial Action and Resistance*. Ellis Horwood Ltd, Chichester.
Schimmel, J. & Husa, W. J. (1956). The effect of various preservatives on microorganisms isolated from deteriorated syrups. *J. Am. Pharm. Ass.*, **XLV**(4), 204–8.
Schmitt, H. G. (1987). The action of antimicrobial substances. *Parfümerie + Cosmetics*, **1**, 3–9.
Schultdt, P. H. & Wolf, C. N. (1958). Fungitoxicity of substituted s-triazines. *Contributions from Boyce Thompson Int.*, **18**, 377–93.
Schuphan, L., Westphal, D., Haque, A. & Ebing, W. (1981). Biological and chemical behaviour of perhalogenmethylmercapto fungicides: Metabolism and in-vitro reactions of dichlofluanid in comparison with captan. *Am. Chem. Soc.*, Symposium Series **158**, 85–96.
Schweinfurth, H. A. & Günzel, P. (1987). The tributyltins: Mammalian toxicity and risk evaluation for humans. *Proc. Oceans*, **87**, 1421–37.
Scott, C. R. & Wolf, P. A. (1962). Antibacterial activity of a series of quaternaries prepared from hexamethylenetetramine and halohydrocarbons. *Appl. Microbiol.*, **10**, 211–16.
Seal, K. J. (1988). The biodeterioration and biodegradation of naturally occurring and synthetic plastic polymers. *Biodet. Abstracts*, **2**(4), 295–317.
Sharma, K. D. (1989). Studies on antimicrobial activity of β-hydroxynaphthaldehyde as a possible protectant against biodeterioration. *Int. Biodet.*, **25**, 123–9.

Shere, L. (1948). Some comparisons of the disinfecting properties of hypochlorites and quaternary ammonium compounds. *Milk Plant Monthly*, **37**, 66–9.
Springle, W. R. (1989). Enzymatic liquefaction of emulsion paints and cellulosic thickeners. Chemspec Europe 89 BACS Symposium, pp. 33–7.
Stonehill, A., Kops, S. & Borick, P. M. (1963). Buffered glutaraldehyde, a new chemical sterilizing solution. *Am. J. Hosp. Pharm.*, **20**, 458–65.
Tannenbaum, M. & Bricker, C. E. (1951). Microdetermination of free formaldehyde. *Anal. Chem.*, **23**, 354–7.
Taylor, W. S. (1965). Fungus resistant paints. USP 3, 199, 990.
Upsher, F. J. & Roseblade, R. J. (1984). Assessment by tropical exposure of some fungicides in plasticized PVC. *Int. Biodet.*, **20**(4), 243–52.
Valcke, A. R. (1989). Suitability of Propiconazole (R 49362) as a new-generation wood preserving fungicide. IRG Document No: IRG/WP/3529.
Valcke, A. R. & Goodwine, W. R. (1985). *Azaconazole, a New Wood Preservative*. American Wood Preserver's Association, 6 pp.
Walker, J. F. (1975). *Formaldehyde*. R. E. Krieger Publishing Co., New York.
Wallhäußer, K. H. (1984). *Praxis der Sterilisation, Desinfektion, Konservierung*. Georg Thieme Verlag, Stuttgart–New York, p. 360.
Warner, P. L., Kornegay, G. B. & Redl, G. (1980). Antibakterielle Verbindungen, Verfahren zu ihrer Herstellung und daraus hergestellte Mittel. DE-OS 3016110.
Wee, Y. C. (1988). Growth of algae on exterior painted masonry surfaces. *Int. Biodet.*, **24**, 367–71.
Whiteley, P. (1966). The occurrence and prevention of mould and algal growth on paint films. *Soc. Chem. Ind. Monogr.*, **23**, 162–9.
Wien, R., Harrison, J. & Freeman, W. A. (1948). Diamidines as antibacterial compounds. *Brit. J. Pharmacol.*, **3**, 211–18.
Willingham, G. L. & Mattox, J. R. (1990). Phenoxyalkanols as stabilizer for isothiazolinones. USP 505201.
Wolf, P. A. & Sterner, P. W. (1972). 2,2-Dibromo-3-nitrilopropionamide, a compound with slimicidal activity. *Appl. Microbiol.*, **24**(4), 581–4.
Wölfel, L., Mach, F. & Chattopadhyay, S. P. (1985). Comparative cytological studies on the effect of cetyltrimethylammonium bromide on bacterial cells *Zentralblatt für Mikrobiologie*, **140**, 631–9.
Wollmann, H., Haufe, F. & Haufe, U. (1963). Perkutane Toxizität von Benzylalkohol beim Meerschweinchen. *Pharmazie*, **22**, 455.
Xu, H. H. & Schurr, K. M. (1990). Genotoxicity of 22 pesticides in microtitration SOS chromo-test. *Toxicity Assessment: Int. J.*, **5**, 1–14.

INDEX

Index of Compounds

Subject Index